The New Asian Innovation Dynamics

Technology, Globalization and Development Series

Series Editor: Anthony P. D'Costa, Professor of Indian Studies, Copenhagen Business School, Denmark.

The series examines technological change in the larger global context by identifying where the markets are, who is specializing in what areas of technology and why, what is the nature of global trade in technologies, and what are some of the social and political challenges posed by such change. Relationships between the OECD and rapidly growing Asian and Latin American economies are brought out by this series. At the same time, the series will also address the issue of why some regions such as Africa and the Middle East lag behind. The question of whether technology is important for development, and how various governments have targeted technology development, whether through education, skill development, or R&D also adds to the diversity of the series. The series includes comparative and regional studies, including single-country case studies of large economies such as India, China, Japan, South Korea, and Brazil. While no single approach or methodology is used exclusively, scholars within the series rely on political economy approaches that are sensitive to institutional and historical realities when analyzing specific countries/regions or technologies.

Titles include:

Anthony P. D'Costa (*editor*)
THE NEW ECONOMY IN DEVELOPMENT
ICT Challenges and Opportunities

Govindan Parayil (*editor*)
POLITICAL ECONOMY AND INFORMATION CAPITALISM IN INDIA
Digital Divide, Development and Equity

Govindan Parayil and Anthony P. D'Costa (*editors*)
THE NEW ASIAN INNOVATION DYNAMICS
China and India in Perspective

Technology, Globalization and Development Series.
Series Standing Order ISBN 1–4039–3591–2

You can receive future titles in this series as they are published by placing a standing order. Please contact your bookseller or, in case of difficulty, write to us at the address below with your name and address, the title of the series and one of the ISBNs quoted above.

Customer Services Department, Macmillan Distribution Ltd, Houndmills, Basingstoke, Hampshire RG21 6XS, England

---

The New Asian Innovation Dynamics

China and India in Perspective

Edited by

Govindan Parayil
Vice Rector of the United Nations University, Tokyo
Professor of Science, Technology and Innovation (on leave)
University of Oslo, Norway

and

Anthony P. D'Costa
Professor of Indian Studies
Copenhagen Business School, Denmark

First published 2009 by
PALGRAVE MACMILLAN

Palgrave Macmillan in the UK is an imprint of Macmillan Publishers Limited, registered in England, company number 785998, of Houndmills, Basingstoke, Hampshire RG21 6XS.

Palgrave Macmillan in the US is a division of St Martin's Press LLC, 175 Fifth Avenue, New York, NY 10010.

Palgrave Macmillan is the global academic imprint of the above companies and has companies and representatives throughout the world.

Palgrave® and Macmillan® are registered trademarks in the United States, the United Kingdom, Europe and other countries

ISBN-13: 978—0—230—20945—9   hardback

This book is printed on paper suitable for recycling and made from fully managed and sustained forest sources. Logging, pulping and manufacturing processes are expected to conform to the environmental regulations of the country of origin.

A catalogue record for this book is available from the British Library.

Library of Congress Cataloging-in-Publication Data

The new Asian innovation dynamics : China and India in perspective / edited by Govindan Parayil and Anthony P. D'Costa.
    p. cm. – (Technology, globalisation and development series)
    Includes bibliographical references and index.
    ISBN 978-0-230-20945-9 (alk. paper)
    1. Technological innovations—Economic aspects—China. 2. Technological innovations—Economic aspects—India. I. Parayil, Govindan, 1955-II. D'Costa, Anthony P., 1957-HC430.T4N49 2008
    338'.0640951—dc22

                                                                        2008030436

10   9   8   7   6   5   4   3   2   1
18   17   16   15   14   13   12   11   10   09

Transferred to Digital Printing in 2011

Contents

Notes on the Contributors

Editors

Govindan Parayil is Vice Rector of the United Nations University, Tokyo and Professor of Science, Technology and Innovation (on leave) at the Centre for Technology, Innovation and Culture, University of Oslo, Norway.

Anthony P. D'Costa is Professor of Indian Studies, Asia Research Centre, Copenhagen Business School, Frederiksberg, Denmark.

Contributors

Albert J. Abma is with the Science & Society Group, Faculty of Natural Sciences and Mathematics, University of Groningen, The Netherlands.

Cong Cao is a researcher with the University of Oregon and the Neil D. Levin Graduate Institute of International Relations and Commerce, State University of New York, New York.

Cristina Chaminade is Associate Professor in Innovation Studies at CIRCLE, Lund University, Sweden.

Joanna Chataway is Professor in the Department of Development Policy and Practice at the Open University, Milton Keynes, UK.

Kalpana Chaturvedi is at the Research and Enterprise Support, University of Surrey, Guildford, UK.

Lars Coenen is Assistant Professor at CIRCLE, Lund University, Sweden.

Peter Gammeltoft is Associate Professor, Department of International Economics & Management, Copenhagen Business School, Frederiksberg, Denmark.

Menno P. Gerkema is with the Science & Society Group, Faculty of Natural Sciences and Mathematics, University of Groningen, The Netherlands.

Julie Marie Kjersem is with the International Marketing Department at Novo Nordisk, Copenhagen, Denmark.

Tomoko Kobayashi is with the Human Geography Division, Department of General Systems Studies, University of Tokyo, Japan.

Xielin Liu is Professor at the Graduate University of the Chinese Academy of Science, Beijing, China.

Nannan Lundin is with the Swedish Environmental Research Institute, Stockholm, Sweden.

Sylvia Schwaag Serger is Senior Advisor Asia at the Swedish Institute for Growth Policy Studies, Östersund, Sweden.

Denis Fred Simon is Provost and Vice President for Academic Affairs, Neil D. Levin Graduate Institute of International Relations and Commerce, State University of New York, New York.

Richard P. Suttmeier is Professor of Political Science (Emeritus) at University of Oregon, USA.

Jayan Jose Thomas is Assistant Professor at the Madras School of Economics, Chennai, India.

Henny J. van der Windt is with the Science & Society Group, Faculty of Natural Sciences and Mathematics, University of Groningen, The Netherlands.

Jan Vang is Assistant Professor at the Copenhagen Institute of Technology, Aalborg University, Denmark.

Nicolien F. Wieringa is with the Science & Society Group, Faculty of Natural Sciences and Mathematics, University of Groningen, The Netherlands.

Foreword

The fastest growing centres for science and technology (S&T) are located in Asia. While Japan, South Korea, Singapore, and Taiwan are still ahead of other Asian countries in key areas of S&T and innovation, China and India get the biggest headlines because of the spectacular developments that have taken place there over the last decade or more. Together, their potential in science and technology is enormous.

While China and India do not as yet have a well-articulated national innovation system, they are clearly poised to become S&T superpowers. They are responding to the deepening trans-national flows of labour, capital, production, services, knowledge, information, and culture that characterize our globalizing world. Their innovation dynamics and their competitiveness in the global knowledge economy are increasingly felt, as witnessed by the host of multinational companies and Western and other governments and public bodies, including many universities, which are flocking there to take stock of, exploit, and/or benefit from the new S&T and innovation potential.

This volume explores the drivers and constraints of the Chinese and Indian innovation dynamics. It is the result of an effort by a network of researchers, initially based in the Nordic countries, under the auspices of NIAS-Nordic Institute for Asian Studies to explore the *New Asian Dynamics in Science, Technology and Innovation*, the title of its first international conference in Gilleleje, Denmark, in September 2006.

The aim of the conference was to discuss, from different disciplinary perspectives, the whole spectrum of challenges relating to the development of basic and applied science and commercialized technology in different economic, political, social and cultural contexts in Asia. There were both empirically oriented contributions on specific aspects and sectors, and papers that discussed broader questions from comparative perspectives.

However, this book travels beyond the scope of the traditional conference volume. The editors, Professor Govindan Parayil, United Nations University and Oslo University, and Professor Anthony P. D'Costa, Copenhagen Business School – both members of the network behind the conference in Gilleleje – decided that the volume should focus on China and India, exclusively. They not only invited contributions from some of the conference participants, but also from additional authors to present a balanced picture of the field under investigation. They also create a common framework of analysis that incorporates the analytical approaches and insights of the individual contributions. Therefore, the volume is able to present a string of subtle, empirically based and well balanced analyses that promote new and coherent insights into

China's and India's innovation dynamics and the perspectives for the rest of the world.

The newness of what happens in China and India lies in the reformulation of the approaches pursued to link innovations within science and technology policies, internationalization strategies through foreign direct investment and trade, and a variety of institutional arrangements involving the state, business, multinational and national firms, and public research institutes. It is patently obvious that other OECD economies, particularly the smaller Nordic and Asian economies, cannot remain passive to these ongoing changes in competitive advantage in the Asian region. This volume offers further insights into the processes that have challenged the rest of the world to rethink and explore the 'new Asian innovation dynamics'.

This volume would not have come about without the conference in Gilleleje, Denmark, in 2006. Therefore, it is appropriate to thank the sponsors of the conference for their timely and generous support: Asia House Foundation, Copenhagen; The EAC Foundation, Copenhagen; The Swedish School of Advanced Asian and Pacific Studies (SSAAPS); The Social Sciences Research Council, Denmark; The Nordic Council of Ministers; and the Nordic NIAS Council through NIAS.

The organizing committee of the Gilleleje conference included: Prof. Thommy Svensson, SSAAPS – Swedish School of Advanced Asia-Pacific Studies (Chair); Prof. Jan Annerstedt, Copenhagen Business School; Dr. Ester Barinaga, Associate Professor, Royal Institute of Technology, Stockholm; Prof. Hans C. Blomqvist, Vice Rector, Swedish Business School in Vaasa, Finland; Dr. Jørgen Delman, Director, NIAS – Nordic Institute of Asian Studies, Copenhagen; Prof. Govindan Parayil, Centre for Technology, Innovation and Culture, University of Oslo; and Prof. Paul Midford, Norwegian University of Science and Technology (NTNU).

The research network behind the conference are now moving to the next phase with a project that will focus on the role of innovative Asian hub cities in global, national and local innovation systems.

I am hopeful this book will receive wider attention among policy makers and academics in both developed and developing countries who are interested in innovation dynamics in larger Asian economies.

Jørgen Delman
Director, NIAS-Nordic Institute of Asian Studies
Copenhagen, Denmark
26 May 2008

Acknowledgements

Chapter 10 is a revised and expanded version of an article that first appeared in *Physics Today* and it is reprinted with permission from Cong Cao, Richard P. Suttmeier, and Dennis Fred Simon, 'China's 15-year science and technology plan' *Physics Today*, December 2006, pp. 38–43. Copyright 2006, American Institute of Physics.

1
China, India, and the New Asian Innovation Dynamics: An Introduction

Anthony P. D'Costa[1] and Govindan Parayil

1.1 Introduction

In recent decades China and India have been attracting headlines in the business press and electronic media for their stellar economic growth and business performance. China's huge trade surplus, based in part on high-technology exports, and India's pre-eminent position as the world's leading offshore destination for software service exports place these countries in enviable positions. In addition to unprecedented growth rates, they are undergoing considerable changes in their production structure and export profile. Both China and India are also emerging as global centres for research and development (R&D) and as knowledge-based exporters of services (UNCTAD, 2004). They, along with other emerging Asian economies, are now active participants in the global economic system. They produce not only basic manufactured goods such as textiles and consumer electronics but also form partnerships with multinational firms to innovate in advanced information and biotechnology sectors (UNCTAD, 2004).

The objective of this volume is to identify the innovation dynamics of China and India in the science- and technology-based industrial sectors. Both China and India are emerging as key players in science-based technology (see Niosi and Reid, 2007).[2] Indirectly, the volume addresses whether or not China and India will drain knowledge-intensive economic activities from the OECD and how the smaller OECD economies of Europe might cope with this presumed inevitability (see WTO, 2005). Moreover, China and India are increasingly integrated with the Asian regional and global economies, albeit in different ways, and as such the institutional architectures behind their innovation efforts deserve scrutiny. China's high-technology exports and India's software exports are both reflections of this broader science and technology effort. Simultaneously, some Asian countries, such as Japan, rely on South East Asia for manufacturing production, but retain significant R&D and design capabilities at home, while South Korea, with its aggressive policy, is working hard to narrow its technology gap with Japan.

Other strategies include Taiwan's growing investment links with mainland China, Singapore's growing dependence on foreign professionals for innovative activities, and multinationals from everywhere tapping into Asian science and technology capabilities and human resources. In all cases, the state is involved in fostering national innovative capability. We refer to these attributes in the wider Asian context as the 'new Asian innovation dynamics' (NAID).

The newness lies in the various approaches linking innovations with science and technology policies, internationalization strategies through foreign direct investment (FDI) and trade, and a variety of institutional arrangements involving the state, business, multinational firms, and public research institutes. Given a wide variety of national histories and institutional contexts, we anticipate a range of policies and strategies among governments and businesses under international economic integration. In addition, the dynamics of innovation look different when viewed from the perspective of a local or multinational firm, the industry, or the national economy. It is this gamut of activities, actors, and contexts that we label 'NAID'. Our intent is not to capture regional economic dynamics in their totality, but to identify the principal forces and actors at work that are contributing to innovation strategies in China and India. The study assumes that upgrading economic activities is not inevitable and we argue that strategic choices made by multinational firms, local businesses, and governments are key to understanding Asia's new innovation dynamics.

In this context, OECD economies, and in particular the smaller Scandinavian and Asian economies, cannot remain passive to the ongoing changes in competitive advantage in the Asian region. Nor can large OECD economies ignore the emerging opportunities in China and India. Active collaboration of Asian businesses, with their foreign counterparts, and purposive government interventions for innovation, promise to benefit both OECD and developing countries in reorganizing their economies.

While this volume is chiefly concerned with innovatory activities in high technology, we believe that the benefits of innovation must accrue to vulnerable groups that are displaced in the OECD and excluded in the non-OECD economies. Innovation must be socially relevant, broadly enhancing productive employment, public welfare, and support systems for those who cannot make the adjustment when comparative and competitive advantages shift. It behooves states in both sets of countries to be aware of and work to minimize such transitional social costs.

Following multiple sets of literature on innovation dynamics from innovation and business strategy studies, technology policy, and political economy, this volume focuses on the following issues:

- The forms and magnitudes of the shift in the division of labour in knowledge-intensive activities in global and Asian regional contexts,

- The challenges and opportunities in China and India for high-technology industrial and services upgrading,
- The involvement of principal actors such as multinational companies and Asian businesses and the particular industrial and business sectors undergoing global shifts,
- The main institutional arrangements such as those between the government, public research institutes, universities, and industries; associated innovation policy instruments; and business models influencing changing competitiveness of firms, regions, and countries, and
- Science, technology, and innovation policies and competitive strategies in China and India in the larger Asian context.

This opening chapter presents some basic issues surrounding innovation dynamics in China and India and thematically introduces the chapters ahead. First, a brief comparison of selected techno-economic indicators is made for the two countries. Next, two standard analytical frameworks – the national innovation system (NIS) and the triple helix model (THM) – are briefly introduced to discuss the critical institutional architectures that lie behind national and regional innovation systems. These are the interactions between the state, national businesses, multinational enterprises, universities, and public and private research institutes. The actual workings of these institutions are examined variously by the different authors for different knowledge-intensive sectors in the rest of the volume. Section 1.4 identifies some of the broader challenges faced by China and India. The following section briefly presents some broad policy areas for both developing and smaller OECD economies where issues of competitiveness, employment, growth, and worsening inequality have become politically charged issues. The final section outlines the individual chapters.

1.2   China and India compared

Innovation dynamics are broadly influenced by interactive national and global factors. While the precise causal mechanisms are unclear, scholars and practitioners have identified national endogenous factors such as education and investment in research and development as critical (Cypher and Dietz, 2004: 377–96). Of course, these variables are significantly influenced by national historical-institutional contexts, which in turn are shaped by external factors. In the contemporary innovation context, knowledge creation and its dissemination is no longer a national affair. With increasing international economic integration active engagement of national firms and research institutions create new learning opportunities while maintaining open channels for knowledge transfer via multinational investments and foreign talent.

Selected techno-economic indicators for China and India are presented in Table 1.1. The economic structure of both countries has changed

Table 1.1 China and India: key techno-economic indicators

	China	India
1. School enrollment, tertiary (% gross)	20.3	11.4
2. GDP (current $, trillion)	**2.7**	**0.91**
3. GDP growth (annual %)	10.7	9.2
4. Industry, value added (% of GDP)	47.0	27.7
5. Services, value added (% of GDP)	41.1	54.7
6. Exports of goods and services (% of GDP)	**36.8**	20.3
7. Imports of goods and services (% of GDP)	**32.9**	23.3
8. Internet users (per 1,000 people)	85.1	54.8
9. High-technology exports (% of manufactured exports)	30.6	5*
10. Foreign direct investment, net inflows (BOP, current US$)	79.1	6.6
11. Share of GDP spent on R&D (%)	1.3	0.8
12. Science and technology workers (million)	2.25	1.1E
13. Science, Medicine, Engineering postgraduates (million)	0.50	0.35EE
14. Number of PhDs awarded in science-related subjects	23,500**	5000-6714
15. US patents granted to Chinese applicants	424#	341
16. Multinational R&D centres	750	150

Notes: Items 1–10 obtained from the World Bank.
Data pertain to 2005 and bold numbers pertain to 2006, unless otherwise noted.
* = 2000, # = 2003, ** = 2004, E = engineers, EE = only engineers, BOP = Balance of Payments
Sources: World Bank, World Development Indicators database, April 2007.
Wilsdon and Keeley (2007: 19–20) for China, Bound (2007: 11–12, 36) for India.

dramatically, with a significant reduction in agriculture for China. However, there is a notable difference. China's manufacturing sector is far ahead of India's, while India's service sector is even further ahead of China's. This difference is highlighted by the two countries' different production and export profiles. This is further illustrated by China's far greater share of high-technology exports compared to India's. What the table does not indicate is the large share of high-technology services exported by India. Its $31 billion exports of software and related services represent nearly a third of India's total exports (NASSCOM, 2007).

The export pattern in both countries could be partly explained by an expanding science and technology ecosystem in the two countries. This is evident from the tertiary education infrastructure and its output, as illustrated by enrollments, science and technology workers, professional postgraduates, and PhDs in science-related subjects. Although China is far ahead of India due to an early and more centralized emphasis on science and technology development, the emphasis by both countries on technical education and high-technology economic activities has set in motion the generation of an absolutely large pool of science and technology talent. Consequently, not only are they leveraging this resource for export purposes, but foreign companies themselves are clamouring to deploy that talent for their global operations. For example, numerous multinational R&D centres have been established in both China and India (see Table 1.1, UNCTAD, 2004). This economic integration both via national exports and multinational investments is a major driver of innovation dynamics in the two countries. Additionally, the return flow of expatriate talent from abroad adds to the national technical and entrepreneurial talent. The routes by which national efforts are translated to innovative capability and how multinational firms both utilize and contribute to innovative dynamics are first examined by a brief discussion of the institutional arrangements in the next section and further elaborated in the individual chapters.

1.3 NIS and THM: framing innovation dynamics in China and India

In this chapter we present mainly the national innovation system (NIS) and the triple helix model (THM) as frameworks to bring out the underlying processes responsible for the generalized shifts in knowledge-intensive activities. We do not review the literature on NIS, THM of state–business–university interactive dynamics, and cluster analysis at this point.[3] The individual chapters in this volume undertake that task in various ways.

In one sense knowledge-intensive activities are no different from other economic activities. In a market context the production of goods and services is driven by competition and profits, irrespective of whether they are labour-, resource-, capital-, or knowledge-intensive. Similarly, initial economic conditions, institutional legacies, business strategies, and government policies influence the expansion of economic activities. For example, it is difficult to imagine India's competitiveness in the pharmaceutical and IT sectors without reference to past national patent policies, tertiary education, investment in government research institutes (GRIs), and industry and science parks such as India's Software Technology Parks (D'Costa, 2003a, Chapter 4 this volume). Beijing Municipality's Zhongguancun Science Park, with forty universities and 130 research institutes, is also a good example of state involvement.[4]

The decision to transnationalize production is also dependent on a wide range of economic, institutional, and political factors not the least of which are the changing relative costs of production and dispersion of technological capabilities in the global economy. However, there are at least three crucial differences between knowledge-intensive and other kinds of economic activities. The first is the high skill and education content of the professionals and workers in knowledge-based activities. This makes the active participation of universities and research institutions necessary in research and development (R&D) and creating science and engineering talent. The second is the qualitative differences in the dynamics of innovation in high-technology sectors.[5] There are far more presumed externalities (economic and non-economic) associated with knowledge-intensive activities, hence they are targets of state intervention for national development, especially on matters related to R&D policy, human resource development, education policy, and so on. The third is the uncertainty associated with new knowledge generation and the innovation process itself. Hence, there are inherent difficulties in correctly predicting emerging technologies and their trajectories, and figuring out how to foster them institutionally.

It is a fact that OECD economies have enjoyed superior educational and technological infrastructures for a longer time than developing countries. Hence, their persistent near-monopoly control over core technological areas can be assumed to be sustained in the medium term. However, what may be unsettling this status quo is Asia's rise in skill- and knowledge-based activities due to the deployment of an absolutely large size and growing relative abundance of human capital, churned out of an expanding technical education infrastructure. This is certainly true of large countries such as China and India for whom a small percentage of highly trained people can translate into millions of knowledgeable professionals (Figure 1.1).

In 1980 China and India together accounted for about 10 per cent (7 million) of the global population with tertiary education. By 2000 the figure stood at 18 per cent (35 million). Though China has higher graduate enrollments, India's graduate enrollment was not insignificant at 10 million in 2004, of which 11 per cent was in engineering (National Council of Applied Economic Research 2005, in Bound 2007: 11). India's current stock of young talent is roughly 14 million, which is roughly one and a half times that of the USA and double China's (Bound, 2007: 11). Even at the doctoral level in science and engineering (S&E) both China and India are rapidly increasing their pool. China, for example, increased its PhDs awarded from 2,556 in 1991 to 12,465 in 2003 (National Science Board, 2006: A2-123). Corresponding figures for India are 8,383 to 13,733. Although the USA continues to lead countries in doctoral degrees in S&E, with nearly 41,000 PhDs in 2003, a large share of this is earned by foreign students, many of them from India and China. For example, 30% of S&E doctorates earned by foreigners in the USA during the period 1983–2003 were earned by students of Chinese and Indian origin (National Science Board, 2006: Figure O-32). When all of Asia is included the share is close to 67 per cent.

The second feature of knowledge-intensive activities is their systemic evolution and development. Increasingly, there is convergence among different knowledge-based activities, thus altering the scope of technology trajectories. For example, the biomedical sub-sector of biotechnology involving gene therapy entails not only researchers in the biological and medical sciences but also in information and nanotechnologies. This is complemented by a

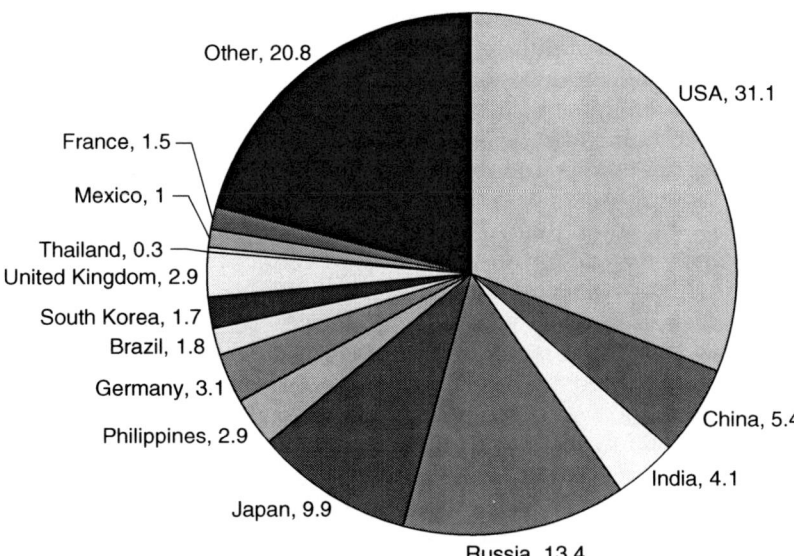

Distribution of population 15+ years (tertiary education),
1980 (73 million)

Other, 20.8

France, 1.5

Mexico, 1

Thailand, 0.3
United Kingdom, 2.9

South Korea, 1.7

Brazil, 1.8

Germany, 3.1

Philippines, 2.9

Japan, 9.9

Russia, 13.4

USA, 31.1

China, 5.4

India, 4.1

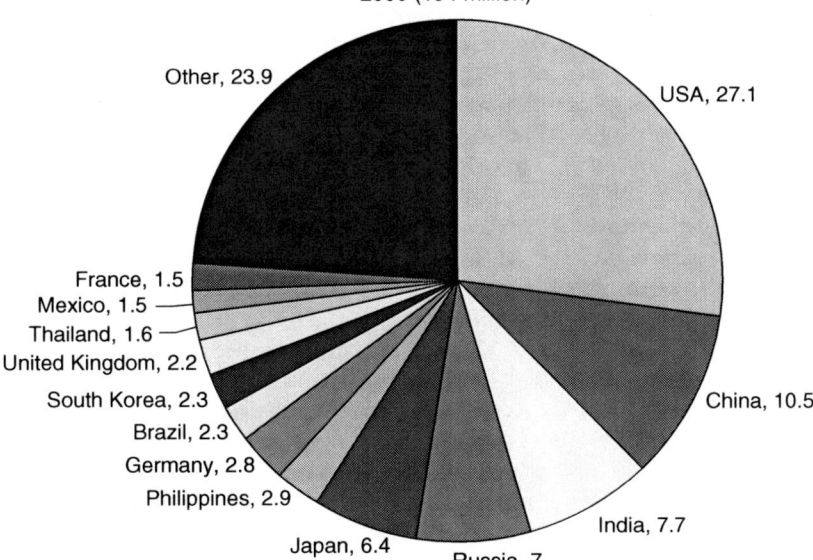

Distribution of population 15+ years (tertiary education),
2000 (194 million)

Other, 23.9

France, 1.5
Mexico, 1.5
Thailand, 1.6
United Kingdom, 2.2

South Korea, 2.3

Brazil, 2.3

Germany, 2.8

Philippines, 2.9

Japan, 6.4

Russia, 7

USA, 27.1

China, 10.5

India, 7.7

Figure 1.1   Distribution of population 15+ years, 1980 and 2000
Source: National Science Board 2006: p. o-11.

system of clinical trials, production, and final demand management involving a whole spectrum of actors such as doctors, hospitals, health ministries, intellectual property rights (IPR) intermediaries, pharmaceutical companies and so on. Thus, innovation cannot be seen as a simple linear progression from simple to complex activities but rather as a compounded and cumulative product of a number of interrelated processes. These involve not only the depth of scientific and engineering research in new knowledge but also the institutional arrangements by which such knowledge is generated, disseminated, and protected to create a system of innovation (see Allarakhia and Wensley, 2007). Within these overlapping process loops, most innovations ultimately depend on market demand and social acceptability. Thus, there is a significant demand pull dimension in the equation, which is also influenced by institutional dynamics.

While NIS and THM frameworks emphasize different institutions, the main actors are the state (GRIs), private national business, multinational enterprises, and the university system. The precise institutional mix and their pathways of interactions in different country and regional contexts are ill-defined. For example, in Beijing and Shenzen, China the role of universities and GRIs has been substantially different (Chen and Kenny 2005). Notwithstanding the challenges of institutional coordination, governments, businesses, and universities are expected to work together to create and diffuse new commercial knowledge. Foreign firms, for strategic reasons, find NIS attractive if research capability and abundant availability of highly skilled technical professionals are integral to it.[6] This is amply demonstrated by offshoring practices of American IT firms in India and increasingly by MNC-driven basic research in China and India. Host governments also eye multinationals for their technologies. Thus, until recently China has pursued a 'technology for market' policy, trying to leverage its large domestic economy to obtain knowledge from multinationals (Cao, Suttmeier, and Simon, Chapter 10).

Whatever the mode of entry of firms and interaction among institutions, the goal of governments is to foster geographical clusters of businesses such that an ecosystem of research institutes, businesses, R&D-oriented multinational enterprises, and universities synergistically propel the region to innovate (Vang, Chaminade and Coenen, Chapter 8). The associated spillover effects, inter-sectoral linkages, and university-based spin-offs associated with dynamic clusters suggest economic expansion and growing sophistication of firm capabilities. This is possible through feedback mechanisms and integrative learning at the local level (Kjersem and Gammeltoft, Chapter 7). In addition, both the region and its institutions are expected to be globally networked with similar institutions elsewhere, especially global centres of innovation such as Silicon Valley in the USA, Bio Valley in Switzerland, Cambridge, UK, and high-tech electronics in the Tokyo–Yokohama corridor in Japan. However, the extent to which knowledge-intensive production and

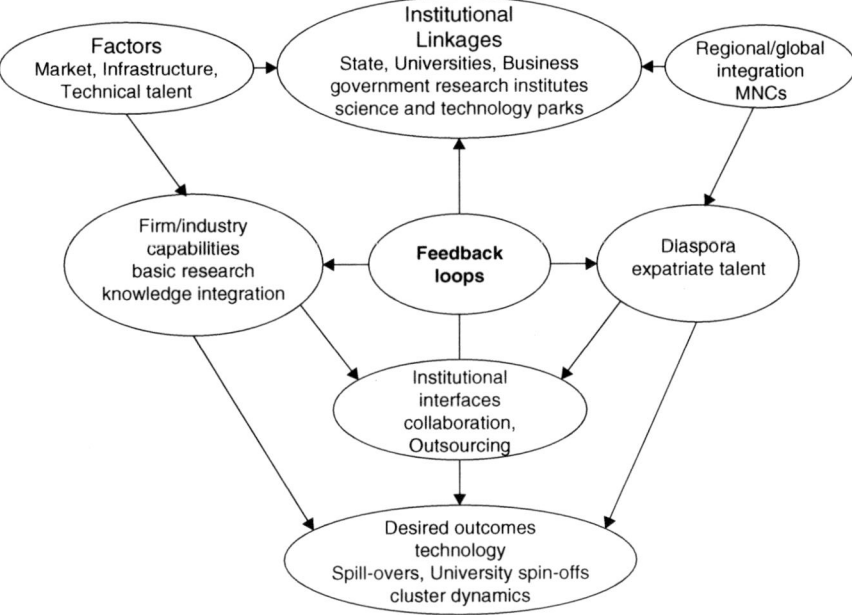

Figure 1.2   Innovation dynamics in Asia
Source: Compiled by authors.

services exhibit such characteristics in specific countries is, in the end, an empirical matter. Developing countries such as China and India are still in their early phases of development in research-based activities, albeit on a steep learning curve.

A stylized version of innovation dynamics associated with the NIS/THM/Cluster frameworks is presented in Figure 1.2. There are some basic drivers of innovation milieu such as large, growing markets, industry-specific infrastructure, and the availability of a large pool of technical talent. These are best utilized under an institutional architecture that links the principal actors and stakeholders such as relevant government ministries and departments, multinational firms, universities, government research institutes, and industry. Foreign direct investment, joint-ventures, offshore outsourcing arrangements through subsidiaries or through contract research are some of the principal forms of external institutional interfaces. Expatriate technical talent can also contribute to innovation dynamics through start-ups, partnerships, research, linkages to foreign markets, and commercial and technical knowledge transfer.

Institutional interfaces are best served under conditions of increasing firm technology and research capabilities such that an emerging international division of labour between basic research and knowledge integration becomes feasible (Chaturvedi and Chataway, Chapter 6). Given the state of knowledge, the innovation milieu, and factors present, Asian economies are likely to enjoy increasing advantages in basic research, while OECD economies specialize in knowledge integration (Abma et al., Chapter 11). The desired outcomes of a well-functioning NIS/THM are technology spin-offs, increasing research capabilities, and synergy among the cluster of institutions.

1.4   Challenges to innovation in Asia

Most developing countries, including China and India, do not yet have a well-articulated national innovation system (NIS) (see Bound, 2007: 20; Wilsdon and Keeley, 2007: 32). The task of setting one up is far too ambitious in terms of operationalization, resources, and institutional coordination. The move away from the top-down statist, plan-based systems of economic management in both countries is in the right direction but most public institutions, including government research institutions (GRIs) and the university system, still do not operate independently. Their success in big science and mega projects for national security such as missile and nuclear technologies has not been matched by effective linkages to business sectors for civilian applications of defence-related technologies.[7] Furthermore, market demand, though increasing, is still, on the average far below OECD norms in per capita terms, constrained by growing inequality and social polarization.

There is a consensus around what kind of actors are involved in fostering knowledge-based activities. However, it is unclear as to what the internal make-up of these institutional actors is, how they might relate to each other, for how long, and the conditions under which inter-institutional interactions take place. It is also not clear whether the state should always prevail, though intuitively we know that knowledge-based activities demand far greater effort in terms of infrastructure development and policy directions, which most private sectors in developing countries find daunting. Only a few governments have successfully fostered innovations. If, as some argue (Saxenian, 2006, 1994), geography matters, making the innovation milieu of Silicon Valley 'sticky', then how do developing countries such as China and India, who have a growing supply of talent, create the national institutional basis for innovation-led growth?

One answer to this barrier is to make strategic policy choices regarding the sources of innovation. However, various challenges remain. Should developing countries rely fully on foreign sources for the transfer of technology or leverage MNCs through FDI to secure technology. Singapore followed the latter path (Parayil, 2005). Avoiding MNCs in this age of globalization is nearly impossible and both China and India have leveraged their large markets for foreign technology and FDI.

However, the consequences of the structural power of MNCs and large OECD markets such as the USA should not be underestimated (D'Costa, 2004a, 2002). For example, the Indian IT industry is heavily dependent on the US market (nearly 65 per cent of exports) and closed off from the second largest IT market in Japan (with barely 3 per cent of exports) (D'Costa and Kobayashi Chapter 9). Hence, the locking in of innovation capabilities due to a lack of geographical and product diversification cannot be ruled out (D'Costa, 2004b). Structural power also works through higher international prices for drugs, which compel companies in developing countries to cater to lucrative OECD markets. This could unwittingly foreclose the development of essential drugs for poor communities in China and India (Thomas Chapter 5). There are other problems associated with too great a reliance on MNCs or export markets for stimulating innovation. For example, high licensing fees, royalty payments, repatriation of profits and dividends associated with FDI entail outflows of foreign exchange.

China, whose foreign exchange reserves today are around $1 trillion, has not been satisfied with the strategy of foreign 'technology for market' approach as there have been few technological spillovers in the domestic market (Schwaag Serger, Chapter 3).[8] Betting on its large market, China is thus toying with alternative approaches such as 'indigenous innovations' and its own technical standards. If the MNC outsourcing of knowledge-based activities is premised on internal resource reallocation, then one could argue that MNCs retain core activities, which rely on home country ecosystems and thereby structurally limit technology spillovers in the host economies.[9] At the same time, as WTO norms and related Trade Related Intellectual Property Rights (TRIPS) kick in, Asian countries are expected to be prohibited from reverse engineering technologies and may become mired in legal battles for copyright infringements or patent violations (Chaturvedi and Chataway, Chapter 6; Thomas, Chapter 5).

MNCs also introduce distortions in labour markets by inducing rapid wage growth (D'Costa, 2006). The ability of MNCs to provide higher salaries, better working conditions, and the possibility of global assignments attracts disproportionately high calibre local talent. While labour turnover may be high in general, such professionals tend to circulate among MNC employers (Schwaag Serger, Chapter 3). Similarly, in India there is poaching of Indian IT talent by both MNCs and large domestic firms leaving smaller domestic firms settling for second best talent.[10] This is no doubt a market outcome and may not be an issue if the high industry growth is matched by growth in high-quality educational infrastructure. But there is evidence of quality problems as illustrated by talent shortages in both China and India (see Bound 2007:22; NASSCOM 2004). There is also considerable talent flight in India – from non-IT sectors to the IT sector due to higher remuneration (D'Costa 2003a). Of particular concern is the unattractiveness of pursuing PhDs in technical fields due to lucrative IT industry jobs (see Table 1.1 for the low number of PhDs awarded in India). Over the long haul this is likely to result in shortages of well-qualified faculty in the very areas that contribute to an innovation ecosystem (Bound, 2007:16, 23, 36).[11]

The structural power of MNCs and global market dynamics constraining innovative capability are often complemented by the general weakness of NIS/THM architecture in most Asian countries. Even India's IT industry growth does not rest on strong synergistic inter-firm collaboration and business–university relations (D'Costa, Chapter 4; McManus, Li, and Moitra, 2007; Lema and Hesbjerg, 2003). Government research institutes are neither administratively nor organizationally geared towards working with the private sector (Sridharan, 2004, 1995; Parthasarathy, 2004). In general, there is a lack of an innovation ecosystem in Asian countries (Bound, 2007: 14, 19). South Korea, which leapt into high-technology sectors by following on the heels of Japan, suffers from weak university R&D (Lim, 2006). China's science base is still weak, which is partly reflected in its difficulty in luring back overseas Chinese talent (Cao, Suttemier and Simon, Chapter 10), while Singapore's doctoral programs are small and hence cannot be a major source of university-based innovations.

The difficulty in establishing an effective NIS or THM is heavily circumscribed by conditions of 'peripheral capitalism' whereby leapfrogging in some sectors is possible without fundamentally changing the structure of the economy. This translates into twenty-first-century globalized sectors coexisting with production units still using obsolete technologies and organizational practices. These sectors are disjointed, with the former group experiencing increasing returns while the latter languishes in diminishing returns (Parayil, 2005). While the supply side of innovations is no doubt critical, and targeted investments can promote them, without a matching demand base it becomes hard to sustain the innovative momentum. Inequality, income polarization, and uneven development are rampant in India and China (D'Costa, 2003a, 2003b; Basu, 2006; Khan and Riskin, 2000). As a recent Forbes listing put it, India now has more billionaires than Japan, with India adding '14 new billionaires since last year to bring its total on the list to 36 with a combined wealth of US$191 billion, . . . , contrasting sharply with the 400 million Indians who still live on less than a dollar a day' (International Herald Tribune, 2007). 80 per cent of India's population lives on $2 a day (UNDP, 2003). This lack of wider demand can act as India's Achilles heel in the pursuit of sustained economic growth and innovative capability.

The unequal distribution of economic benefits during the recent economic boom in India thanks to its high-technology industries has been realistically captured by Amartya Sen in his statement: 'A democratic country can hardly want to maintain a divisiveness that makes it part California and part sub-Saharan Africa' (Sen, 2007a). The reference to sub-Saharan Africa is prompted by the fact that there are more malnourished people in India than in sub-Saharan Africa.[12] Thus the biotechnology industry in large poor countries

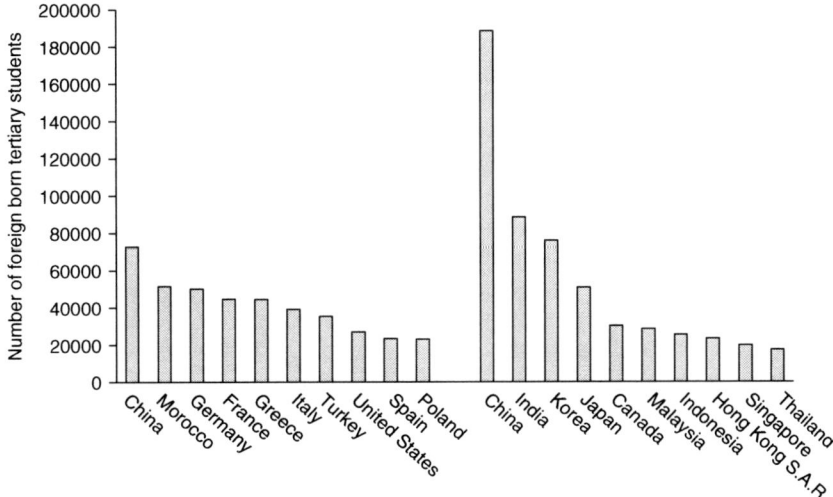

Figure 1.3 International tertiary students in OECD Europe and outside of Europe, 2003

Source: OECD 2006.

such as India and China has much commercial and scientific promise but it is also laced with social and ethical dilemmas (Thomas, Chapter 5). In the absence of serious distributive policies a few highly dynamic innovation clusters are likely to remain as enclaves and cater to better-off social classes.

The strength of China, India, and other Asian economies is in human capital endowments (D'Costa, 2008). However, here too there are challenges in either keeping some talent at home for domestic innovation or generating more. As production becomes globally decentralized the demand for professionals also becomes globalized. With demographic shifts in OECD economies, resulting in low fertility rates and ageing populations, the demand for young technical professionals is high and is expected to grow in the future (D'Costa, 2004c). Already there are shortages of technical labour in many markets as evidenced by considerable international movement of people with tertiary education (Figure 1.3). A large number of students from the developing world travel to OECD countries to study and a sizeable fraction remains in the OECD economies. This suits rich receiving countries such as the USA well as they secure ready-made talent trained elsewhere. The recent call made by Bill Gates to admit more foreign workers under the H1 B (employment) visa programme to the USA is a testimony of the insufficiency of local talent availability even in the USA.

Large economies such as China and India do not necessarily suffer from brain drain. However, China, because of its one-child policy, has an ageing society and thus emerging shortages of talent. While India could reap its demographic dividend with a much larger and younger population compared to China, talent shortages are still expected in certain markets (D'Costa and Kobayashi, Chapter 9).

Given shortages of skilled workers, a number of countries will be challenged to sustain their innovative activities (see Brown, 2001). While China and India are likely to witness shortages in specific product and service markets within knowledge-based activities, smaller OECD countries in Scandinavia and rapidly ageing Japan will be stretched thin (D'Costa, 2007). For example, Denmark established a specialized Information Technology University in Copenhagen and welcomes highly trained researchers and scientists.[13] But to maintain international competitiveness Danish firms have begun R&D operations in both China and India (Schwaag Serger, Chapter 3; Kjersem and Gammeltoft, Chapter 7). Singapore has aggressively pursued multi-pronged policies of recruiting foreign talent and outward and inward foreign investments. European countries have been slow on this front due to their reluctance to destabilize the social contract of a welfare state and because high-quality workers are still available regionally. But there are other large OECD economies such as Japan and Germany, which have been ageing rapidly but have not been very open to foreign workers. These countries may face even graver problems as they have a sizeable part of their GDP attached to high technology activities.

Japan's reliance on Chinese workers currently may not be sustainable for the long haul since the demand for skills in China itself is expected to expand rapidly (D'Costa and Kobayashi, Chapter 9). Japan, Germany, and other European countries will not only have to rely on foreign talent but also diversify their sources for such workers (D'Costa, 2008; D'Costa, in Wiegand, 2007). China, India, and other developing countries will also have to continue to invest in technical educational infrastructure to meet their own and global demand, which of course would reduce investment funds for universal education and physical infrastructure, a necessary part of an innovative society.

There are many other challenges to the internationalization of knowledge-intensive activities such as high labour turnover and costs and thus inflationary pressures. Suffice to say, cultural and institutional problems in interfacing between foreign and domestic partners and institutions are routine (Kjersen and Gammeltoft, Chapter 7; D'Costa and Kobayashi, Chapter 9). High growth places undue pressure on the existing infrastructure (D'Costa, Chapter 4). China has a far better record in building infrastructure ahead of demand, while India remains mired in political indecisiveness and decrepit infrastructure. There are growing fears that poor infrastructure in India may stifle growth (Hamm and Lakshman, 2006) as well as a scarcity of highly skilled knowledge workers, known as the 'Bangalore bug' (Rajan and Subramanian, 2006). Given the multiple challenges faced by China, India, and other Asian countries the question is how best to participate in the globaliza-tion of knowledge-intensive activities. We take up this issue below by looking at some broad policy options for strengthening the innovation system in a global context.

1.5 Broad outlines of innovation and social policies

No doubt there are considerable tensions between domestic sources of inno-vation and MNC-driven internationalization of innovation, especially since technology spillovers, as in the Chinese case, have been seen as limited. Scepticism about MNCs may be also driven by techno-nationalist sentiments (Cao, Suttmeier and Simon, Chapter 10). While India has recently embraced internationalization enthusiastically and successfully in the IT sector, many structural challenges remain (D'Costa, 2003c). Despite the massive advantage of China and India over most countries in raw human capital, both suf-fer from a weak institutional architecture behind their innovation systems (D'Costa, Chapter 4). More importantly, the growing disparity in incomes and wealth in China and India may drag down their innovation potential as innovation is driven by a wide and deep user base. Under these conditions, forging and strengthening the interactions among THM institutions will be necessary (Vang, Chaminade and Coenen, Chapter 8). At the same time, social and technology policies aimed at enhancing income and employment for the poor will go a long way in sustaining broad-based development.

It is obvious that human capital today is a critical factor in innovation. However, in order to utilize its full capabilities an institutional makeover is necessary. This includes, but is not limited to, greater autonomy of GRIs and more collaboration between industry and university. Both China and India have talented people but they also have legacies of a top-down approach to research. China is trying to transition from a state-dominated research sys-tem to an enterprise-based innovation system (Liu and Lundin, Chapter 2). Chinese universities have been allowed to form commercial spin-offs suggest-ing a degree of enterprise autonomy regarding investment and technology decisions. However, firm-level entrepreneurialism will be difficult in China with a cultural propensity that does not encourage independent decisions in a hierarchical environment. India poses different kind of legacy challenges. Unlike their Chinese counterparts, Indian universities legally cannot initiate commercial enterprise.[14] The paradox is that Indian firms have been quite entrepreneurial but rarely collaborative as implied by a functioning THM framework. Sharing of specialized expertise is integral to tackling complex-ity, suggesting that the incentive structure industries face must encourage inter-institutional collaboration.

Private Indian firms have demonstrated their mettle in initiating many new businesses with foreign partners. But they remain risk averse. For example, in biotechnology India secures pre-competition and contract research, both of which are considered less risky (Abma et al., Chapter 11). Similarly, the Indian IT industry has found it convenient to stay with customized services for western markets, thereby failing to seriously tap the large and dynamic East Asian markets, especially Japan (D'Costa, 2004a). They have not demonstrated their willingness to cooperate with each other or with the established GRIs, who seem to be stymied by legacy problems of their own such as overt security concerns or jurisdictional conflicts over private and public domains.

Both China and India have responded well to the global demand for technical talent, although the quality of education remains an issue. In India both state and private sectors have invested in educational and technological infrastructure. However, from an NIS/THM point of view they are still inadequate in terms of research orientation. Barring a few institutions of higher learning, the Indian educational system is essentially geared towards creating raw technical talent. There is low-key research activity, as evidenced by few engineering PhDs, a weak publications profile, and little original research (Dahlman and Utz, 2005). Although China has a better record than India on these indicators, it too has shortcomings, especially with respect to research-based innovation. Incentives for Indian firms to undertake research are low, as illustrated by the small number of Indian firms engaged in research (Bound, 2007: 33). Creating a research culture is not easy since resources, people, and an ecosystem of institutional interfaces are necessary. At the minimum, and for the long term, the reorientation and reorganization of university systems should be designed to attract talent for research, which could then serve specific industry needs.

It is evident that redesigning the institutional architecture must be high on the agenda since an ecosystem to foster research is a major weakness. Talent creation, which both China and India generally enjoy, must be specifically consistent with this. One of the advantages China, India, and other Asian countries possess, in the absence of a home-grown research ecosystem, is an expatriate talent pool. For example, Indians comprise 14 per cent of the three million foreign-born science and engineering graduates in the USA, of which 300,000 are doctorates (National Science Board, 2006). Over the years the migration of students to pursue technical education in OECD economies but especially to the USA, provides a ready-made brain bank for sending countries (D'Costa, 2008). Since creating an ecosystem is challenging, one avenue could be to tap into overseas talent for national R&D capability. This would call for an open and vibrant research and innovation ecosystem. Of course, there is a tension between spending to attract expatriate talent versus bringing existing domestic talent up to global standards. However, there is already increasing evidence of a reverse flow of talent in a number of countries, including China and India.[15]

In revamping the educational infrastructure, there is of course no reason why foreign students could not study in Indian or Chinese universities. Since innovation is generally an open-ended process with a considerable exchange of knowledge and mutual cultural appreciation, it would serve the host economies well to support a research ecosystem that draws on national and foreign talent. A considerable number of foreign students, especially from China and India, already study in the USA (D'Costa and Kobayashi, Chapter 9). Within Asia, there is some evidence of students from South-East Asia and Japan studying in India. In Japan students from China, Taiwan, and Korea are present and local prefectural governments in Japan have begun recruiting Chinese students to fill talent shortages (D'Costa and Kobayashi, Chapter 9).[16] These cross-national exchanges at the university level are likely to challenge developing country universities to benchmark themselves to global standards, encouraging institutional collaboration at various levels, domestically as well as internationally.

Open systems would also entail commercial and technological cooperation between India and China. The recent establishment of the India–China Business Alliance is an illustration (Hindu Businessline, 2006). While they compete in some areas, there are complementarities between the two countries. China's strength in high-technology manufacturing and India's success in the global software industry along with their large markets to each other are important for rising bilateral trade. Large Chinese high-technology firms are already visibly present in India, with Huwaei, ZTE, and Haier all having a presence in Bangalore (Liu and Lundin, Chapter 2). Large Indian IT companies such as Satyam, Infosys, TCS, and WIPRO are also located in Beijing, Dalian, and Shanghai in China (D'Costa and Kobayashi, Chapter 9). Not only is the Chinese market attractive for Indian IT companies but Chinese engineers have a cost advantage over Indian engineers. Finally, Chinese firms are tied to the Japanese IT industry and Indian firms would like to get a piece of the large Japanese market by using Chinese professionals and geographically being close to Japan. There are also several areas in biotechnology where both countries could work together, given the importance of low-cost drugs and treatment in their countries (Thomas, Chapter 5).

Since employment is the principal preoccupation of most OECD governments and since rich countries are unable to compete in low-wage goods and services, yielding knowledge-intensive activities to Asian competitors is often seen as a national crisis. However, the OECD firms can insert themselves in the current new Asian innovation dynamics to strengthen their NIS/THM. One avenue is to reorganize research and education through international outsourcing of R&D (UNCTAD, 2005). In the medium term, a knowledge-based division of labour between China and India with large talent pools and smaller European economies specializing in R&D, particularly in integrating knowledge at different levels seems appropriate (Abma et al., Chapter 11).

The two-way process of enhancing innovation systems implies engaging China, India, and other Asian countries in large-scale basic research on a contract basis. This also means that the European and Asian OECD economies will have to revamp their research system to attract talent from Asia and away from the USA, the 'world's largest skill magnet' (Lowell in Ackers, 2005: 100). To facilitate talent mobility, more relaxed immigration policies will have to be designed as the demand for high skills in IT, biotechnology, agriculture, life sciences, and nanotechnology will increase. In smaller OECD countries limited home markets and scarcity of skilled workers, compounded by restrictive immigration policies and declining fertility rates, will dampen innovation possibilities.

Knowledge-intensive activities require scientists and engineers. However, for talent-rich but impoverished India and China, public spending on social welfare is inadequate. In their quest for techno-nationalism, the bias toward middle-class aspirations often outstrips the immediate, more mundane forms of political commitment for social transformation. In an integrating world economy, such social outcomes are largely left to the devices of trickle-down economics. Even China has dismantled many of its social protections. What this suggests is that without broader social development, their innovative capability will remain narrow and exclusivist. In addition to renewed public commitment for social policies successful Chinese and Indian businesses ought to take a leadership role on the education front, especially in the area of universal education system at the primary and secondary levels. The plea made by Amartya Sen in his keynote address to NASSCOM, India's software association, to contribute to India's universal education is noteworthy (Sen, 2007b). Without such an emphasis the enclave nature of internationalized high-technology activities, which excludes the majority of the population, cannot be ethically or politically sustained.

Public health is another area where Asian innovation dynamics can have a major impact. Considerable resources are devoted globally for new drugs and treatment. However, since the incentives are market-driven, there is an income bias that limits the availability of treatment in poor societies. Consequently, in the absence of any meaningful social programmes, developing countries face dual healthcare systems – one for the affluent and the other for the less fortunate. The current patent rules are flawed when it comes to life-saving drugs, making an alternative reward system necessary (Stiglitz, 2007). States also have a responsibility to promote research that directly enhances the welfare of the people, especially in generic drugs and treatments for tropical diseases. Finally, as new research and treatments pose ethical and moral dilemmas, such as stem cell research and genetic engineering, more civic participation and public debates are crucial. At one level India is in an advantageous position due to its openness to public discussion on innovation and social policies, while in China such debates may be suppressed for political and pragmatic reasons. Clearly on matters of social policy involving science and technology an ecosystem that also encourages debate and dialogue will be inherently valuable.

Science, technology, and innovation are in the end social activities influenced by politics. However, the social and thus political nature of these commercially-driven activities calls for a more thoughtful approach to generating and sustaining them – whether in terms of national priorities, resources allocated, or policy directions. Governments, businesses, and universities as well as civic societies have important roles to play to foster innovations in a changing global economy and also have a responsibility for the well-being of their citizens. If the cards are played right there is no reason why leading Asian countries cannot leverage the new innovation dynamics in their favour and sustain a more inclusive form of knowledge-led development.

1.6 Chapter outlines

Chapter 1 by D'Costa and Parayil introduces some of the key themes surrounding the new Asian innovation dynamics. It identifies the principal economic and business forces and actors at work at the national, regional, and global levels in both China and India. It introduces the importance of interactions among various institutions behind technology and innovations and the ways by which strategic choices made by multinational firms, local businesses, and governments influence innovation dynamics. It ends with policy options for overcoming innovation challenges faced by both OECD and non-OECD economies in an era of globalization.

In Chapter 2, Liu and Lundin sketch the contours of the Chinese innovation system and its progressive transformation in the last two decades. The change from a research institute-dominated innovation system to an enterprise-centred system has strengthened industry–science linkages. But with globalization the Chinese innovation system has become an open one, producing both new opportunities and daunting challenges. In such a dynamic and open innovation system, Liu and Lundin argue, the Chinese national strategy of indigenous innovation will be influenced by global linkages and partnerships.

Schwaag Serger takes up the issue of global linkages in Chapter 3 by examining foreign R&D activities in China. She analyzes their impact and role on China's innovation system and shows that foreign R&D activities have been rising rapidly in recent years. She argues that foreign firms going beyond product development or adaptation for the Chinese market to establish strategic or global R&D centres in China are beneficial for the local economy. However, so far the benefits of positive spillover from foreign firms' R&D activities have not been fully realized and thus there is growing domestic criticism of preferential policies for foreign R&D investments. Overall, Schwaag Serger argues the increase in foreign R&D in China is symptomatic of a more profound structural change within the broader innovation dynamics in Asia and beyond.

In Chapter 4, D'Costa disputes that there is strong synergy among local institutions, as implied by the sheer physical presence of multinational high-technology firms in R&D, numerous universities, engineering colleges, management and public policy institutes, and government research institutes. Relying on the triple helix model, D'Costa suggests that Bangalore's (and India's) IT industry is based on an extensive growth model, which does not encourage thick institutional linkages such as those encapsulated by the interactions between industry, academia, and government. The chapter establishes the empirical basis for such an outcome without foreclosing the possibility of generating a new trajectory for software growth. A new pathway could be forged by leveraging India's technical talent both at home and abroad for R&D in a revamped university and government research system, along with strong university–industry linkages. Intensive growth driven by science, technology, and innovation, social inclusion, and export market diversification towards East Asia are expected to place India on a more solid innovation platform.

Thomas in Chapter 5 examines the prospects of innovation in India and China by analyzing the pharmaceutical and biotechnology industries. The context for examining these industries is the globalization of pharmaceutical R&D and post-TRIPS intellectual property regimes. According to Thomas, globalization is constraining the future supply of affordable medicines for the poor and yet China and India are providing commercial opportunities to multinational companies to reduce costs through contract research and new drug markets in India and China. Thomas argues that rather than competing by cutting wage costs, India and China need to take the lead in developing innovative products that would also benefit the poor in the developing world and not just the middle classes who benefit from globalization.

Overlapping with Thomas, Chaturvedi and Chataway in Chapter 6 examine the experience, growth, and evolution of the Indian pharmaceutical industry. As an emerging high-technology sector that demands considerable scientific and technological inputs for innovations, they look at firms in the pre-TRIPS patent regimes as a prelude to investigating the current strategies of leading Indian firms. Suggesting that firms have responded well to policy changes, Chaturvedi and Chataway argue that favourable public policies for an 'enabling environment' have been complemented by entrepreneurial initiatives by Indian firms in clinical research, custom manufacturing, and R&D-led services. Thus firm-level strategies are critical not only in 'making the policy work' but, more importantly, in the 'making of the policy itself'.

Following Schwaag Serger's understanding of China becoming a global R&D centre, Kjerstem and Gameltoft in Chapter 7 argue that multinationals who offshore knowledge-intensive activities to China will often encounter unforeseen difficulties in the exchange and protection of knowledge between geographically and organizationally dispersed R&D units. They analyze the emerging phenomenon of foreign investments in R&D in China approached both from the supply side of the evolution of the Chinese system of science, technology and innovation (ST&I), and from the demand side of foreign firms investing in high-tech activities in China. The latter is discussed through a case analysis of the Chinese R&D labs of Novo Nordisk, GN Resound, and BenQ Siemens.

Echoing D'Costa, Chaminade et al., in Chapter 8, analyze the role of universities and government research institutes (GRIs) in initiating and sustaining the development of regional innovation systems in Asian countries. They specifically focus their discussion on the software industry cluster in Bangalore, India. By reviewing innovation systems research, they argue that the specific context in which interaction between university and industry takes place is ignored, especially in developing countries. This chapter aims to reduce this gap by making an empirically-based analysis of the role universities can play in initiating, sustaining, and deepening Bangalore's regional innovation system for the IT service and software industry.

Taking off from the triple helix model in which technical professionals become critical to the innovation process, D'Costa and Kobayashi in Chapter 9 examine the Japanese software industry in light of impending labour shortages. Drawing on the understanding that the international mobility of talent contributes to epistemic communities they investigate the pattern and magnitudes of Chinese and Indian talent flows to Japan. They explain the divergent patterns of engagement of the two countries but conclude that flows of talent from China and India will not only increase but that such flows will be beneficial for all three countries in enhancing their innovative competencies.

In Chapter 10, Cao et al. revisit China's innovation policies. They provide a detailed exposition of the Medium to Long-Term Plan for the Development of Science and Technology (2006–2020) introduced by the Chinese central government in 2006 to turn China into an 'innovation-oriented society' by the year 2020. According to Cao et al., the plan commits China to the development of capabilities for 'indigenous innovation' (*zizhu chuangxin*) and to 'leapfrogging' into leading positions in new science-based industries by the end of the plan period. China plans to invest 2.5 per cent of its burgeoning GDP in R&D by 2020, up from 1.41 per cent in 2006. In this chapter, Cao et al. review the objectives of the plan and some of the subsequent implementing measures, and explore whether state directed innovation is likely to be a winning strategy for twenty-first-century technological development.

Based on the records of international R&D companies in China and India, Abma et al. in Chapter 11 show that outsourcing R&D has both negative and positive effects on the innovation cycle. The production of basic scientific knowledge can improve, but activities concerning the integration of scientific knowledge into marketing and corporate strategy are weaker. At this time, according to Abma et al., the outsourcing of R&D still seems an experiment for most multinational companies but is likely to lead to new transnational production strategies. On a global scale, continental specialization in R&D (and management of science and knowledge integration) is possible, while Asia can become a centre for knowledge production. This future scenario has major implications for technology policy and the education of scientists in OECD countries.

The final chapter by the editors concludes by synthesizing the key themes running through the individual chapters. It relies on the new Asian innovation dynamics discussed in each chapter and raises a number of questions for further research. Are China and India following similar or divergent innovation trajectories? Are they complementary or competing? What are the key common innovation patterns in these two emerging giants, especially in terms of sectoral developments, their successes and failures? Finally, what are the global opportunities and challenges that China and India face and how could they leverage their human resources and institutional systems to sustain their innovation capabilities?

Notes

1. We thank Janette Rawlings for her editorial support.
2. There are, of course, many differences between India and China. Pertinent to the discussion of innovation India's IPR regime is considered better than China's, although China is working to improve it. China is ahead of India in economic reforms and thus international economic integration by a decade and has ten times the FDI compared to India. China has far more university spin-offs than India, while Indian firms have demonstrated a greater entrepreneurial development in the IT industry. China remains state dominated yet maintains a non-ideological position when it comes to the economy, while the Indian state parlays varying interests and retains considerable control over science and technology policy issues.
3. We recognize the difference between national innovation system (NIS) and regional innovation system (RIS). We do not dwell on the differences here but leave it for individual authors to address them in their chapters if relevant to their theoretical approaches. For a critical discussion of the differences between the two see Acs et al. (2000).
4. D'Costa, Field Visit, Zhongguancun Science Park, Beijing, November 2007.
5. We use knowledge-intensive and high-technology interchangeably.
6. We recognize that NIS/THM operates at the national level, but often policies are sector-based and outcomes are at the regional/city levels. However, as there is greater interconnectedness between different branches of formal knowledge a sector becomes wider than is normally associated. Thus creating talent for the IT industry involves government, business, and universities at the national level, which could be broadened to include policies for high technology sectors and R&D in the life sciences.
7. For example, both countries have developed satellite and other communications technologies but there have been few applications of these technologies in various social sectors such as agriculture (weather forecasting).
8. Of course, it does not imply that spin-offs are impossible. A former Texas Instruments manager in India established Ittiam (I think therefore I am) in 2001 in Bangalore, after having worked for TI for many years (Bound, 2007: 35). In China the Chinese Academy of Sciences and Beijing University have successfully

spawned high-technology firms. It is evident that private entrepreneurship in India is strong, while China's state-dominated system is also able to foster new enterprises.

9. D'Costa, Field Visit to Bangalore, February 2005.
10. But see Liu and Buck (2007), who do find spillovers from MNC FDI in China. But the recent decision by Intel to build a chip fabrication plant in Dalian, China reveals structural dependence of non-OECD economies on foreign technology (Flynn, 2007: C3). By the time Intel's Chinese plant comes into operation the chip making technology will have already fallen behind by two generations of what Intel will produce in its home economy.
11. D'Costa, Personal Interview with Chancellor of Vellore Institute of Technology, Tokyo, February 2006.
12. India's impoverished masses can also promote knowledge-based life sciences because of lower clinical trial costs and the promise of large markets for specific drugs. For example, there are 32 million diabetics in India alone (Bound, 2007: 43). The effect of lower trial costs is lower drug development costs (Chaturvedi and Chataway, Chapter 6). One estimate put Indian costs at $50 million versus $1 billion elsewhere in the developed world (Bound, 2007: 34).
13. D'Costa, Personal Interview with Rector and Pro-rector, IT University, Copenhagen, February 2006.
14. This is a situation that needs urgent remedy according to India's National Knowledge Commission (NKC), which makes policy recommendations to the Indian Prime Minister on revamping India's educational system, among other issues. See NKC's recommendation on GRIs and patents at http://www. knowledgecommission.gov.in/downloads/news/news10.pdf (accessed 27 March 2007).
15. The Indian government recognizes the importance of its overseas diaspora by giving 15-year visas to persons of Indian origin (PIO). It recently instituted the Overseas Citizen of India (OCI), which provides a life-long visa. Taiwan and South Korea have successfully attracted back their overseas students (Saxenian 2004, Lim 2006). With economic dynamism in China and India we anticipate increasing rates of return migrants.
16. D'Costa, Research Visit, Kitakyushu Science Park, Japan, June 2005.

References

Ackers, L. (2005) 'Moving People and Knowledge: Scientific Mobility in the European Union', *International Migration*, 43(5): 99–131.

Acs, Z.J., J. de la Mothe and G. Paquet (2000) 'Regional Innovation: In Search of an Enabling Strategy', in Zoltan J. Acs (ed.), *Regional Innovation, Knowledge and Global Change*, London and New York: Pinter, pp. 37–49.

Allarakhia, M. and A. Wensley (2007) 'Systems Biology: A Disruptive Biopharmaceutical Research Paradigm', *Technological Forecasting & Social Change*, 74(9): 1643–60.

Basu, K. (2006) 'Globalization, Poverty, and Inequality: What is the Relationship? What Can be Done?', *World Development*, 34(8): 1361–73.

Bound, K. (2007) *India: The Uneven Innovator*, London: Demos.

Brown, P. (2001) 'Skill Formation in the Twenty-First Century', in P. Brown, A. Green and H. Lauder (eds), *High Skills: Globalization, Competitiveness, and Skill Formation*, Oxford: Oxford University Press, pp. 1–55.

Chen, K. and M. Kenny, (2005) 'Universities/Research Institutes and Regional Innovation Systems: The Cases of Beijing and Shenzen'. Paper Presented at Universities as Drivers of the Urban Economies in Asia, World Bank and Social Science Research Council, Singapore, 24–5 May.

Cypher, J.M. and J.L. Dietz (2004) *The Process of Economic Development*, 2nd edition, New York: Routledge.

D'Costa, A.P. (2008) 'The International Mobility of Technical Talent: Trends and Development Implications', in A. Solimano (ed.). *International Mobility of Talent and Development Impact*, Oxford: Oxford University Press, pp. 44–83.

D'Costa, A.P. (2007) 'The Looming Labour Shortage: Can India Meet Japan's Technical Worker Needs?', *Ekonomisuto* (*The Weekly Economist*, in Japanese), January.

D'Costa, A.P. (2006a) 'ICTs and Decoupled Development: Theories, Trajectories and Transitions', in G. Parayil (ed.), *Political Economy & Information Capitalism in India: Digital Divide, Development and Equity*, Basingstoke: Palgrave Macmillan, pp. 11–34.

D'Costa, A.P. (2004a) 'The Indian Software Industry in the Global Division of Labour', in A.P. D'Costa and E. Sridharan (eds), *India in the Global Software Industry: Innovation, Firm Strategies and Development*, Basingstoke: Palgrave Macmillan, pp. 1–26.

D'Costa, A.P. (2004b) 'Export Growth and Path-Dependence: The Locking-in of Innovations in the Software Industry', in A.P. D'Costa and E. Sridharan (eds), *India in the Global Software Industry: Innovation, Firm Strategies and Development*, Basingstoke: Palgrave Macmillan, pp. 51–82.

D'Costa, A.P. (2004c) 'Globalization, Development, and the Mobility of Technical Talent: India and Japan in Comparative Perspectives', UN University, World Institute of Development Economics Research, Helsinki, Research Paper Series (WIDER RP2004/62). Accessed at http://www.wider.unu.edu/publications.

D'Costa, A.P. (2003a) 'Uneven and Combined Development: Understanding India's Software Exports', *World Development*, 13 (1): 211–26.

D'Costa, A.P. (2003b) 'Catching Up and Falling Behind: Inequality, IT, and the Asian Diaspora', in K.C. Ho et al. (eds) *Asia Encounters the Internet*, London: Routledge, pp. 44–66.

D'Costa, A.P. (2003c) 'The Indian Software Industry in the Global Division of Labour', in A.P. D'Costa and E. Sridharan (eds), *India in the Global Software Industry: Innovation, Firm Strategies and Development*, Basingstokes: Palgrave Macmillan, pp. 1–26.

D'Costa, A.P. (2002) 'Software Outsourcing and Policy Implications: An Indian Perspective', *International Journal of Technology Management*, 24(7/8): 705–23.

Dahlman, C. and A. Utz (2005) *India and the Knowledge Economy: Leveraging Strengths and Opportunities*, Washington, DC: The World Bank.

Flynn, L.J. (2007) 'Intel, Already With Operations in China, Appears Ready to Build a Chip Plant There', *New York Times*, 23 March.

Hamm, S. and N. Lakshman (2007) 'The Trouble with India: Crumbling Roads, Jammed Airports, and Power Blackouts could Hobble Growth', *Business Week Online*, 19 March.

Hindu (2007) 'Education, Healthcare need Greater Government Investment: Amartya Sen', http://www.thehindu.com/2007/02/14/stories/2007021406991400.htm. Accessed on 14 February 2007.

Hindu Businessline (2006) 'India–China Business Alliance Launched, March 13', www.moneycontrol.com. Accessed on 15 March 2007.

International Herald Tribune (2007) 'Viewpoint: Following the Money in Asia', http://www.iht.com/articles/2007/03/12/bloomberg/sxpesek.php. Accessed on 3 December 2007.

Khan, A.R. and C. Riskin (2000) *Inequality and Poverty in China in the Age of Globalization*, New York: Oxford University Press.

Lema, R. and B. Hesbjerg (2003) *The Virtual Extension: A Search for Collective Efficiency in the Software Cluster in Bangalore*, Roskilde: University of Roskilde, Public Administration and Public Economics & International Development Studies.

Lim, C. (2006) 'Korean National System of Innovation and FDIs'. Paper presented at the International Conference of the European Alliance for Asian Studies, Institute of Asian Affairs and the German Asia-Pacific Business Association, Hamburg, Germany 17–19 March.

Liu, X. and T. Buck (2007) 'Innovation Performance and Channels for International Technology Spillovers: Evidence from Chinese High-tech Industries', *Research Policy*, 36(3): 355–66.

McManus, J., M. Li and D. Moitra (2007) *China and India: Opportunities and Threats for the Global Software Industry*, Oxford: Chandos Publishing.

NASSCOM (2007) 'Indian IT Industry: NASSCOM Analysis', http://www.nasscom.in/upload/5216/IT%20Industry%20Factsheet%20-%20Sep%2007.pdf. Accessed on 17 December 2007.

NASSCOM (2004) *The IT Industry in India: Strategic Review 2004*, New Delhi: NASSCOM.

National Knowledge Commission (2007) 'Knowledge Commission Recommends Enactment of Law on Patent Rights', http://www.knowledgecommission.gov.in/downloads/news/news10.pdf. Accessed on 27 March 2007.

National Science Board (2006) *Science and Engineering Indicators 2006*, vol. I, Washington, DC: US National Science Foundation.

Niosi, J. and S.E. Reid (2007) 'Biotechnology and Nanotechnology: Science-based Enabling Technologies as Windows of Opportunity for LDCs?', *World Development*, 35(3): 426–38.

OECD (2006) *Trends in International Migration*, SOPEMI 2006. Paris: OECD.

Parayil, G. (2005) 'From "Silicon Island" to "Biopolis of Asia": Innovation Policy and Shifting Competitive Strategy in Singapore', *California Management Review*, 47(2): 50–73.

Parthasarathy, B. (2004) 'Globalizing Information Technology: The Domestic Policy Context for India's Software Production and Exports', *Iterations: An International Journal of Software History*, May: 1–38.

Rajan, R. and A. Subramanian (2006) 'India Needs Skill to Solve the "Bangalore Bug"', *Financial Times*, 17 March, p. 11.

Saxenian, A. (2006) *The New Argonauts: Regional Advantage in a Global Economy*, Cambridge, MA: Harvard University Press.

Saxenian, A. (2004) 'The Silicon Valley Connection: Transnational Networks and Regional Development in Taiwan, China, and India', in A.P. D'Costa and E. Sridharan (eds), *India in the Global Software Industry: Innovation, Firm Strategies and Development*, Basingstoke: Palgrave Macmillan, pp. 164–92.

Saxenian, A. (1994) Regional *Advantage: Culture and Competition in Silicon Valley and Route 128*, Cambridge, MA: Harvard University Press.

Sen, A. (2007a) 'Can Life Begin at 60 for the Sprightly Indian Economy?', *Financial Times*, London 14 August.

Sen, A. (2007b) 'IT and India', Keynote Address at the NASSCOM 2007 India Leadership Forum in Mumbai on 7 February 2007, The Hindu Online edition at: http://www.hindu.com/nic/itindia.htm. Accessed on 8 March 2007.

Sridharan, E. (2004) 'Evolving Towards Innovation? The Recent Evolution and Future Trajectory of the Indian Software Industry' in A.P. D'Costa and E. Sridharan (eds), *India in the Global Software Industry: Innovation, Firm Strategies and Development*, Basingstoke: Palgrave Macmillan, pp. 27–50.

Sridharan, E. (1995) 'Liberalization and Technology Policy: Redefining Self-Reliance', in T.V. Sathyamurthy (ed.), *Industry and Agriculture in India since Independence*, New Delhi: Oxford University Press, pp. 150–88.

Stiglitz, J. (2007) 'Dying in the Name of Monopoly', www.businessday.co.za/articles/topstories.aspx?ID=BD4A407148, Accessed 3 April 2007.

UNCTAD (2005) *World Investment Report: TNCs and the Internationalization of R&D*, New York: UNCTAD.

UNCTAD (2004) *World Investment Report: The Shift Toward Services*, New York: UNCTAD.

UNDP (2003) *Human Development Report 2003: Millennium Development Goals*, New York: Oxford University Press.

Wiegand, B. (2007) 'Hjernernes nye Kredsløb', *Mandag Morgen* MM 06, February.

Wilsdon, J. and Keeley, J. (2007) *China: The Next Science Superpower?* London: Demos.

World Bank, World Development Indicators Database, April 2007, www.worldbank.org.

WTO (2005) *World Trade Report 2005*, Geneva: WTO.

2
The National Innovation System of China in Transition: From Plan-Based to Market-Driven System

*Xielin Liu and Nannan Lundin**

2.1 Introduction

The 'National Innovation System' (NIS) is a relatively new concept in China. It was introduced in China in the mid-1990s when the first independent evaluation of the science and technology (S&T) capacity of China was conducted by the International Development Research Centre (IDRC) of Canada. Associated with the rapid economic growth and the ongoing structural reforms as well as the increased importance of S&T capacity for the Chinese economy, NIS has become an increasingly useful tool in the analysis of innovative capacity and providing inputs for policy making in China. In the recently released 'National Guidelines for Medium- and Long-term Plans for Science and Technology Development (2006–2020) of China' (MOST, 2006), S&T is considered the key driving force for sustainable economic growth and for the transformation of China into an innovation-oriented nation through the construction of an enterprise-centred NIS with strong indigenous innovation capacity. Accompanied by the strong policy messages, some specific targets were also set up such that the research and development (R&D) expenditure needs to increase rapidly so that the R&D to GDP ratio will rise from the current level of 1.3 per cent in 2005 to 2.0 per cent by 2010 and 2.5 per cent by 2020.

Even though the concept of NIS and its implications are defined in different ways (see, inter alia, Freeman, 1987; Nelson et al., 1993; Lundvall, 1992), it provides a useful framework for identifying key innovation performers and important elements in innovation activities. Nevertheless, it is also important to keep in mind that there are shortcomings in such an approach that need to be recognized. First, in addition to a clear mapping of key performers in the NIS, such as enterprises, research institutes and universities, the linkages and interactions among these key actors, or the dynamics inside the

* The valuable comments and suggestions on an earlier version of this chapter by Anthony D'Costa and Govindan Parayil are gratefully acknowledged.

system must be taken into account. Secondly, in the context of globalization, the analysis of NIS needs to be associated with the increasing global integration of economies and markets, which implies both new opportunities and challenges for the development of NIS in China. Finally, country-specific characteristics need also to be highlighted when applying the generalized NIS framework, as 'most public policies influencing the innovation process or the economy as a whole are still designed and implemented at the national level' (Edquist, 2006).

In the Chinese context, its innovation system has been a largely plan-based one. Government research institutes (GRIs) played a dominant role and the government acted as the core coordinator, from idea generation to final user of new products through issuing National Science and Technology Plans both annually and in every five years. Industrial enterprises played a trivial role in such a system (Liu and White, 2001). Even though the system was efficient for implementing some targeted R&D programmes, such as the development of nuclear bombs and missile systems, it was an inefficient system in terms of innovation in general.

Since the 1980s, China launched its economic reform and open-door policy towards a market-oriented economy of greater openness. Industrial enterprises with diversified ownerships have gradually become the main player of innovation during the reform process. A large number of state-owned enterprises (SOEs) were transformed into equity share or privately owned enterprises. The market mechanisms thus, to a large extent, replaced the government plans as driving and coordinating forces of the Chinese innovation system. In a sense, the system has been heading towards a market-driven innovation system.

Furthermore, China's economic growth is to a great extent related to its openness in terms of international trade and foreign direct investment (FDI) as results of the open-door policy. China has benefited from the globalization in many aspects, such as accelerated structural changes, improved international trade performance and job creation. The field of S&T is not an exception and the NIS of China has been transforming into an open NIS. In addition, innovation activities in China are closely linked with the global development, facing both new opportunities and daunting challenges.

In this chapter we present the transformation process of the NIS in China. As China's political and economic systems are unique, the innovation system and policy making need to be grounded in this uniqueness. Therefore, we aim to put emphasis on key elements of the NIS in terms of actors and linkages as well as on country-specific factors, which are important determinants of both strength and weakness of the Chinese NIS.[1]

In section 2.2 we provide a short review of how the Chinese innovation system has evolved over the last twenty years. In section 2.3 key linkages in the system will be illustrated and discussed. The enterprise system will be presented in more detail in section 2.4. Key country-specific factors in the

Table 2.1 The relative importance of key actors in R&D expenditure (%)

Performers	1990	1996	2000	2005
Research institutes	50	41	29	21
Universities	12	13	9	10
Enterprises	27	37	60	68

Source: China Statistical Yearbook on Science and Technology, 2004, 2006.

Chinese innovation system, such as the role of government, foreign direct investment, and the globalization of innovation are presented in section 2.5. The chapter concludes in section 2.6 with some general observations relating to the evolving innovation dynamics in China.

2.2 Evolution from research institutes to an enterprise-centred system

The key performers of innovation activities in China are government research institutes, universities and enterprises. The most crucial element in the S&T structural reforms is to adjust the specific role played by these key performers. The purpose of the adjustment is to obtain a better balance between improving the market orientation of S&T activities and boosting strategic and long-term innovation capacity.

Since the beginning of economic reforms, the Chinese innovation system has undergone significant changes, in terms of the relative importance of key actors and the mechanisms that drive the development of the innovation system. As shown in Table 2.1, the share of R&D expenditure by research institutes in the total R&D expenditure nationwide has gradually decreased from 50 per cent in 1990 to 21 per cent in 2005, while the corresponding share for enterprises has increased from 27 per cent to 68 per cent during the same period. This change is driven by a combination of the restructuring of research institutes, the expansion of the higher education sector and the strengthening of innovation capacity of enterprises. In particular, the privatization and the opening up of the manufacturing sector for foreign firms have largely contributed to this transformation.

2.2.1 Planned innovation system

In its previously entrenched socialist regime, the Chinese innovation system was dominated by a linear model of innovation with a clear-cut of division of labour. The government functioned as the key coordinator of the system and government research institutes (GRIs) played a dominant role in all innovation activities.

Between the 1950s and the 1980s, GRIs were established at different administrative levels with various goals and orientations. The most important of these were at the national level, such as the Chinese Academy of Science (CAS). Most basic scientific research was done by the CAS and some large research universities such as Beijing University and Tsinghua University. There were also hundreds of industrial research institutes under a wide range of industrial ministries in different regions, focusing on applied research and developmental tasks. For example, in each province, there was an Academy of Agriculture Science and the CAS has several regional branches in Beijing, Shanghai and other major Chinese cities. These regional GRIs conducted R&D tasks, which were defined as relevant for the region's development.

In addition, the higher education sector played a complementary role for the GRIs. Most universities at that time were not involved in research, except the large research universities such as Beijing and Tsinghua. Many specialized universities focused on industry-specific technical education. For example, there were universities specialized in light industries, telecommunication, metallurgy, printing and so on.

In general, the role played by industrial enterprises in the innovation system was limited and they functioned as manufacturing and/or sales units. Most of them did not do any R&D, while only some large SOEs had their own R&D laboratories and their work focused mainly on experimental issues. Hence, the innovation system was constructed and driven largely according to a linear and hierarchical model, in which there was a clear-cut division between theoretically oriented research and innovation activities. The resource allocation, therefore, followed a plan-based model with top-down decision-marking mechanism.

Following this clear-cut division of labour in knowledge creation and product manufacturing, a key question was how to introduce new technologies and products into the market? This was the task for the government. The main policy tools of the government were the Annual and the Five-Year Economic and Science & Technology (S&T) plans. Even at the government level, there was a sophisticated division of labour in policy making. For example, the State Planning Committee (the current State Development and Reform Commission) was the central body for allocating production targets for enterprises and also responsible for introducing new technologies into the economic system. The Ministry of Science and Technology (MOST) was in charge of the annual and the five-year plans in the field of science and technology.

For a long time, S&T was considered to be of strategic importance to overcome shortages of goods and services as well as to strengthen China's military position. High priority was given to a few large national projects, which involved thousands of scientists and engineers from a large number of government research institutes, universities, enterprises and hospitals across the country. The successes of building nuclear bombs, artificial insulin production and other major discoveries were the results of this planning regime and they reinforced the impression that great successes in S&T could be achieved, albeit with huge costs.

Despite the success in a few prioritized fields, the innovation system as a whole was less efficient. The enterprises were output-oriented so that they cared primarily about output, not quality and inputs. In other words, efficiency and profitability were not sufficiently taken into account in the absence of market incentive in the planned system (Liu and White, 2001). At the same time, IPR was hardly used as an incentive to encourage innovation activities. On the other hand, many imitation activities, such as the production of generic drugs in the biomedical industries, were the key activities in the absence of an IP regime. Research institutes and universities were funded by the government and produced project reports with limited industrial use.

From the 1950s through the 1960s and 1970s, China imported foreign technologies on a large scale from the former Soviet Union, Germany and Japan. The imported technologies laid the foundation for the Chinese chemical, automobile, steel, textile and several other industries. For many industrial GRIs, from 1949 to the early 1980s, their main tasks were to assimilate the imported technologies. In order to replace the imported technology and to save foreign currency, incremental innovations through reverse engineering on imported technology were also implemented. Many new industrial sectors were established in the 1970s around the same time when South Korea initiated its new growth path targeting automobiles, shipbuilding, ICT and steel. However, the Chinese enterprises in these sectors have been lagging behind Korean enterprises for many years because of a high degree of foreign technology dependency and low level of absorptive capacity. Even though Korean firms also relied on foreign technology during their catch-up phase, they managed to successfully utilize these technologies to enhance their own competitiveness in the export markets. In contrast, many Chinese enterprises got stuck in a pattern of lag and low-level technological trajectory.

2.2.2 Transition to an enterprise-centred system

In the earlier plan-based innovation system, there was little space for curiosity-driven research. The basic research share of GDP, on average has been around 0.07 per cent during the period 2000–04 in China. Compared to other developing or transition economies, China had a much lower basic research share of GDP level than Argentina (0.10 per cent in 2002) and Russia (0.17 per cent in 2002). Compared to more developed economies, this share has been very low. For instance, the share of basic research in GDP was 0.49 per cent for the US, 0.35 per cent for South Korea and 0.39 per cent for Japan in 2002 (OECD, 2005).

Since the economic reform was initiated in 1978, the S&T system of China was soon exposed to some form of rudimentary market competition. The objectives of the reform were twofold: (i) to introduce a competition-based funding systems and (ii) to establish a new governance system of S&T institutions in order to more efficiently commercialize R&D results. The key initial changes were to reform the funding system and to make the governance of S&T institutions more flexible. It meant that government reduced the direct funding for GRIs, and the funding of GRIs were expected to be increasingly diversified and to come from sources other than the government. While this change aimed to enhance incentives for innovation and to accelerate commercialization, it imposed increased pressure on scientists and led to short-term research projects for pursuing more immediate economic returns.

In order to accelerate the process from research to commercial products, the government also encouraged GRIs and universities to set up their own spin-offs and encouraged scientists to leave their research position and engage in commercial activities. Furthermore, a new institution called Technical Market was introduced. This new specialized market was supposed to facilitate technology transactions between suppliers and users of technology. Moreover, special economic zones were also established across China to support the development of high-tech enterprises.[2]

In the 1990s, after more than ten years of reform, there was still a great gap between the research activities of GRIs and the needs of industrial sectors. In the meantime, the government research system underwent a significant change as most of the industry-specific ministries were abolished. The new structural challenge was how to deal with the industrial GRIs, which were previously affiliated to several ministries. Towards the end of 1998, the State Council decided to transform 242 GRIs at the national level into technology-based enterprises or technology service agencies. The transformation process was led by the State Council and implemented by the respective ministries through various administrative re-constructions and regulations. This important structural change implied that the dominance of GRIs in the Chinese innovation system was changed and, instead, the industrial enterprises were on the way to becoming the core of the innovation system. Since 2000, the enterprises performed more than 60 per cent of total R&D in China (as shown in Table 2.1). However, GRIs and universities are still the key players in frontier science and technological research. They still attract a larger number of talented scientists than do enterprises.

2.3   Industry–science linkages

The intensity and efficiency of industry–science linkage are important indicators of innovation capability at the system level in a country. As there was a functional division of labour in knowledge creation and diffusion for a long time in China, strong barriers existed between the knowledge creation by GRIs and universities and the utilization of knowledge by enterprises. But since the introduction of economic reforms in China, under the strong competition pressures and supported by various institutional changes, the industry–science linkages have been improving greatly in the past 20 years. In this section, we illustrate industry–science linkages in the Chinese innovation system in the following three aspects: university spin-off, R&D outsourcing, and co-publication.

2.3.1   GRIs and university spin-off

GRIs and universities were allowed and encouraged to set up their own spin-offs so that they could commercialize their technology and research results directly. In this way, GRIs and universities could be more integrated in the economic activities. Spin-off companies could also provide GRIs and universities with some financial resources, which could compensate for budget cuts from the government. Up to 2004, there were around 2,400 spin-off enterprises established and generated around US$9.7 billion revenue (Ministry of Education, 2005).[3] Although the number of spin-off firms in China is small compared to that of the Chinese industrial sector, it is very important for high-tech industries in China. Spin-off companies provided many scientists from GRIs and universities with opportunities to access new market opportunities. The policy to encourage spin-offs also gave birth to many successful domestic high-tech companies, such as Lenovo (from the CAS) and Beida Founder (from Peking University), which are now leading companies in the Chinese ICT industry. Most of the Chinese biotechnology companies are also spin-offs. For example, Shenyang Sunshine Pharmaceutical Co. Ltd., Beijing Shuanglu Pharmaceutical Co. Ltd., and Anhui Anke Biotechnology Co. Ltd. were all founded by former researchers from research institutes (Liu and Lundin, 2006a).

However, since 2000, as the government has continuously strengthened its support for research and higher education, many GRIs and universities no longer consider development of spin-off companies as one of their primary functions.

2.3.2   S&T outsourcing by industrial enterprises

As an integral part of the establishment of science–industry linkage, GRIs and universities began to conduct contract research for the industrial sector. This type of activities has been beneficial for the industrial sector, as most of Chinese enterprises, especially small and medium-sized enterprises (SMEs), have limited innovation capabilities. Outsourcing of S&T research to GRIs and/or universities has become important development strategy for industrial enterprises. For instance, the share of universities' S&T funds from industrial enterprises was about 38 per cent of their total research funds in 2004. However, the share of government research institutes' S&T funds from industrial enterprises remains relatively low and was at a level around 6 per cent of their total S&T funds in 2004. It is because GRIs are still heavily relying on direct funding from the government (China Statistical Yearbook on Science and Technology, 2005).

2.3.3   Joint publications

The majority of published scientific papers in China are submitted by individual researchers from either the higher education sector or research institutes. As another indicator of industry–science linkage, the number of joint scientific publications by researchers from universities and the industrial sector is still relatively small. For IPR and other reasons, the S&T staff from industrial enterprises is typically reluctant to publish papers. But recently, researchers from universities and to an increasing extent engineers and researchers from industrial enterprises have become co-authors of science and technology publications. For instance, the number of co-authored papers, by researchers from universities as the first author, together with engineers/researchers from the industrial sector has rapidly increased from 867 articles (1.7 per cent of the total number of scientific papers published) in 2000, to 7,421 pieces (7.4 per cent of the total) in 2003 (Chinese Institute of Information, 2005). This intensified interaction and co-operation may promote innovation capacity in both sectors as well as enhance the mutual understanding of their different, but closely related innovation activities.

2.4   The enterprise system

For a long time, enterprises have typically operated as manufacturing units with few, if any, R&D activities or formal R&D centres. Their production capability was maintained and upgraded mainly through technology imports and enterprises spent more money on technology imports than on their own R&D during the period before 1998. Since the 1980s, SOEs were given more autonomy to invest and innovate based on their own strategic decisions. In addition, enterprises with different ownerships such as private and foreign enterprises have also, to a large extent, engaged in innovation activities. This wave of privatization and competition has given enterprises strong incentives to invest in product development and innovation. Table 2.2 shows the pattern of how large and medium-sized enterprises gradually increased their R&D inputs and R&D intensity. Nevertheless the R&D intensity is still quite low, compared to that of developed countries.

Table 2.2   R&D expenditure and technology import (unit: US$100 million)[3]

	Expenditure on R&D	Share of total sales, %	Expenditure on technology imports	Share of total sales, %
1995	17.1	0.46	43.4	1.17
2000	42.7	0.71	29.6	0.49
2005	152.7	0.76	36.2	0.18

Source: China Statistical Yearbook on Science and Technology, 2004, 2006.

In terms of output, the innovation capability of Chinese enterprises is still relatively low. Their innovation capability is mostly focused on incremental innovation with little radical innovation, which can be observed from the patenting activities of the enterprises. Patents registered in China are classified into three categories: invention, utility model and (appearance) design. The classification of patents differs from the international standard. For instance, design refers to new appearance and utility model refers to functionality modification or improvement, without substantial technological contents. The invention patents are thus presumably more R&D intensive than the other two types of patents. Chinese enterprises have relatively high levels of patenting activity in utility model and design, which account for the largest increase in the total number of patents, but low in invention patents. However, since 2000, the number of invention patents granted has also increased more rapidly than before (see Table 2.3). Furthermore, the patenting activities differ significantly between domestic and foreign firms in China. For instance, even though both domestic and foreign firms have rapidly increased their patent applications, the largest increases in both applications for invention patents and invention patents granted are from foreign firms. Moreover, the technological sophistication and the claims required in patent applications by foreign firms, in general, are more advanced than in domestic firms.

The growth of small firms is a relatively new phenomenon in the Chinese enterprise system. The market was opened up for non-state-owned small firms only after the 1980s. As most of them started their business by taking advantage of market opportunities, their innovation capability is still low, despite the rapidly rising importance accorded to entrepreneurship and small firms in economic growth. If China is to make rapid strides in building up innovation capacity, it is crucial to bring small firms into this process. Firstly, small firms play an important role in job creation in China and absorb a large number of new entrants into the labor market as well as those former large SOE employees who were laid off during the structural reforms. The enhanced innovation capacity will not only increase the potential of small firms to grow, but also help them to create better jobs. Secondly, while FDI and globalization of R&D have been highly MNC-dominated phenomena, in recent

Table 2.3 Patents granted in China – by type of patents (unit: number)

	1995	2000	2005
Total number of patents granted	45,064	105,345	214,003
Share of invention patent, (%)	7.5	12.0	24.9
Share of utility model patent, (%)	67.6	52.0	27.1
Share of design patent, (%)	24.9	36.0	38.0

Source: China Statistical Yearbook on Science and Technology, 2004, 2006.

Table 2.4   Comparison of key S&T indicators of small and large S&T-based firms (%), (2004)

	Small S&T-based enterprises				Large S&T-based enterprise			
	R&D/ Sales	Export of new products/ sales	Tech import/ sales	Patent/ 100 persons	R&D/ Sales	Export of new products/ sales	Tech import/ sales	Patent/ 100 persons
SOE	1.2	0.3	0.2	0.5	0.9	1.6	0.3	0.1
Joint venture: HTM	1.0	4.2	0.2	0.4	1.0	23.0	0.4	0.4
Joint venture: Foreign	1.6	4.2	0.6	0.4	1.3	6.4	1.2	0.7
Foreign	1.4	6.6	0.2	0.8	1.0	24.4	0.2	0.3
Private	1.6	3.2	0.1	0.7	0.7	5.9	0.1	0.9

Source: Lundin et al., 2006a.

years it has been observed that small foreign firms are making greater efforts to enter the Chinese market and to participate in the process of globalization of R&D.

At present, the share of R&D conducted by small firms is still low in China (accounting for 14 per cent of total R&D expenditure in the business sector in 2004, which is slightly lower than the OECD average of 17 per cent). However, their innovative potentials, indicated in terms of R&D intensity and patent output, should not be underestimated. As shown in Table 2.4, the comparison of key S&T indicators of large and small S&T-based firms, across various ownerships suggests that small firms have, in general, a higher R&D intensity (measured by R&D/sales ratio) than large firms. Small foreign firms are particularly active in invention-related patent applications. However, due to various resource and institutional constrains, small firms have also limited access to foreign technology and low capability to enter foreign markets. Furthermore, there are also substantial differences across ownerships among both small and large and medium-sized enterprises in their inputs and outputs of innovation activities (Lundin et al, 2006a).[5]

2.5   Key characteristics of the innovation system of China

Similar to the transition of the Chinese economy, the innovation system of China is also undergoing transition from a plan-based to a market-driven open system. In this process, there are important country-specific characteristics, resulting from both historical reasons as well as the uniqueness of the Chinese economy, in terms of its size, social and political contexts, and its rapid integration into the global economic system. In this section, we

Table 2.5 National R&D programmes of China (in US$ million)[6]

	1996	2001	2003
973 Basic Research Program	N.A.	7.2	9.7
863 National High Tech R&D Program	5.4	30.3	54.3
Key Technologies R&D Program	6.3	12.8	15.1
Torch Program (for high technology)	0.6	0.6	0.6
Spark Program (for rural SMEs)	0.5	1.2	1.2
Key S&T Diffusion Program	0.2	0.2	0.2

Source: MOST, China Science and Technology Development Report, 2006.

highlight several country-specific characteristics, which influence the shaping of the new innovation system and the new dynamics inside the system as well as the pace by which the transition process unfolds.

2.5.1 Role of the Chinese government

The central government plays an important role in the development of the Chinese innovation system even though the market force has been an increasing countervailing force. For example, the government agencies at different levels control land, large investment projects, infrastructure development, and market accesses to some strategic industrial and service sectors such as automobile industry and financial services. In terms of innovation and national R&D programs, various long-run and short-run plans are important instruments, through which government policy influences S&T development in China.

2.5.1.1 National R&D programmes

China has developed a system of national R&D programmes to support innovation activities. Table 2.5 gives a brief overview of the main programs controlled by the Ministry of Science and Technology (MOST).

In addition, there are the National Innovation Fund (INNOFUND) for S&T-based SMEs (about US$4 million per year) and the National Science Foundation of China (NSFC) mainly for basic research, which had a budget around US$420 million in 2006 (NSFC, 2006). The national programs are very important not only for funding. Universities and GRIs give very high priority to governmental projects and many talented scientists are involved in these projects as main researchers. Furthermore, many other regional and industrial funds often follow the national projects, in terms of topics and fields of study.

2.5.1.2 Policy instruments

The Chinese government has adapted various important policy instruments to encourage innovation activities and to promote transfer and commercialization of R&D results. One of the important policies is to establish special enterprise zones and incubator facilities to promote high-tech industries in China. This policy started in the end of 1980s by following the Silicon Valley model. There are 53 high-tech development zones at the national level. The first one, Zhongguancun high-tech zone, was established in Beijing in 1988. The high-tech zone policy covers the following aspects:

- To establish well-functioning infrastructural facilities so that the high-tech zones could serve as platform for innovation activities and interactions.
- To provide preferential treatments for high-tech firms in the form of a broad range of tax incentives.
- To create a new governance model, which is characterized by 'small government, but big service' to reduce transaction costs.
- To establish clusters in order to promote active interactions and close co-operation among the firms.

In the past two decades, these high-tech zones have expanded rapidly in terms of their size and scope of activities and therefore played an important role in promoting the development of high-tech industries in China. Up to now, more than 90 per cent of high-tech firms and incubators are located in these high-tech zones and most of them are spin-offs from universities and GRIs, new private and FDI firms. In 2004, the total value-added of these high-tech zones reached US$66.4 billion, accounting for about 8.8 per cent of GDP. Their exports were about US$82.4 billion, making up about 12 per cent of the total exports of Chinese industrial sectors (MOST, 2006).

The first business incubator in China was established in 1987 in Wuhan. By 2005, more than 490 incubators had been created across the country and most of them are located in Beijing, Shanghai and Shenzhen. ICT and biomedical industries are the two most favoured fields, but the number of incubators specialized in ICTs is much larger than the biomedical sector.

Regarding the ownership of IPR, the following important steps have been taken to facilitate the commercialization of R&D results:

- Firstly, inspired by the Bay–Dole model from the US, the first step taken by the Chinese government is to allow IPR resulting from government-funded R&D projects to be commercialized.
- Secondly, the ownership of IPR resulting from government-funded R&D projects could be transferred to the university or GRI who conducted the projects, instead of being government-owned intangible assets.
- Thirdly, since 1998 individual inventors involved in government-funded R&D projects have been allowed to obtain a royalty of up to 35 per cent of the license fee when the research results are transferred.

2.5.1.3 Long-term Plan: indigenous innovation

'National Guidelines for Medium- and Long-term Plans for Science and Technology Development (2006–2020) of China' is the current long-term S&T policy framework of China. The most interesting element of the new plan is the declared intention to strengthen 'independent' or 'indigenous' innovation. However, the interpretation of 'indigenous innovation' is still an issue of both political and economic debate in China, due to the underlying tension between indigenous innovation effort and the increasing openness of the innovation system.

Why indigenous? There are, at least three different factors behind this concept. Firstly, the economic growth of China has been strongly dependent on foreign technology and foreign invested firms. Since 2000, foreign-invested enterprises accounted for more than 85 per cent of all high-tech exports (China Statistics Yearbook on high-tech technology industry, 2004–2006). In recent years, there has been an increasing frustration among domestic actors, caused by the factor that 'market for technology' policy has not resulted in the immediate and automatic knowledge and technology spillovers from foreign to Chinese enterprises that policy makers had hoped for. Secondly, a culture of imitation and copying is common not only in product development and design, but also in the field of scientific research. Hence innovations from domestic knowledge bases and intellectual property rights are acutely needed in China. Thirdly, the high growth rate of the Chinese economy during the last twenty years will not be sustainable without a change in the development strategy. China needs, for example, more energy-efficient and environment-friendly technologies, new management skills and new organizational practices to ensure sustainable growth in the near future. To summarize, the concept of 'indigenous innovation' expresses the need for enhancement of both the absorptive and innovative capacity of domestic firms as well as for generating stronger spillovers.

There are three main policies selected to pursue the indigenous innovation strategy. First, the government plans to increase R&D by 2020 to 2.5 per cent of GDP (from the current level of 1.3 per cent). Since GDP growth is projected to increase at a similar pace as in the past two decades, the increase of R&D to GDP ratio implies a huge increase of R&D expenditure in absolute terms. Today China has the second-largest budget on R&D in terms of purchasing power parity, trailing only the USA, but has surpassed Japan (OECD, 2006).[7] Secondly, various fiscal policy instruments to enhance innovation capability are assumed. The new tax policy will make R&D expenditure 150 per cent tax deductible, thus effectively constituting a net subsidy, as well as accelerated depreciation for R&D equipment worth up to approximately US$37,500 (MOST, 2006). In addition, the public procurement of technology will be adopted to promote indigenous innovation activities. The purpose of current public procurement practice is to cut costs rather than promote indigenous innovation. The new public procurement policy aims to give priority for indigenous innovative products in public procurement in terms of price and volume in various forms.

2.5.2   Foreign direct investment (FDI), high-tech industries and R&D

Since the open-door policy was implemented in China, FDI firms have become increasingly important in production as well as in R&D in China. During the period 1998–2004, the number of large and medium-sized FDI firms has been increasing steadily. While the shares of value-added and exports of FDI firms in the Chinese industrial sector have reached a relatively high level (40 per cent and 76 per cent, respectively in 2004), the shares of R&D expenditure and employment are still relatively low (29 per cent and 34 per cent respectively in 2004). Apart from the large number of new FDI establishments of capital-intensity processing manufacturing units, R&D-related activities in these new FDI firms are still relatively limited. It implies that, FDI firms' production in the Chinese industrial sector has been relatively capital-intensive, but not really R&D-intensive manufacturing.

Beyond the manufacturing sector, the internationalization of the high-tech industries is of significant importance to China's upward mobility in global innovation. But such a move also has some controversial characteristics. On the one hand, the increased trade volume shows the international competitiveness of the high-tech industries of China. But on the other hand, the dominance of FDI firms and the large share of processing of imported materials as well as the reliance on foreign technology raise the following questions: Are China's high-tech industries really high-tech? And are the high-tech industries in China really Chinese? Nevertheless, there are also substantial cross-industrial variations in the high-tech industries. As a well-known fact, the ICT sectors are the most internationalized high-tech industries, in which value-added, technology imports and exports are dominated by FDI firms. Regarding R&D expenditure, the share of FDI firms in the computer and office equipment industry has the most remarkable increase and FDI firms in the medical equipment and instruments industry have also noticeably increased their contribution to R&D investment at the industry level (see Table 2.6).

In addition to the relative importance of FDI firms at the industry level, another important and controversial question is, are FDI firms more R&D-intensive than domestic firms? While the R&D intensities across different ownerships have increased during the period 1998–2004, so far domestic firms, both state-owned and private, have higher R&D intensity than FDI firms. The implications of these observations are:

- Domestic firms in China are strengthening their innovation capacity through increased R&D investments. This is achieved not only by the

Table 2.6 Importance of FDI firms across high-tech industries, 1998 and 2004 (share in the high-tech industries, %)

	Number of FDI firms	Share of number of LMEs	R&D expenditure	Tech Import	Export	Employment
1998						
Pharmaceutical products	83	16	20	4	19	11
Electronics & telecommunication	349	52	41	77	86	42
Computer & office equipment	70	59	37	94	94	51
Medical equipment & instrument	28	20	11	41	40	14
2004						
Pharmaceutical products	158	21	22	20	21	16
Electronics & telecommunication	1145	72	42	93	93	73
Computer & office equipment	336	86	82	98	98	91
Medical equipment & instrument	105	38	27	33	88	36

Source: Lundin et al., 2006b.

increased R&D investments in the SOEs, but is also driven by an increased number of entrepreneurial and S&T-based private firms.

- The lower R&D intensities in FDI firms may be explained by two types of FDI activities in China. For the first, some of the FDI firms' activities are still capital- or labour-intensive manufacturing in the high-tech industries. For the second, while some foreign firms are increasing their R&D effort in China, the R&D activities are still based in the OECD home countries where the firms originate.
- Even though the R&D intensities in the high-tech industries have increased over time, they are still at a much lower level compared to the high-tech industries in the OECD countries.

As shown in Table 2.7, the R&D intensities in most of the high-tech industries, except in the aerospace industry, were not, on average, substantially higher than in the manufacturing sector. In an international comparison to the US and Japan, the difference is remarkable. From a long-term perspective, the R&D intensities need to, and will be further boosted, driven by continued indigenous R&D efforts and intensified competition between domestic and FDI firms when the technology gaps between them are being narrowed. Furthermore, the narrowed technology gap can also facilitate strategic alliances

Table 2.7   R&D intensity in the high-tech industries (%)

	R&D/value-added 2001	R&D/value-added 2003	R&D/value-added U.S. (2001)	R&D/value-added Japan (2001)
Manufacturing average	3.4	2.0	8.7	9.9
High-tech average	5.0	4.4	27.2	26.3
Aerospace	15.0	15.8	14.4	22.3
Pharmacy	2.6	2.7	14.8	22.9
Computers and office machines	4.1	2.5	36.7	30.7
Electronic, telecommunications	5.8	5.4	37.2	18.6
Medical equipment and meters	2.5	3.0	36.8	30.2

Source: China Statistics Yearbook on high technology industries, 2004, 2005.

among firms with various ownerships and thereby boost R&D investments in both domestic and FDI firms.

In terms of innovation output, one of the largest differences between domestic and foreign applications is in the structure of applications. For domestic firms, the majority of their patent applications are utility model or design, although the number of invention applications has also been increasing. For foreign patent applications, the invention application is the main category. The number of invention applications by domestic firms exceeded for the first time their foreign counterparts in 2003. However, the foreign firms still outperformed their Chinese counterparts significantly in terms of the number of granted invention patents in the past years (see Figure 2.1).

There is no straightforward explanation as to why the number of domestic patents granted lags behind the number of foreign patents granted. The awareness of protection of intellectual property right has indeed increased among domestic firms and innovation activities have become an important strategic reaction for both domestic and foreign firms to enhance their competitiveness. However, the gap (see Figure 2.1) may suggest that, despite the increased awareness, the technology gaps, in terms of degree of novelty and technological sophistication between domestic and foreign firms, make the catch-up process of domestic firm somehow difficult.

Among foreign patent applicants, the multinational enterprises from Japan and the USA are the most active, while German, Korean and French companies are also applying for a large number of patents in China (Table 2.8). The distribution by field of technology reflects to a large extent the competitive strengths of these multinationals in the Chinese market.

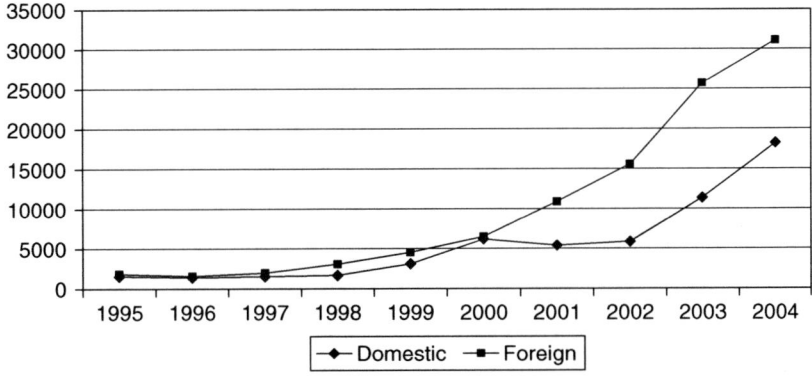

Figure 2.1 Domestic and foreign invention patents granted
Source: China Statistical Yearbook on Science and Technology, 2005.

Table 2.8 Top ten foreign enterprises in applications for invention patents in China (2003)

Ranking	Country	Enterprise	Number of applications
1	Japan	Matsushita Electric Industrial Co., Ltd.	1,817
2	South Korea	Samsung Electronics Co., Ltd.	1,560
3	Japan	Canon Co., Ltd.	820
4	Japan	Seiko Epson Corp.	781
5	South Korea	LG Electronics Corp.	624
6	Japan	Toshiba, Inc.	583
7	United States	IBM Corporation	581
8	Japan	Sony Corp.	560
9	Japan	Mitsubishi Electric Co., Ltd.	556
10	Japan	Sanyo Electrical Motors Co., Ltd.	541

Source: Ministry of Science and Technology Indicators, 2005.

2.5.3 Globalization of R&D and China

In recent years the number of R&D centres of multinational enterprises in large cities such as Beijing and Shanghai has increased rapidly. The purpose of these establishments is mainly twofold: to take advantage of abundant and relatively cheap R&D human resources in China and to locate R&D units near their (existing) manufacturing units in China. According to von Zedtwitz (2006), there were 199 foreign R&D facilities in China in the beginning of 2004 (Figure 2.2). The number has increased rapidly since then and currently possibly amounts to 250–300.

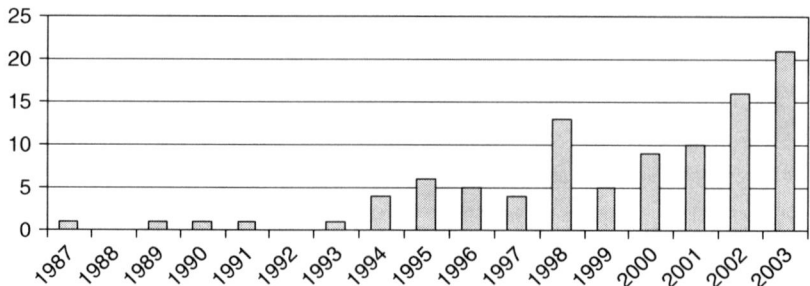

Figure 2.2   Number of new establishments of foreign R&D labs in China, 1987–2003
Source: von Zedtwitz (2006).

The globalization of R&D in China can also be observed from the co-operation between foreign enterprises and Chinese universities and research institutes. This new type of co-operation is in an initial and immature stage and it is still very difficult for foreign enterprise to find original ideas and sufficiently innovative projects through this kind of co-operation. At the current stage, foreign enterprises do not buy ready-made projects or research, rather they utilize the existing R&D research capacity and facilities (which were often purchased by the support of governmental funding and of very high standard) to carry out research projects, which are defined by the foreign enterprises themselves and modified during the working process to adapt to local conditions.

Nevertheless, the mutual benefits generated through such co-operative efforts should not be underestimated. It will provide local universities and research institutes with additional funding and more advanced equipment. More importantly, it will also generate positive demonstration and spill-over effects to the universities and allow them to get more informed about the international research frontier. Finally, it can be an efficient way for foreign firms to identify research units and personnel with high research capacity.

Compared to the level of inward foreign direct investment (which reached US$72 billion in 2005, UN World Investment Report, 2006[8]), China's outward direct investment (ODI) and cross-border Mergers and Acquisitions (M&A) are still very limited. By the end of 2005, China's aggregated ODI reached US$57.2 billion, accounting for only 0.59 per cent of global ODI and ranked 17th in the world among outward investors in 2005 (MOC, 2006). However, this low level of ODI may change, associated with both the increased openness of the Chinese economy, new government policies and the relaxing of financial controls as well as the efforts to diversify China's huge foreign exchange reserves.[9] In 1999, the Chinese government launched the 'Go Out' policy and China's Ministry of Commerce predicts that ODI will maintain an average annual growth rate of over 22 per cent in the years to come and will exceed US$60 billion by 2010.

Table 2.9   Selected M&A deals by Chinese firms, 2001–05

Chinese bidder	Target foreign firm/Unit	Industry	Bid value (US$ million)
Holly group	Philips Semiconductors, CDM hand-set reference design (US), 2001	Telecom	180
TCL International	Schneider Electronics AG (Germany), 2002	Electronics	8.5
TCL international	Thomson SA, Television manufacturing unit (France), 2003	Electronics	N.A.
BOE Technology Group	Hyundai display technology, (South Korea), 2003	Electronics	1,305
Shanghai Auto Industry Corporation (SAIC)	Ssangyong Motor (South Korea), 2004	Automotive	474
Lenovo group	IBM, PC Division (US), 2004	IT	1,620
Nanjing Automotive	MG Rover Group (UK), 2005	Automotive	87

Source: Wu (2005), The Boston Consulting Group (2005) and various press reports.

Even though at the current stage, Asia and Latin America account for 90 per cent of China's ODI I order to target the acquisition of energy and natural resources, the new 'Go Out' strategy is also a measure to promote and facilitate the internationalization of Chinese firms in S&T-intensive sectors. It aims at encouraging successful Chinese firms to strengthen their technological capacity and build brand recognition as well as to counter intensified competition in the Chinese market by investing abroad.

Recently, some Chinese enterprises, particularly in the electronics and ICT sectors, have initiated their international R&D activities, by either acquisition of foreign enterprise/units or through setting up R&D organizations in OECD countries. The high-profile M&A deals involving Chinese enterprises in the high-tech sectors have caused huge attention worldwide. In these M&A deals, the access to R&D centres of western sellers is one of the key elements. For example, in the TCL–Thomson deal, it included Thomson's R&D centres in Germany, Singapore and the US. Similarly, in the Lenovo–IBM deal, Lenovo took over IBM's R&D centres in Japan and the US (See Table 2.9).

In a recent report from Boston Consulting Group (BCG, 2006), among the top 100 emerging global companies from developing economies, 44 are Chinese firm, 18 of which are in the ICT sector and a few from the automobile sector. Even though the number of such Chinese firms is very small and the scale of their international R&D activities is still small, a new generation of Chinese firms seem to emerge as important players in S&T-intensive (instead of labour-intensive) segments of the global market. The innovation capacities of these Chinese firms and their ability to tap into the global network have therefore generated large interests, from both research- and policy-making perspectives. In other words, will these emerging Chinese multinationals become global players in the near future?

2.6   Conclusion

Similar to the Chinese economy, which is in the process of transforming from a planned economy to a more market-oriented economy, great changes have also taken place in the Chinese innovation system during the past twenty years. The innovation system has become dynamic with great potentials. The structural adjustments in various forms have made enterprises the core of the national innovation system. In the industrial sector, SOEs have undergone reforms of governance. Many large non-state-owned enterprises have emerged, such as Huawei, Lenovo and Haier, who are not only the driving force for innovation capacity building in the Chinese market, but are also on the way to entering the global market. SMEs have also become more important players in the Chinese economy as well as in the innovation domain, driven by competition and entrepreneurial spirit. The increased openness of the innovation system, spurred by FDI in both high-tech manufacturing and purely R&D-oriented activities, has also created significant incentives for structural changes and generated mutual learning opportunities among domestic and foreign enterprises.

However, the Chinese innovation system is still in an early stage of transition. The emerging innovation-oriented enterprise system is still weak in terms of innovation capacity and innovation activities are mostly focused on incremental innovations. In addition, because of low indigenous innovation capacity and technological gaps, the cross-sector and cross-ownership spillovers are still limited. Hence GRIs and universities are still very important in R&D activities as well as in terms of R&D human resources. Furthermore, in this transition process, the specific Chinese characteristics, such as the need for continuing structural reforms, the changing role played by the government, and also the tension between indigenous capacity building and increased openness of the Chinese economy, are serious challenges for the future development of NIS. From an S&T policy point of view, there are at least three important aspects that need to be taken into serious considerations. For the first, from the planned to the market-driven system, the government will still have strong influence on the emerging innovation system through various policies, strategies and investments. However, the role played by the government in this process will be more like a conduit than a planner. In other words, the role played by the government is not only to provide financial incentives, but also to create an innovation-friendly environment. Secondly, while the importance of enterprises is increasing, there is still a strong need for strengthening the higher education sector and human resources. The balance between the short-run leapfrogging and the long-run strategic capacity building is crucial for the future development of the NIS in China. Finally, knowledge and technology diffusion through commercialization and industrialization of S&T/R&D results remain a key challenge as the barriers in such processes are associated with both inadequate innovation capacity and insufficient market opening mechanisms.

In such contexts, there are two major forces that will jointly shape the future development of the Chinese innovation system: One is the national strategy of indigenous innovation, which focuses on how to promote domestic innovation capability building. The second is an open innovation approach, which is based on knowledge creation and technology acquisition through global linkages and partnership.

Notes

1. When discussing economic and S&T development in China, the significant regional disparities are an important aspect that should be taken into account. However, the regional dimension of NIS is beyond the scope of this chapter and needs to be discussed in more details in a separate study.
2. More detailed information on spin-offs and special economic zones are given in section 2.3.1 and section 2.5.1.2.
3. Total revenue of 2,355 spin-offs amounted RMB 80.7 billion and converted into US dollars using the annual average exchange rate at the value of 8.28.
4. The nominal values of R&D expenditure and technology import in RMB were converted to US dollars using the annual average exchange rates in 1995 (US$1 = 8.31 RMB), 2000 (US$1 = 8.28 RMB) and 2005 (US$1 = 8.19 RMB).
5. The industrial enterprises in China can be divided into the following ownership categories: SOE, joint venture with enterprises from Hong Kong, Taiwan and Macau (HTM), joint venture with foreign enterprises, wholly foreign-owned and private enterprises.
6. The nominal values of government appropriation in RMB were converted to US dollars using the annual average exchange rates in 1996 (US$1 = 8.31 RMB), 2001 (US$1 = 8.28 RMB) and 2003 (US$1 = 8.28 RMB).
7. However, to make an international comparison of this measure of China to other OECD countries is not a straightforward task, particularly when taking account of the potential overestimation caused by the conversion from Chinese currency into USD PPP. See a more detailed discussion in Schaaper (2005).
8. This large increase is also due to the fact that, in 2005 for the first time data on Chinese inward FDI included inflows to financial industries. In 2005, non-financial FDI alone was US$60 billion. The FDI into financial service surged to US$12 billion, which was driven by large-scale investment in China's largest state-owned banks.
9. In 2005, China's foreign currency reserves increased by US$209 billion and reached US$819 billion. It exceeded those of Japan and has become the world's largest in 2006.

References

Boston Consulting Group (2006) *China's Global Challengers: the Strategic Implications of Chinese Outbound M&A*, BCG report, May.

Chinese Institute of Information (2005) *China Science Paper and Citation Analysis*, Beijing.

Edquist, C. (2006) 'System of Innovation: Perspectives and Challenge', in J. Fagerberg, D. Mowery and R. Nelson (eds), *The Oxford Handbook of Innovation*. Oxford: Oxford University Press, pp. 181–208.

FIAS (2005) 'South Multinationals: a Growing Phenomenon', Washington, DC: FIAS. Accessed at rru.worldbank.org/PapersLinks/Open.aspx?id=6686.

Freeman, C. (1987) *Technology Policy and Economic Performance: Lessons from Japan*, London: Pinter.

IBM Global Business Service (2006) *Going Global: Prospects and Challenges for Chinese Companies on the World Stage*, New York: IBM Institute for Business Value.

Lazonick, W. (2006) 'The Innovative Firm', in J. Fagerberg, D. Mowery and R. Nelson (eds), *The Oxford Handbook of Innovation*. Oxford: Oxford University Press, pp. 29–55.

Liu, X. L. et al. (2006) *Chinese Report of Regional Innovation Capability*, Beijing: Chinese Science Press.

Liu, X. L. and N. Lundin (2006a) *Globalisation of the Biomedical Industry and the System of Innovation in China*, Stockholm: SNS.

Liu, X. L. and N. Lundin (2006b) 'China's Development Model: an Alternative Strategy for Technological Catch-up', mimeo.

Liu, X. and S. White (2001) 'Comparing Innovation Systems: a Framework and Application to China's Transitional Context', *Research Policy*, 30: 1091–114.

Lu, F. and K.D. Feng (2005) *The Policy Choice of Developing Indigenous IPR Chinese Automobile Industry*, Beijing: Peking University Press.

Lundin, N., F. Sjöholm, J.C. Qian and P. He (2006a) 'The Role of Small Enterprises in China's Technological Development', Working Paper No. 695, Research Institute of Industrial Economics (RIIE).

Lundin, N. F. Sjöholm, J.C. Qian and P. He (2006b) 'Technology Development and Job Creation in China', Working Paper No. 697, Research Institute of Industrial Economics (RIIE).

Lundvall, B.A. (ed.) (1992) *National Systems of Innovation: Towards a Theory of Innovation and Interactive Learning*, London: Pinter.

Ministry of Commerce (MOC) (2006) *China Commerce Yearbook*, Beijing.

Ministry of Commerce of the People's Republic of China (MOC) (2006) 'China's Overseas Investment in 2005 Hits New High'. Accessed at http://english.mofcom.gov.cn/aarticle/counselorsreport/asiareport/200611/20061103606164.html.

Ministry of Education (MOC) *Statistics of Science and Technology in High Education*, 2000–2005.

Ministry of Education (MOC) (2005) *Statistics of University's Industry in 2004 in China*.

Ministry of Science and Technology (MOST) (2005) *The Yellow Book on Science and Technology Vol. 7: China Science and Technology Indicators 2004*, Beijing: Scientific and Technical Documents Publishing House.

Ministry of Science and Technology (MOST) (2006) *China Science and Technology Development Report*, Beijing: Chinese S&T Literature Press.

Mowery, D. C. and B.N. Sampat (2006) 'Universities in National Innovation Systems', in J. Fagerberg, D. Mowery and R. Nelson (eds), *The Oxford Handbook of Innovation*. Oxford: Oxford University Press, pp. 209–39.

Mu, Q. and K. Lee (2005) 'Knowledge Diffusion, Market Segmentation and Technological Catch-up: the Case of the Telecommunication Industry in China', *Research
Policy*, 34: 759–83.

National Bureau of Statistics (NBS) (2004) *China Statistical Yearbook on High Technology
Industry, 2004*, Beijing: China Statistical Press.

National Bureau of Statistics (NBS) (2004) *China Statistical Yearbook on Science and
Technology, 2004*, Beijing: China Statistical Press.

National Bureau of Statistics (NBS) (2005) *China Statistical Yearbook on Science and
Technology, 2005*, Beijing: China Statistical Press.

National Bureau of Statistics (NBS) (2006) *China Statistical Yearbook on Science and
Technology 2006*, Beijing: China Statistical Press.

Nelson, R.R. (ed.) (1993) *National Systems of Innovation: a Comparative Study*, Oxford:
Oxford University Press.

OECD (2005) *OECD Science, Technology and Industry Scoreboard 2005*, Paris: OECD.

OECD (2006) *OECD Science, Technology and Industry Outlook 2006*, Paris: OECD.

Schaaper, M. (2005) 'An Emerging Knowledge-based Economy in China? Indicators
from OECD Databases', *STI Working Paper 2004/4*, Paris: OECD.

UNCTAD (2006) *World Investment Report 2006 – FDI from Developing and Transitional
Economies: Implication for Development*, United Nations Conference on Trade and
Development. Accessed at http://www.unctad.org/Templates/webflyer.asp?docid=
7431&intItemID=3968&lang=1&mode=downloads.

von Zedtwitz, M. (2006) 'Chinese multinationals: new contenders in global R&D'. Conference presentation. Available at http://goingglobal2006.vtt.fi/programme.htm.

Wu, F. (2005) 'The Globalization of Corporate China', *NBR Analysis*. 16(2): 5–29.

3
Foreign Corporate R&D in China: Trends and Policy Issues

Sylvia Schwaag Serger

3.1 Introduction

From being an essentially closed and planned economy until the late 1970s, China has become one of the most important trading nations and recipients of foreign direct investment (FDI) in the world (UNCTAD, 2005a, 2005b). Whereas previously FDI into China consisted mainly of acquisition or green-field investments in production, extraction and distribution facilities, today increasing numbers of multinational corporations (MNCs) are investing in research and development (R&D) in China (Gassmann and Han, 2004). US firms' R&D expenditures in China, for example, increased more than tenfold between 1998 and 2002, growing from US$52 million in 1998 to US$646 million in 2002 (National Science Foundation, 2006). Confirming this trend, in two recent surveys MNCs ranked China as one of the most attractive locations for future R&D investments (see A.T. Kearney, 2006; UNCTAD, 2005b).

Foreign corporate R&D in China can be explained partially by a general trend we are witnessing towards the internationalization of R&D, as firms increasingly locate R&D outside their home country.[1] The internationalization of R&D, in turn, is the latest step in a progression whereby companies initially located distribution, production, and finally different R&D activities outside their home countries (Gassmann and Han, 2004). At the same time, however, the increase of foreign corporate R&D activities in China reveals a fundamental shift in the international economic geography. R&D, knowledge and human capital are no longer the exclusive domain or privilege of developed countries. Nor is their location as path-dependent as one may have previously assumed. Rather, knowledge is becoming an increasingly internationalized, and even mobile asset with developing countries actively competing for resources such as corporate R&D activities and highly skilled labour (UNCTAD, 2005c).

In addition to increases in the shares of R&D expenditures by US, European and Japanese firms outside their home country, the growing importance of

50

patenting and publishing activities involving two or more countries and rising shares of foreign students in national university systems all indicate that knowledge is becoming more internationalized (see, for example, Eurostat, 2006). Thus, between 1994 and 2002, US firms' R&D expenditures increased more rapidly abroad than at home (National Science Foundation, 2006). At the same time, R&D expenditure by foreign companies in the US, as a percentage of total industrial R&D expenditure, increased (National Science Foundation, 2006). Sweden, the UK, Finland, Japan and Germany are other examples of countries where the share of R&D investments funded by foreign firms has been increasing (Karlsson, 2006).

While it can be argued that knowledge is becoming increasingly internationalized, this does not mean that it is location-neutral. A conducive local context remains vital for transforming knowledge into innovation. Thus, globalization does not reduce the importance of local combinations and agglomerations of physical and social capital and competencies as a determinant of innovative capacity and competitiveness (Saxenian, 2006; Scott and Storper, 2003).

For many years the Chinese government actively encouraged and promoted foreign corporate R&D in China, viewing it as a way of upgrading domestic technology and skills by importing, and hopefully internalizing, foreign know-how. Recently, there has been growing scepticism over the benefits of foreign corporate R&D for China's innovation system, with some observers within China even arguing that it may have hurt China's innovative capacity. Some academics and policy makers criticize foreign firms' presence and their behaviour in China, claiming that they charge unduly high licenses for their patents, 'crowd out' domestic firms in the market for highly skilled labour, and thwart technology transfer and knowledge spillovers (Lin, 2006; Breidne and Schwaag Serger, 2006). Foreign firms are also seen as dominating standards and technology platforms and reducing Chinese companies to the role of producers with low profit margins.

The increasing tendency for companies to locate R&D in China is also arousing concern abroad. Governments and public opinion in foreign countries worry that, after production and purchasing, now R&D will also move to China. In particular, there are growing concerns in developed countries that multinational companies will increasingly set up R&D in China at the expense of Europe and the USA. As journalist Abe De Ramos (2003) formulated the question in a 2003 issue of *CFO Magazine*, 'USA companies are beginning to outsource technology research and development to India and China. Will a meltdown in tech jobs follow?' While, so far, we do not see the rise in firms' R&D in China and India directly coinciding with a decline of R&D employment in Europe or the United States, the question illustrates concerns and reactions triggered by these developments in media and policy arenas.

In this chapter, I examine what brings foreign corporate R&D to China and what the development means for China and the rest of the world.

The analysis shows that Chinese government requirements explain some of the foreign R&D activities in China, which is sometimes referred to as the 'market-for-technology strategy' (Long, 2005). At the same time, a significant and growing share of R&D activities is drawn to China by a unique combination of China's market size, strategic importance and attractive human capital. Turning to the question of effects and spillovers of foreign R&D for China, I argue that foreign R&D reveals both the strengths and significant weaknesses of China's rapidly evolving innovation system.

Overall, the increase in foreign R&D in China is symptomatic of a more profound structural change. Firstly, knowledge resources are increasing global. Secondly, they are growing more rapidly in the East and, particularly, in developing countries, than in the West. It is important to point out that so far, we see no indications of R&D in China and India growing at the expense of Europe and the USA. Nonetheless, the dominance of the USA, Europe and Japan as obvious centres and magnets of knowledge is waning. This development has far-reaching policy implications for international economic relations and for future prosperity and economic development in both developed and developing countries.

In the first section I examine the nature, quantity and location of foreign firms' R&D operations in China. I then identify drivers explaining China's rising importance as a location for foreign corporate R&D activities and assess their relative importance. In the third section, I assess the domestic debate in China regarding the benefits of foreign corporate R&D before drawing conclusions on policy implications and possible future developments.

The findings of this chapter are based on a combination of reviews of existing studies on foreign R&D in China, analyses of press clippings and annual company reports, and interviews and surveys. Between June 2005 and June 2007, I interviewed approximately 60 senior executives and other experts on foreign R&D in China, such as representatives of chambers of commerce, employers' organizations, trade associations, universities and colleges, government authorities, international organizations, academics, and journalists.

The aim of the interviews was to supplement the lack of reliable statistics and to gain a realistic picture of the nature and drivers of foreign R&D operations in China. To assess how much innovative R&D is conducted by foreign companies in China, I have cross-referenced press clippings (for example, on the establishment of R&D centres) with companies' annual reports and websites, and interviews.

3.2 Foreign corporate R&D: what kind, how much, where?[2]

The establishment of R&D centres by foreign companies is a recent, but rapidly growing, phenomenon. Up to the late 1990s, there were relatively few R&D activities by foreign companies in China. In the past five years, foreign corporate R&D in China has increased dramatically (von Zedtwitz, 2004; Schwaag Serger, 2006). Furthermore, while adaptive R&D continues to dominate foreign firms' R&D activities in China, large multinational companies, many of them technology leaders in their fields, are increasingly locating innovative R&D in China. I use the term 'innovative' to differentiate between R&D activities devoted merely to adapting products to the Chinese market (adaptive R&D) and operations with a scope and nature that exceeds the domestic Chinese market. Centres with innovative R&D functions are also sometimes referred to as 'global R&D centres'.

There are three ways for foreign companies to establish R&D operations in China: as wholly independent R&D laboratories; as R&D departments or activities within either a branch of a Chinese operation or a joint venture with Chinese partners; and as cooperative R&D with Chinese research universities or institutes (von Zedtwitz, 2004). According to Gassmann and Han (2004), the more sensitive the technology is for the company, the more likely that its Chinese operations are wholly foreign-owned. Foreign R&D centres are mainly greenfield establishments. This contribution focuses primarily on wholly foreign-owned R&D centres.

For several reasons, it is difficult to assess accurately R&D activities by foreign companies in China. First, the figures of R&D employees in foreign companies in China only reflect part of the actual R&D activity. Many foreign companies have R&D cooperation with, or buy R&D services from, Chinese companies. Secondly, R&D is one of firms' most strategic, and therefore sensitive, activities, which means that managers are not always keen to disclose how much or what kind of R&D they have and where. A third reason for why it is difficult to gain a clear picture of foreign corporate R&D activities in China is that foreign companies are offered incentives (financial and otherwise) to establish R&D operations in China. Chinese authorities sometimes require companies to set up local R&D in return for being allowed to manufacture or sell in China. As one person interviewed by the author said, 'the Chinese demanded [that we carry out R&D in China], so we hired a few engineers'. As a result, some R&D activities exist more on paper than in reality. Gassman and Han (2004) observe that preferential treatment and government incentives for foreign R&D facilities may induce some foreign companies to register their activities as R&D even if they would not otherwise be classified as such. Companies may do so to establish goodwill with Chinese authorities since they strongly encourage foreign technology transfer (Gassman and Han, 2004).

The Chinese Ministry of Commerce stated that by late 2005 there were more than 750 foreign-established or foreign-invested R&D centres in China. More recently, the official number has been as high as 1,000 (Science and Technology Daily, 2006). For the reasons mentioned above, the number of operative centres actually carrying out R&D is likely to be considerably smaller. According to von Zedtwitz (2004), there were 199 operative foreign R&D facilities in China at the beginning of 2004. According to my estimates, the number has increased rapidly since then, possibly amounting to around 350–450 by 2006.[3]

It is even more difficult to assess how many of these companies carry out *innovative* or *global* R&D, i.e. R&D which is of relevance to the firms' global R&D operations, rather than only being aimed at adapting products to the Chinese market. The distinction is obviously somewhat arbitrary, since it is difficult to draw a clear line between innovative and adaptive R&D. Nonetheless, it is useful to attempt to make such a distinction since there are differences in the drivers and wider implications of the two types.[4] Adaptive R&D can be argued to be location-specific, determined by the need for proximity to market or production. Innovative or global R&D, on the other hand, refers to activities which are not tied to a specific national market and which, in theory, could be carried out elsewhere in the world. The trend towards establishing global or innovative R&D centres in China is a more complex and strategically more relevant phenomenon than when companies' R&D consists merely of adapting products to the Chinese market. It raises the question what factors explain companies' decision to establish innovative or global R&D in China, as opposed to other countries.

Numerous companies are choosing China as one of a select group of countries for setting up a global R&D centre. Nokia's research centre in Beijing, for example, is one of the company's eight research laboratories in the world, the others being located in Finland (Helsinki and Tampere), Germany (Bochum), Hungary (Budapest), Japan (Tokyo) and the USA (Cambridge and Palo Alto).[5] Unilever lists its research centre in Shanghai as one of six global R&D sites.[6] Of Fujitsu's seven R&D laboratories, two are in China (Beijing and Shanghai), three in the USA, one in the UK, and one at the headquarters in Kawasaki, Japan.[7]

A further complicating factor, when trying to assess how many companies carry out innovative, or strategic, R&D in China, is the discrepancy between R&D sites declared by a company or the Chinese press to be innovative, and the absence of these sites in the same company's listing, on its homepage or in its annual report, of its global R&D centres. After correcting for this discrepancy, we found around 40–50 large multinational companies with up to 70 facilities performing innovative R&D activities in China (see Appendix 1).

While adaptive R&D activities can be found outside Beijing and Shanghai, and are quite frequently located close to the companies' production facilities, when it comes to innovative R&D operations, with very few exceptions, they tend to be located in Beijing or Shanghai and not necessarily in close vicinity to companies' manufacturing plants. These two cities are the most popular destinations for foreign R&D activities because they offer a combination of highly qualified human resources, well-developed infrastructure, a concentration of industrial and science parks as well as top-class universities and research institutes (Gassmann and Han, 2004). While proximity to government seems to determine location in telecommunications, in other sectors availability of graduates (for example, engineering schools) and access to scientific, fashion or key consumer communities seem to be more important (von Zedtwitz, 2004).

The extent to which foreign companies locate innovative or global R&D functions in China differs significantly according to industry. Initially, telecommunications and IT or personal computer companies have been at the forefront, whereas life-science companies were less likely to locate such functions in China (Asakawa, 2005; Gassmann and Han, 2004). Starting in 2006, a number of chemical and pharmaceutical companies have announced plans to set up global R&D in China (Tremblay, 2006). Both Novartis and Astra Zeneca announced plans to invest US$100 million in R&D facilities in China (press clippings and company homepages).

Whereas R&D investments were initially concentrated within high-technology industries and activities, lately, a number of foreign-owned or foreign-invested global product design centres have sprung up in the Shanghai area. Philips, Sony, GM, Omron and Motorola have established design centres in China, and several companies report concrete plans to do so in the near future. Companies are attracted to China because it offers good and inexpensive designers (Business Week, 2005b). Some are also starting to view the Chinese market as strategically important, not only because of its size, but because it is a dynamic and rapidly changing country that is assuming an increasingly significant role as a global trendsetter (see also section 3.4.1).

3.3 Foreign corporate R&D in China: what are the drivers?

Based on interviews as well as existing studies and surveys, I have identified three principal drivers for the location of R&D centres by foreign firms in China.[8] The first driver is proximity to market and production. Many foreign centres are set up to adapt products and services to the strategically important Chinese market and/or to be near production facilities, which are already in China. The second reason for companies to locate R&D to China is political or institutional conditions. Examples of this driving force include 'local content' rules, or national standards (von Zedtwitz, 2004). There are also national regulations that may require foreign companies interested in setting up production facilities to set up R&D facilities as well as fiscal incentives. The third factor attracting R&D to China is the supply of knowledge resources in China. While all three factors play a role in explaining foreign companies' R&D activities in China, the relative weight of each factor has been changing over time.[9]

3.3.1 Proximity to market and production

In principle, all companies with R&D facilities in China had manufacturing, purchasing and/or distribution activities there before they set up research or product development. Multinational companies initially established production in China *both* because of low production costs and because of an attractive domestic market. Eventually, as companies gained experience from operating in China, some of them started to increase the range of their activities there by adding R&D to sales, purchasing and production activities. Dramatic growth rates in the number of PC and Internet users, for example, make China an attractive market for companies such as Google and Hewlett Packard, explaining why telecommunications and IT companies have dominated foreign R&D activities in China. Analysing company-level data on science and technology activities, Motohashi (2006) finds that 'the major motivation of foreign R&D in China is "market driven" instead of "technological driven" or "human resource driven"'. However, he acknowledges that the motivations differ depending on where in China companies establish R&D. Thus, market-driven R&D is observed primarily in Guangdong, whereas foreign companies' R&D operations in Beijing are more technology driven, due largely to a concentration of scientific institutions in China's capital. He finds that 'Shanghai, with both a large industrial as well as strong science sector, is in-between.'

3.3.2   Human capital

Human resources have grown significantly in China in recent years – in terms of both quantity and quality. China's increasing research strength, combined with well-equipped laboratories and a large supply of relatively inexpensive scientists and engineers, is attracting the attention and investments of many R&D-intensive companies. Thus, for example, a growing number of pharmaceutical companies are choosing China as a location for carrying out clinical trials. The labour cost for a PhD-level researcher in Shanghai is estimated to be around one-fifth of the cost for a similar resource in Silicon Valley and the costs for conducting clinical trials in China are about one-quarter of the costs in the USA (Business Week, 2005a). Interviews confirm that well-qualified, motivated and relatively inexpensive engineers, doctors and other scientists constitute an important pull factor in companies' considerations to establish R&D activities in China. As mentioned above, foreign companies' innovative R&D operations are strongly concentrated in the Beijing and Shanghai area. Both offer a large supply of highly skilled labour, which is explained by a concentration of internationally renowned universities and research institutions.[10]

When asked to reflect upon their experiences from carrying out R&D in China, R&D managers of three foreign firms indicated a high level of satisfaction with the quality of human capital. They acknowledged that Chinese employees, on average, may lack some management capabilities or the ability to think 'out of the box' in comparison to employees in other regions. However, several executives pointed out that they do not expect this to remain a shortcoming of Chinese employees. Evidence of this can be seen in the number of companies beginning to promote local Chinese employees to replace foreigners in key management positions (The Wall Street Journal, 2006).

3.3.3 Policies and government incentives

Perhaps the most important reason why foreign companies initially located R&D in China was because they had to. Thus, in addition to its abundant labour supply, low manufacturing and transport costs, and a large domestic market, foreign investors have been drawn to China by a combination of FDI-friendly policies and 'persuasion'. Since China began to open its borders to foreign companies, it has pursued a policy of requiring companies interested in producing or selling goods and services in China to transfer technology, sometimes referred to as the 'market for technology strategy' (Cao et al., 2006; Gassmann and Han, 2004). As an example, China used its market as leverage for requiring technology transfer when automobile companies competed for licenses to establish joint venture establishments in China in the late 1990s when there was speculation that this would be the last license to be issued for a long time (Gassmann and Han, 2004).

Furthermore, companies are establishing R&D operations in China because significant tax rebates and other financial incentives are on offer. In addition to preferential policies for FDI in general, a number of policies are targeted at attracting technology-intensive activities of foreign companies.[11] Examples of the policies targeting foreign technology-intensive activities are exemptions from customs duties and VAT on the import of equipment and technologies for self-use. Gains from technology transfer activities can be exempt from business and enterprise income taxes. Some R&D and wage expenses can be used to offset enterprise income taxes. Science parks and high-tech development zones advertise tax rebates and other benefits on their websites for companies willing to establish R&D activities on their premises.[12]

Some of China's policies for attracting foreign R&D are argued to be in conflict with WTO rules. Officially, the requirements for foreign firms to establish R&D in China have now been removed (Schwaag Serger, 2006). However, in practice, many companies are still 'encouraged' or pressured to locate R&D in China (see Walsh, 2003; Long, 2005). Overall, there are currently no signs that China intends to phase out preferential policies for attracting foreign R&D.[13] The 11th five-year plan for utilizing foreign investment published by the National Development and Reform Commission (NDRC) in November 2006 explicitly calls for the design of preferential policies to encourage foreign firms to establish R&D centres in China.

Finally, domestic technical requirements and standards provide a further explanation why companies such as Motorola, Microsoft, Ericsson, Sony Ericsson, and Nokia were among the first to set up extensive R&D operations in China (von Zedtwitz, 2004). China is pursuing a policy of developing national standards in high-tech fields, particularly IT, telecommunications and biotechnology. Examples include initiatives to develop standards in wireless local area networks (WAPI) technology, and third-generation mobile telephony (TD-SCDMA). As explained in Suttmeier and Yao (2004) and Kennedy (2006), this policy is driven both by an ambition to promote the development of internationally successful Chinese high-tech firms and by the desire to appropriate a greater share of the gains from globalization and innovation. Thus, there is currently a widespread perception that foreign companies are earning unfair returns or profits, which according to some constitute 'monopoly rents', from owning and excessively charging Chinese companies for patents and royalties. China's market size lends its efforts to establish its own standards considerable weight. As a result, China's standards policy provides a further explanation for why foreign firms seeking to compete in certain high-tech sectors, both within China and abroad, are increasingly establishing R&D within China's borders.

3.3.4   Challenges of locating R&D to China

Gassmann and Han (2004) identified a number of barriers that foreign companies face when managing R&D in China. They categorized these barriers as stemming from either complexity or unpredictability, or a combination of both. Barriers experienced by companies within their organization include management and communication difficulties due to language and cultural gaps and high employee turnover rates.[14] Outside the organization, companies identify bureaucracy and insufficient enforcement of intellectual property rights as important obstacles to operating R&D activities in China.

One issue that is often mentioned as an important factor deterring foreign companies from locating their R&D in China is the weak protection of intellectual property rights (IPR) in that country. In my interviews, I found no clear consensus on this matter. While some interviewees listed fear of piracy as a clear concern, others claimed that this was not a significant obstacle. Some executives also expressed their optimism that the issue of IPR would resolve itself over time, as Chinese companies become more innovative and thus acquire a stake in good IPR laws and enforcement. Zhao (2006) argues that, first, weak IPR protection results in the underutilization of 'innovative talents' in countries such as China and India, and, secondly, that multinational enterprises (MNEs) have internal structures and mechanisms that enable them to protect their research results in countries with weak IPR protection, such as China (p. 1185). While the former makes it attractive to locate R&D in countries with weak IPR and thus to be able to tap into the pool of underutilized human capital, the latter makes it possible for firms to do so. According to Zhao, the combination of these two factors explains why multinationals are increasingly locating R&D operations to countries with weak IPR protection.

3.4 Foreign R&D and China's innovation system

Recently, there has been growing criticism of China's strong focus on attracting FDI. China's FDI policies are claimed to be 'tilting' the playing field in favour of foreign and foreign-funded companies (see, for example, China Daily, 2004, and Prasad and Wei, 2005). Some observers claim that India's less aggressive promotion of FDI, and its 'favoring [of] domestic investment over foreign', has allowed the development of domestic companies that can now compete on the international market (The Economist, 2005b). Partially in response to the domestic criticism, the government recently announced the introduction of a unified corporate income tax rate for foreign and domestic companies. According to official estimates, previously domestic enterprise paid on average 25 per cent income tax, whereas foreign firms paid only 15 per cent (Asia Times, 2007).

Economists are also questioning more generally whether China's policies aimed at promoting investment, both domestic and foreign, are beneficial to China's long-term economic development and prosperity. The debate over the benefits of preferential investment policies is connected partly to concerns about growing inequalities and accusations that China's growth has benefited a small minority at the expense of the vast majority of people. Kuijs and Wang argue that

> ... reducing subsidies to industry and investment, encouraging the development of the services industry, and reducing barriers to labor mobility would result in a more balanced growth and a substantial reduction in the income gap between rural and urban residents. (Kuijs and Wang 2006: 1)

Furthermore, China's dependence on foreign companies for the production and export of high-technology products is identified as an example of its 'unhealthy' reliance on external technologies. Companies with either whole or partial foreign ownership accounted for 57 per cent of China's total exports in 2004 and 58 per cent of total imports, in value terms. Their share of national industrial output has increased from 2 per cent in 1990 to 32 per cent in 2004 (Invest in China website). Foreign companies also dominate exports of high-technology goods, accounting for more than 80 per cent of China's total high-tech exports in 2005 (China High-Tech Industry Statistics, 2006).

Criticism of the government's policy of attracting foreign R&D is also growing. Critics are questioning to what extent there are positive spillovers from foreign R&D centres to domestic companies and research institutions. They claim that foreign research centres may actually be starving domestic companies of the best scientists and engineers and criticize the government for putting too much emphasis on attracting foreign technologies, rather than promoting the growth of domestic technologies. Thus, for example, a recent article reported that there were around 1,000 R&D institutions by foreign firms in China and asked: 'Are these 1,000 rivals and 1,000 obstacles, or 1,000 partners and 1,000 opportunities?' (Science and Technology Daily 2006, 'Whether China-based Foreign-funded R&D Institutions Can be Included in China's Independent Innovation System', 27 September; see also Cao, 2004 or Yuan, 2006). In the latest *Medium and Long-Term Plan for Science and Technology Development* issued by the State Council in 2006, the government emphasized the need to strengthen 'independent' or 'indigenous' innovation. Among other things the plan calls for measures to reduce China's dependence on foreign technology and innovation (cf. Breidne and Schwaag Serger, 2006; Cao et al., Chapter 10). These and other indications of rising so-called 'techno-nationalism' are creating concerns among foreign companies in China (Suttmeier and Yao, 2004).

Several studies have examined the impact of foreign R&D activities on host countries in general (Blomström and Kokko, 1998; Reddy, 2005) and on developing countries in particular (UNCTAD 2005c).[15] Foreign corporate R&D can have positive technology or knowledge spillover effects on the host country, although their nature and scope can vary considerably. Spillovers from foreign R&D activities may occur through value chain linkages with domestic firms. Thus, domestic firms may either be suppliers, customers or competitors of foreign R&D centres. In all three cases, the interaction with or presence of a foreign R&D centre can lead to transfer or upgrading of knowledge or technology. Another channel for potential spillovers is through R&D cooperation between foreign R&D centres and Chinese universities, institutes or other organizations. There are examples of this happening with foreign firms setting up joint R&D laboratories with Chinese institutes or universities (Chen, 2006). In addition, Lundin and Schwaag Serger (2007) show that the outsourcing of science and technology activities by foreign firms to Chinese firms increased dramatically between 2000 and 2004, providing an important channel for knowledge transfer. A third, and perhaps the most important, conduit for knowledge spillovers from foreign R&D centres to the surrounding domestic environment is people. Thus,

> The transfer of technology from MNC [Multinational Corporations] parents to affiliates is not only embodied in machinery, equipment, patent rights, and expatriate managers and technicians, but is also realized through the training of the affiliates' local employees... The various skills gained while working for an affiliate may spill over as the employees move to other firms, or set up their own businesses. (Blomström and Kokko, 1998: 13–14)

Positive spillover effects from foreign R&D centres are not guaranteed, however, but depend on a conducive local environment (Blomström and Sjöholm, 1999). Among other things, positive spillovers require a certain minimum level of human capital or 'local capability' (Blomström and Kokko, 1998). Xu (2000) examined technology diffusion from foreign MNCs in 40 different host countries. He found that whether technology diffusion occurred depended on the level of development of the host country. In less developed countries, technology diffusion was found to be limited. According to Xu, this was due to the fact that these countries lacked the human capital necessary 'to absorb the technology diffused by MNEs' (Xu, 2000: 479).

Lack of human capital may offer a partial explanation of why positive knowledge spillovers from foreign corporate R&D centres have been limited in China to date. Several studies point out that while China has a relatively high literacy rate compared with India and also with many other developing countries, there is a shortage of highly skilled labour and, particularly, of people with the set of skills necessary for setting up, developing, managing and working in innovative companies (Farrell and Grant, 2005).

In addition to a shortage of human capital, an additional explanation for limited knowledge spillovers could be a lack of mobility of human capital. Thus, in the case of China, positive knowledge spillovers from foreign R&D centres appear to be limited not so much, or not only, because there is too little human capital but rather because a significant portion of its human capital is working in foreign firms and shows little inclination to move to domestic firms or start their own firms. In several surveys carried out in recent years, Chinese students list foreign enterprises as their favourite place for future employment. Thus, for example, 64 per cent of students interviewed in Beijing in 2004 intended to work in a foreign enterprise after graduation ('Why Chinese Graduates Prefer to Work in Foreign Enterprises', survey carried out by Beijing Century Perspective Marketing Research Company, published on China Test and Study website, 6 July 2004). In another survey, where more than 4,000 students from China's top universities were asked to rank their preferred employers, 13 out of the 20 highest-ranking employers were foreign companies, leading the authors conclude that '[t]here is apparently a lot of working power in China and the Chinese students seem to want to use this power in big multinational companies' (Universum Communications, 2005). Similarly, *China Daily* recently observed that '[w]orking at multinational companies (MNCs) has been the preferred choice for college graduates in China for years' (China Daily, 2006a).

While surveys show a clear preference among university graduates to work for foreign firms, there is little available data to show to what extent those who work in foreign companies in China eventually go to work for domestic firms or start up their own firms. Although there are examples of Chinese employees leaving foreign firms to set up their own companies (Chen, 2006), when Chinese employees leave a foreign firm, many of them tend to do so in order to work for another foreign firm that is offering them better pay or a better position. Thus, while the turnover among local employees in foreign companies in China appears to be high, and is identified as a significant problem by foreign employers, this seems to be due more to a 'circulation' of a group of Chinese employees among foreign enterprises, rather than to a movement, and thus a vehicle for knowledge spillovers from foreign MNCs to domestic companies.

When it comes to domestic firms, students prefer state-owned enterprises (SOEs) over privately owned enterprises. This is partially due to the fact that SOEs offer more job security and benefits in terms of healthcare and pension schemes, something which has become an important consideration for jobseekers (China Through a Lens, 'New Trends in Employment of China's Graduates', 15 February 2006). The dismantling of China's public healthcare, pension and education systems can thus be argued to have introduced a bias against private domestic firms in the labour market. The difficulty for them to attract human talent, with the best graduates preferring to work for foreign companies or SOEs, constitutes an important barrier to the development of innovative, competitive and successful firms in China.[16]

In addition to the limited mobility of human capital between foreign and domestic companies, knowledge spillovers or linkages from foreign R&D centres to Chinese companies and institutions may be limited because of a limited receptive capacity among domestic firms and institutions. Examining potentials for spillovers from foreign technology activities to host countries, Narula and Lall (2004) emphasize the importance of 'absorptive capacity' for generating positive spillovers from foreign R&D activities. Several factors indicate that the absorptive capacity of firms and other key actors in China's innovation system is low. First, Chinese firms still invest comparatively little in R&D compared with European, Japanese or North American firms. Thus R&D expenditure as a share of value-added was only 1.9 per cent of value-added in Chinese manufacturing companies (see Table 3.1). The discrepancy becomes even greater when looking at R&D intensity in Chinese high-tech companies compared with more developed countries.

Secondly, a large share of China's business expenditure on R&D is carried out in large state-owned enterprises (SOEs). Thus, 41 of the 50 Chinese companies with the largest R&D expenditure in 2006 were state-owned.[17] However, for several reasons SOEs' ability both to innovate and to absorb knowledge is often comparatively low. Analysts point to problems of corporate governance, weak management skills and organizational structures, monopoly powers, subsidies and preferential policies as factors explaining low innovative and absorptive capacity (see, for example, Kwan, 2006; Li et al., 2007; and China Daily, 2005, 'SOEs Have Low Innovation Capacity: Official', 18 November).

Overall, therefore, it could be argued that part of the problem appears to lie not only in a lack of technological competence but also in the fact that companies and individuals do not have the management and organizational skills and structures to internalize and make use of know-how that is outside their

Table 3.1 R&D intensity of high-tech industries and total manufacturing in selected countries (% of value-added)

	China 2004	USA 2002	Japan 2002	Germany 2002	France 2002	UK 2002	Italy 2002	Korea 2003
Manufacture	1.9	7.8	10.4	7.7	7.4	6.9	2.3	7.3
Total high-tech industries	4.6	27.3	29.9	24.1	28.6	26.0	11.6	18.2
Pharmaceutical products	2.4	21.1	27.0	–	27.2	52.4	6.6	4.4
Aircraft and spacecraft	16.9	18.5	21.6	–	29.4	23.8	23.4	–
Electronic and telecom equipment	5.6	25.4	20.4	39.2	57.2	23.6	19.4	23.4
Computers and office equipments	3.2	32.8	90.4	18.1	15.8	5.9	8.8	4.4
Medical equipment and meters	2.5	49.1	30.1	14.0	16.1	8.3	6.4	10.7

Source: China High-Tech Industry Statistics 2006
(http://www.sts.org.cn/sjkl/gjscy/data2006/data06.htm).

core competence. A recent article examining innovation in China's IT industry pointed out that domestic companies lagged behind foreign competitors not so much because they did not possess the necessary core technologies, but because product development was too technology-driven and failed to be sufficiently market-oriented.[18]

Finally, weak social capital or institutional conditions constitutes an additional important impediment for knowledge spillovers or linkages between foreign firms' R&D activities in China and the surrounding environment. According to Narula (2005),

> In a system, the efficiency of economic actors – firm or non-firm – depends on how much and how efficiently they interact. The means by which interactions take place are referred to as institutions in the economics literature, though sociologists prefer to speak of social capital. (p. 49)

Social capital is used to capture the notion that economic value creation depends not only on physical capital (tangible assets such as land, machinery and so on) and human capital (knowledge and skills), but also on the value that derives from people's willingness and likelihood to share knowledge and information (for an overview of the concept of social capital see, for example, Woolcock, 1998). Social capital can be defined as shared values, norms and trust that reduce transaction costs.

Social capital is sometimes erroneously equated with networks. Many observers point out that China has strong family ties or networks, sometimes referred to as the 'bamboo networks' or the strength of *guanxi* (relationships), and to their importance in the conduct of business and other affairs in China (cf. Watkins-Mathys and Foster, 2006; Ramasamy et al., 2006). A more accurate indicator of a high level of social capital could be the willingness of people to share information or knowledge with people *outside* their immediate network. In countries with weak social capital, business interactions and knowledge transfer is often limited to family networks. However, strong family networks combined with a low level of trust can lead to a suboptimal utilization of resources, if knowledge and information are only shared with people who belong to one's immediate network rather than being channeled to people who might be able to use it most effectively. An example could be a researcher choosing to sell his or her invention to his or her uncle who has a company but not the know-how necessary for the successful commercialization of the specific invention, rather than seeking out companies or other partners, which may be outside the family network but which possess the technical and other resources relevant to the invention.

Weak social capital provides an important impediment for knowledge spillovers or linkages between foreign firms' R&D activities in China and the surrounding environment. Vang and Asheim (2006), for example, examine spillovers from foreign firms' R&D activities in Shanghai and find that the absence of sufficient social capital, and particularly trust, 'is still limiting the possibility of developing interactive learning environments which is a precondition for improving the absorptive capacity at the firm level' (p. 55). A recent analysis of commercialization of life science in China confirmed that lack of 'trust' and, thus, willingness to interact with people outside individuals' personal networks constituted a fundamental barrier to commercialization and the development of a competitive pharmaceutical industry in China (Nilsson et al., 2006). Finally, the comparatively high level of corruption, which can be seen as a proxy for social capital, indicates or further reinforces a low level of trust, undermining the efficiency of economic interactions (see also Narula, 2005).

Some authors argue that foreign R&D activities can have negative net effects on a host country's innovative capacity, particularly when absorptive capacity among domestic firms is low, or when there are other impediments to knowledge transfer and positive spillovers (see, for example, Lall and Narula, 2004; Zhou, 2005). Thus, the presence of foreign firms and their attractiveness as employer may starve domestic firms of highly skilled manpower. While there may be some evidence of this, in the case of China it is also conceivable that the presence of a critical mass of foreign firms with strategic R&D activities plays a significant role in attracting highly educated and experienced people, including overseas Chinese, to the country who may not have come to China otherwise.

3.4.1 Looking forward

The following statement by a foreign executive with a long record of business experience in China succinctly summarizes the development of foreign companies' R&D in China: 'Initially, firms located R&D here because they had to. But once they were here, they realized how attractive it was for them to do R&D in China.' Thus, government policy of 'encouraging' or effectively demanding technology transfer in return for access to the Chinese market played a key role in initially getting companies to establish R&D centres in China. This seemingly 'bitter pill' was sweetened with generous tax and other incentives for foreign companies setting up R&D operations in China. While government policy still plays an important role in some industries, in recent years its role is waning and being supplemented by a combination of factors (for example, the proximity to production facilities, a large market, and human capital) that constitute a strong argument in favour of foreign companies establishing R&D in China. Unless there is a drastic slowdown in its economic development or a radical reversal of its current investment-friendly policies, foreign firms' R&D activities in China are likely to continue to increase in the future. Furthermore, the trend towards establishing strategic or global R&D activities in China will also strengthen as China's market becomes, if anything, more important, and human capital continues to strengthen both in quantity and quality. The importance of China as a base for global R&D is underlined by a new phenomenon, whereby even firms without significant sales or production in China are considering establishing strategic R&D in China. An example of this is Vodafone, which recently announced the establishment of an R&D centre in Beijing.[19] Another emerging trend is for foreign firms to design products aimed specifically at consumers in developing countries. Examples include a new beverage created by Coca Cola's R&D centre in Shanghai, a low-cost PC developed at Dell's China Design Center, and Philips Mobile Phone Design Center in Shanghai which has a specific mission to develop low-cost mobile phones for emerging markets.[20] Both Coca Cola's beverage and the Dell computer were originally designed for the Chinese market but with a potential to be marketed and sold to consumers in other emerging markets.[21]

However, a change in Chinese government policy or in the public attitude towards foreign firms could rapidly reduce firms' willingness to locate R&D activities there. As mentioned earlier, there are some indications of growing wariness towards foreign firms and possible 'techno-nationalism'. Examples are new public procurement rules which favour domestic firms over foreign companies, even those with significant production and research facilities in China (see also Schwaag Serger and Breidne, 2007). These and other indications of a change in China's previously very welcoming stance towards FDI and foreign corporate R&D are raising concerns among foreign firms (New York Times, 2007). Some journalists argue that the recent attempts to reduce foreign penetration of China's economy are a response to growing economic nationalism, and, particularly, resistance and discrimination against Chinese firms and products in countries such as the US (Business Week, 2007). Regardless of the causes for the apparent rise in 'economic nationalism', if these trends continue they are likely to reduce significantly foreign firms' willingness to establish R&D, and particularly, global or innovative R&D in China.

3.5 Discussion and conclusion

Foreign firms' R&D activities have been growing rapidly in recent years, and this trend can be explained by a unique combination of factors. First, low production costs and the domestic market coincide to make China a highly attractive location for production. Secondly, and in addition to these two factors, China offers a competitively priced and abundant pool of highly skilled labour. The combination of these factors means that companies or industries can either locate or have access to the entire value chain of their product in one, albeit large, country. Finally, China has combined favourable policies with 'technology-for-market' strategies to attract foreign R&D. According to several interviewees, it is the combination of factors that makes China attractive for purchasing, production, distribution *and* R&D.

China's rise as an important global manufacturing *and* knowledge base, as well as a large domestic market, will make it an increasingly strong contender for research and development. The emergence of countries such as China and India as attractive locations for corporate R&D indicates a major shift in international economic relations. R&D, knowledge and human capital are no longer the exclusive domain or privilege of developed countries. Nor is their location as path-dependent as one may have previously assumed. Rather, knowledge is becoming a mobile asset. The increasing mobility of knowledge is partially explained by the fact that large multinational companies account for an increasing share of R&D expenditures. Ford Motor or Siemens, for example, have larger R&D budgets than countries such as Spain, Belgium, Switzerland or Brazil (UNCTAD, 2005b). These multinational companies are increasingly locating R&D where the markets are attractive and human capital is good, both in terms of price and quality.

A significant share of foreign R&D operations consists of product adaptation to the domestic market. However, recently a growing number of foreign firms have also located global, strategic or innovative R&D activities in China. In order to maximize the positive spillovers from foreign firms' R&D activities in China, policy efforts should focus on increasing the ability of domestic companies, consumers and institutions to generate but also to receive, absorb and internalize knowledge and new ideas, products and processes (see also Schwaag Serger and Breidne, 2007). This, in turn, requires policy makers to pay attention to 'soft' aspects of innovation and framework or institutional conditions, rather than focusing only on R&D expenditure, patents and the science and engineering graduates. Many of the obstacles currently preventing or limiting positive spillovers from foreign R&D centres to domestic firms and institutions originate in weak institutional conditions. Examples that could be mentioned are a lack of social capital, particularly trust, and a weakness of pull factors that might induce people currently working in foreign R&D centres from eventually moving to domestic private companies or starting their own firms. Given the lack of affordable and widely available healthcare and education in China today, in addition to concerns about mismanaged and insufficient public pension funds, jobseekers will prefer to work for foreign enterprises, SOEs or the government, all of which have the resources to offer pension schemes, health benefits, as well as relative job security. Thus, China's highly inequitable and insufficient provision of social services creates a strong bias against private firms in the labour market.

Even if China still has a long way to go in becoming a knowledge-based economy, the country already has more knowledge resources, quantitatively speaking, than any other country in the world with the exception of the USA. In addition, strong international knowledge and research environments are developing within some high-technology fields with considerable support from the government. At the same time, China's innovation system still suffers from significant weaknesses that prevent it from reaping greater economic and societal benefits, both from the knowledge it invests in through growing R&D expenditures and from the knowledge it attracts through Chinese returnees and foreign firms' R&D activities (Schwaag Serger, 2006). These weaknesses illustrate some important general points. Knowledge and innovation capabilities cannot be created by decree, and the processes of knowledge creation, utilization, and transformation are not top-down processes. Rather, they depend on a complex set of values, learning processes, networks and interactions, which require – and must be gradually shaped by – social capital (defined as shared values, norms and trust that reduce transaction costs), communication and competition.

One fundamental ingredient for resilient and innovative societies and economies is human capital. Human capital, in turn, depends strongly upon a good education system. It could be argued that, over recent decades, China has prioritized research over education. Thus, total government appropriation on education, as a share of total government expenditure, dropped from 15.4 per cent in 1995 to 12.1 per cent in 2003 (UNICEF, 2006). Whereas China's expenditure on R&D is comparatively high for a developing country, having increased from 0.6 per cent of GDP in 1996 to 1.3 per cent in 2005, its education spending, at around 2.8 per cent of GDP, is comparatively low. For example, India and Brazil spend less money, as a percentage of GDP, on R&D but more on education (see Table 3.2).[22]

In its latest five-year plan, the Chinese government has targeted education as a prioritized area and announced its intention to increase budgetary expenditure on education to 4 per cent of GDP by 2010. In addition to increased

Table 3.2 Government R&D and education expenditure as a
percentage of GDP (latest available years)

	R&D expenditure (% of GDP)	Public expenditure on education (% of GDP)
Japan	3.2	3.6
US	2.7	5.7
EU-15	1.9	5.0
EU-25	1.8	na
China	1.4	2.8
Brazil	1.0	4.1
India	0.8	3.3

Sources: UNICEF (2006), UNESCO (2006), UNESCO UIS Database.

funding, China is currently examining the foundations of its education system. Thus, there is a lively debate within China on the general emphasis of education, which is viewed as focusing too much on teaching people to know and to memorize, rather than to think. There are several examples of local governments, schools and universities working to change their curricula and teaching methods. Overall, recent developments indicate that China is working on improving its education system, both by increasing its funding and its quality, and thus acknowledging its key role in innovation and future prosperity.

Notes

1. For an overview over this development and its policy implications, see Karlsson (2006).
2. For this section, I have summarized and updated an earlier report which I wrote on foreign corporate R&D in China (Schwaag Serger, 2006).
3. I arrive at this number by using von Zedtwitz's figure from 2004 as a point of departure and then conducting a search of Chinese and foreign media articles, press releases and company reports to get an estimate of how many foreign companies have established R&D centres since 2004. I focus particularly on companies with existing production facilities, or other relevant presence, in China, since it is very unlikely for firms without manufacturing or other operations in China to set up R&D there.
4. Reddy (2005) distinguishes between five types of foreign R&D units, depending on the nature of the work they carry out; among others he identifies 'Global Technology Units' or 'Corporate Technology Units'.
5. According to its website, Nokia carries out research on visual interaction systems and adaptive terminals, among other things, at its Beijing research centre (Nokia website, accessed 29 November 2006).
6. Unilever website (accessed 25 March 2007) http://www.unilever.com/ourvalues/sciandtech/How_where/On_the_map.asp.

7. Fujitsu website accessed 29 November 2006.
8. This section is a summary of an earlier paper (Schwaag Serger 2006). An overview of trends and drivers of international R&D in general can be found in Karlsson (ed.) (2006), chapter 2.
9. This analysis is derived by combining results from published studies with findings from interviews carried out with R&D managers and other experts between May 2005 and November 2006.
10. While there are signs of emerging labour shortages in certain regions, aggregate unemployment is still high and thus China is likely to have a large pool of cheap labour for quite some time.
11. Preferential FDI policies include low tax rates or tax exemptions on VAT, corporate taxes and income taxes, exemptions from import tariffs on production inputs imported by foreign-invested enterprises (FIEs), favourable land-use rights, administrative support, subsidized office rents, etc. (see, for example, Hong Kong Trade Development Council 2004, and Hou 2004). Foreign companies establishing themselves in China are exempt from corporate income tax for the first two years that they make a profit. After that, they are subject to 15 per cent corporate income tax on average, which is much less than the normal rate for Chinese companies of 33 per cent (Prasad and Wei, 2005).
12. See, for example, Jiangsu Province Taixing Economic Development Zone, http://www.chempark.com.cn/enwhh/htm/1_guide09.htm, or Xi'an High-Tech Development Zone, http://www.cbw.com/business/invest/xian/policies.htm.
13. This is exemplified by the 'Opinions of Shanghai Municipality on Encouraging Overseas Investment in the Establishment of Research and Development Institutions', formulated and adopted as recently as 2003/04 see http://w2.tdctrade.com/report/reg/reg_040102.htm.
14. Several people I interviewed spoke of a tendency among some Chinese employees in multinational companies to 'circulate' among the community of foreign firms, frequently switching jobs in search of career advancement or higher salaries.
15. A good overview over the literature and relevant issues in FDI-assisted economic development and technology upgrading is also provided in Lall and Narula (2004).
16. In a small survey we carried out we found that job security, healthcare and pensions were important considerations for people with university degrees in their twenties, and that these considerations were one argument for them to view SOEs or the government as attractive employers.
17. Data on R&D expenditure from the China Enterprise Confederation.
18. IT Manager, 'The Problems of "Independent Innovation"', 5 November 2006, p. 12.
19. Vodafone presentation, Demos Conference, London, 16 January 2007.
20. Philips homepage http://www.philips.com.au/About/News/press/article-14451.html.
21. The Economist (2007) 'Orange Gold', 1 March 2007, and International Herald Tribune (2007) 'Dell Introduces a Low-cost PC for China', 21 March 2007.
22. Education expenditure data for India and Brazil from UNESCO UIS Database.

References

Allen, Franklin, Jun Qian and Meijun Qian (2006) 'China's Financial System: Past, Present and Future', *Wharton Financial Institutions Center Working Paper*, No. 05–17, April.

Asakawa, K. (2005) 'Accelerating R&D Investments into India and China', *Columns Back Issues*, No. 4, columns 0137, Research Institute of Economy, Trade and Industry (RIETI).

Asia Times (2007) 'World Bank Downplays Impact of Tax Reform', 13 March.

A.T. Kearney (2006) *FDI Confidence Index*, Global Business Policy Council 2005, vol. 8: 2–3.

Bekier, Matthias M., Richard Huang and Gregory P. Wilson (2005) 'How to Fix China's Banking System', *The McKinsey Quarterly*, 1: 110–19.

Blomström, Magnus and Ari Kokko (1998) 'Multinational Corporations and Spillovers', *Journal of Economic Surveys*, 12(2): 1–31.

Blomström, Magnus and Fredrik Sjöholm (1999) 'Technology Transfer and Spillovers: Does Local Participation with Multinationals Matter?', *European Economic Review*, 43: 915–23.

Breidne, M. and S. Schwaag Serger (2006) 'Stark tro på tillväxt genom teknisk förnyelse i Kinas nya långtidsplan', *Tillväxtpolitisk Utblick*, No. 6, Stockholm: Swedish Institute for Growth Policy Studies (ITPS).

Buderi, R. (2005) 'Microsoft: Getting from "R" to "D"', *Technology Review*, 108(3): 28–30.

Business Week (2005a) 'A New Lab Partner for the US?', by Bruce Einhorn and John Carey, 22 August.

Business Week (2005b) 'China Design', by David Rocks, 21 November.

Business Week (2006) 'Novartis in China: East Meets West in R&D', by Kerry Capell, 6 November.

Business Week (2007) 'China: A Tide of Economic Nationalism', Interview with Beijing Bureau Chief Tiff Roberts, 28 June, http://www.businessweek.com/mediacenter/podcasts/international/international_07_28_06.htm.

Cao, C. (2004a) 'Challenges for Technological Development in China's Industry. Foreign Investors are the Main Providers of Technology', *China Perspectives*, 54.

Chen, Yun-Chung (2006) 'Changing the Shanghai Innovation Systems: The Role of Multinational Corporations R&D Centres', *Science, Technology and Society*, 11(1): 67–107.

China Daily (2004) 'Tempering FDI-inviting Policies', 29 September.

China Daily (2006a) 'Hard Climb', China Business Weekly, 18–24 September.

China Daily (2006b) 'Overseas ventures crucial for Chinese firms', 13 January.

De Ramos, A. (2003) 'The China Syndrome', *CFO Magazine*, October.

Economist, The (2005a) 'The Struggle of the Champions', 6 January.

Economist, The (2005b) 'The Insidious Charms of Foreign Investment', 3 March.

Economist, The (2007) 'Orange Gold', 1 March.

Eurostat (2006) 'R&D and Internationalisation', by Simona Frank, *Statistics in Focus*, 15/2006.

Farrell, D. and A.J. Grant (2005) 'China's Looming Talent Shortage', *The McKinsey Quarterly*, 2005:4: 70–9.

Financial Express (2005) 'Reaping Rich Dividends: China's Experiment With "Storing Brain Power Overseas" is Paying Off', by Pallavi Aiyar, 17 December.

Gassmann, O. and Z. Han (2004) 'Motivations and Barriers of Foreign R&D Activities in China', *R&D Management*, 34(4): 423–37.

Hong Kong Trade Development Council (2004) *Guide to Doing Business in China*, 2004/05 edition.

Karlsson, Magnus (ed.) (2006) *The Internationalization of Corporate R&D: Leveraging the Changing Geography of Innovation*, Stockholm: Swedish Institute for Growth Policy Studies (ITPS).

Kennedy, S. (2006) 'The Political Economy of Standards Coalitions: Explaining China's Involvement in High-Tech Standards Wars', *Asia Policy*, July: 41–62.

Kuijs, L. (2005) 'China's Savings: Where Does All the Money Come From?', *China Economic Quarterly*, 9(3): 40–4.

Kuijs, L. and T. Wang (2006) 'China's Pattern of Growth: Moving to Sustainability and Reducing Inequality', *China and World Economy*, 14(1): 1–14.

Kwan, Chi Hung (2006) 'Who Owns China's State-Owned Enterprises? Toward Establishment of Effective Corporate Governance', Research Institute of Economy, Trade and Industry, *China in Transition*, 28 July 2006, http://www.rieti.go.jp/en/china/06072801.html. Accessed on 28 July 2006.

Li, Y., Y. Liu and F. Ren (2007) 'Product Innovation and Process Innovation in SOEs: Evidence from the Chinese Transition', *Journal of Technology Transfer*, 32: 63–85.

Lin, Zhongping (2006) 'The Influence of MNCs upon China's Independent Innovation Capacity', *China S&T Investment*, May: 40–3.

Long, G. (2005) 'China's Policies on FDI: Review and Evaluation', in Moran et al. (eds), pp. 315–36.

Lundin, N. and S. Schwaag Serger (2007) 'Globalization of R&D and China: Empirical Observations and Policy Implications', Research Institute of Industrial Economics (IFN) Working Paper, Nr. 710, Stockholm, Sweden.

Miesing, P., M. Krieger and N. Slough (2007) 'Towards a Model of Effective Knowledge Transfer Within Transnationals: the Case of Chinese Foreign-invested Enterprises', *Journal of Technology Transfer*, 32: 109–22.

Ministry of Commerce (2005) *Report on the Foreign Trade Situation of China*, Beijing.

Ministry of Science and Technology (2005) *China Science and Technology Statistics Data Book 2005*, Beijing.

Moran, T. H., E. M. Graham and M. Blomström (eds) (2005) *Does Foreign Direct Investment Promote Development?*, Washington, DC: Institute for International Economics.

Motohashi, K. (2006) 'R&D of Multinationals in China: Structure, Motivations and Regional Difference', Draft Version, Research Institute of Economy, Trade and Industry (RIETI).

Narula, R. (2005) 'Knowledge Creation and Why it Matters for Development: the Role of TNCs', in UNCTAD, *Globalization of R&D and Developing Countries*, Geneva: UNCTAD, pp. 43–60.

Narula, R. and S. Lall (2004) 'Foreign Direct Investment and its Role in Economic Development: Do We Need a New Agenda?', *The European Journal of Development Research*, 16(3): 447–64.

National Bureau of Statistics of China (NBS) (2006) *China Statistical Yearbook 2006*, Beijing.

National Science Foundation (2006) *National Science and Engineering Indicators 2006*, Beijing.

New York Times (2007) 'China Stand on Imports Upsets US', by Steven R. Weisman, 15 November.

Nilsson, A. S., H. Fridén and S. Schwaag Serger (2006) *Commercialization of Life Sciences in the USA, Japan and China*, Report A2006:006, Stockholm: Swedish Institute for Growth Policy Studies.

NRCSTD (2005) *Science and Technology Indicators 2004*, Beijing: National Research Center for Science and Technology for Development.

Prasad, E. and S. Wei (2005) 'The Chinese Approach to Capital Inflows: Patterns and Possible Explanations', *Working Paper*, No. 11306, Cambridge, MA: National Bureau of Economic Research (NBER), April.

Ramasamy, Bala, K.H. Goh and Mathew C. H. Yeung (2006) 'Is Guanxi (Relationship) a Bridge to Knowledge Transfer?', *Journal of Business Research*, 59: 130–9.

Reddy, P. (2005) 'R&D-related FDI in Developing Countries: Implications for Host Countries', in UNCTAD, *Globalization of R&D and Developing Countries*, Geneva: UNCTAD, pp. 89–108.

Roland Berger (2007) *Globalization of R&D – Drivers and success factors. Excerpts of the study*, presentation by Robert Ohmayer, 19 April, Stuttgart.

Saxenian, A. (2006) *The New Argonauts: Regional Advantage in a Global Economy*, Cambridge, MA, Harvard University Press.

Schwaag Serger, S. (2006) 'China: From Shopfloor to Knowledge Factory?', Chapter 10 in Magnus Karlsson (ed.), *Internationalization of Corporate R&D: Leveraging the Changing Geography of Innovation*, Stockholm: Swedish Institute for Growth Policy Studies (ITPS).

Schwaag Serger, S. and M. Breidne (2007) 'China's Fifteen-Year Plan for Science and Technology: An Assessment', *Asia Policy*, Research Note, No. 4, pp. 135–64.

Schwaag Serger, S. and E. Widman (2005) *Konkurrensen från Kina – Utmaningar och möjligheter för Sverige*, Report A2005:019, Stockholm: Swedish Institute for Growth Policy Studies.

Science and Technology Daily (2006) 'Whether Foreign-funded R&D Institutions in China Could be Integrated into the Independent Innovation System of the Country', 27 September.

Scott, Allen J. and Michael Storper (2003) 'Regions, Globalization, Development', *Regional Studies*, 37: 579–93.

Suttmeier, R. P. and X. Yao (2004) *China's Post-WTO Technology Policy: Standards, Software, and the Changing Nature of Techno-Nationalism*, NBR (National Bureau of Asian Research) Special Report.

Tremblay, J.-F. (2006) 'R&D Takes Off in Shanghai', *Chemical & Engineering News*, 84(34): 15–22.

UNCTAD (2005a) *Trade and Development Report 2005*, Geneva: UNCTAD.

UNCTAD (2005b) *World Investment Report 2005*, Geneva: UNCTAD.

UNCTAD (2005c) *Globalization of R&D and Developing Countries*, Geneva: UNCTAD.

UNESCO (2006) *Global Education Digest 2006*, Geneva: UNCTAD.

UNICEF (2006) *China's Budget System and the Financing of Education and Health Services for Children*, by H. Mei and X. Wang (edited by Anthony Hodges), published by UNICEF and Office of National Working Committee on Children and Women under the State Council, Beijing.

Vang, J. and B. Asheim (2006) 'Regions, Absorptive Capacity and Strategic Coupling with High-Tech TNCs: Lessons from India and China', *Strategy, Technology & Society*, 11(1): 39–66.

Universum Communications (2005) *The Universum Graduate Survey 2005 – Press Release Asian Surveys*, online publication.

von Zedtwitz, M. (2004) 'Managing Foreign R&D Laboratories in China', *R&D Management*, 34(4): 439–52.

von Zedtwitz, M. (2005) 'China Goes Abroad', in Samuel Passow and Magnus Runnbeck (eds), *What's Next? Strategic Views on Foreign Direct Investment*, published by Invest in Sweden Agency (ISA), pp. 62–9.

Walsh, K. (2003) *Foreign High-Tech R&D in China*, Washington, DC: Henry L. Stimson Center.

Wall Street Journal (2006) 'Firms in China Think Globally, Hire Locally', by Cui Rong, 9 March.

Watkins-Mathys, L. and M. John Foster (2006) 'Entrepreneurship: the Missing Ingredient in China's STIPs?', *Entrepreneurship and Regional Development*, 18(3): 249–74.

Woolcock, M. (1998) 'Social Capital and Economic Development: Towards a Theoretical Synthesis and Policy Framework',*Theory and Society*, 27(2): 151–208.

Xu, B. (2000) 'Multinational Enterprises, Technology Diffusion, and Host Country Productivity Growth', *Journal of Development Economics*, 62: 477–93.

Yuan, J. (2006) 'Perspectives to the Global R&D Strategy of MNCs', *China Information Review* Hotspot focus, 3: 16–19.

Zhao, M. (2006) 'Conducting R&D in Countries with Weak Intellectual Property Rights Protection', *Management Science*, 52(8): 1185–99.

Zhou, Y. (2006) 'Features and Impacts of the Internationalization of R&D by Transnational Corporations: China's case', in UNCTAD 2005c, pp. 109–18.

Appendix 1 Foreign firms with strategic/global R&D centres in China

Company	Location, year of establishment	Home country	Sector	Source
3M	Shanghai (2005)	USA		Roland Berger (2007)
ABB	Beijing (2005), Shanghai (2005)	Switzerland	industrial and farm equipment	company website
Agilent Technologies	Agilent Labs Beijing (2000), Shanghai, Chengdu (2005)	USA	industrial equipment	company website; telephone interview (Jan 2007)
Alcatel-Lucent	Bell Labs Research Center Beijing (2000), Research and Innovation Center Shanghai, Qingdao, Nanjing (2003), Chengdu (2006)	France/USA	network and other communications equipment	company website; telephone interview (Jan 2007)
AMD	Beijing, Shanghai (2006)	USA	semiconductors	company press release Aug. 22, 2006; *Electronic News Aug. 23, 2006*
Astra Zeneca	Clinical Research Unit East Asia, Shanghai (2002), Innovation Center China, Shanghai (2007)	UK	pharmaceuticals	company website
Ciba Specialty Chemicals	Ciba Specialty Chemicals Research and Development Center, Shanghai (2005)	Switzerland	chemicals	company press release
Cisco	China R&D Center, Shanghai (2005)	USA	network and other communications equipment	*Express India*, Oct. 13, 2005
Coca Cola	R&D center, Shanghai	USA	beverages	*The Economist*, "Orange Gold", March 1, 2007
Degussa	Shanghai (2004)	Germany	chemicals	Company website (press release April 23, 2004)
Dell	China Design Center, Shanghai (2002)	USA	computers, office equipment	company website

Company	Center/location (year)	Country	Industry	Source
DoCoMo	Research Lab Beijing (2003)	Japan	telecom	company website
Du Pont	Shanghai (2005)	USA	chemicals	Tremblay 2006
Eli Lily	Shanghai (2003)	USA	pharmaceuticals	*Asia Times*, Aug. 23, 2005
Ericsson	6 R&D centers	Sweden	network and other communications equipment	company website, interviews (Beijing, 2005 and 2006)
Ford	Research and Engineering Center, Nanjing (2007)	USA	automotive	Company website http://media.ford.com/newsroom/release_display.cfm?release=25646
France Telecom	R&D center Beijing (2004)	France	telecom	company information
Fujitsu	Beijing, Shanghai	Japan	Computers, office equipment	company website
General Electric	China Technology Center, Shanghai	USA		Chen 2006
General Motors	Pan-Asia Technical Automotive Center (PATAC), Shanghai, 1997 Global Design Studio, Hybrid Technology Research Center, Shanghai (2008?)	USA	motor vehicles and parts	company website; *Fast Company*, Issue 114, April 2007; New York Times, Oct. 29, 2007, http://www.nytimes.com/2007/10/29/business/worldbusiness/29cnd-auto.html
Google	Beijing, 2006	USA	internet	Company presentation (Beijing 2006)
GlaxoSmithKline (GSK)	GSK Research and Development Center Shanghai, 2007	UK	pharmaceuticals	company website
Hewlett Packard	HP Lab, Beijing, 2005 (?), Software Solution Center Shanghai (2000)	USA	Computers, office equipment	company website and Chen 2006

(Continued)

Appendix 1 Continued

Company	Location, year of establishment	Home country	Sector	Source
Hitachi	Beijing (2000), Shanghai (2004)	Japan	electronics, electrical equipment	Company website, http://www.hqrd.hitachi.co.jp/global/world.cfm
Honeywell	Shanghai (2004)	USA	aerospace and defense	Tremblay 2006
IBM	Beijing (1995)	USA	Computers, office equipment	company website
Infineon	Xian, 2004	Germany	semiconductors	company website
Intel	Intel China Research Center, Beijing, 1998, Shanghai, 2005	USA	semiconductors	*Business Week*, Nov. 6 2006
Matsushita/ Panasonic	Panasonic Beijing Laboratory, Beijing, 2001; Suzhou, 2002	Japan	electronics, electrical equipment	Company presentation (2005), http://www.w3.org/2005/08/SSML/Papers/PBL_Position_Paper_for_W3C_Workshop.htm
Microsoft	Microsoft Research Asia, Beijing, 1998, Microsoft Advanced Technology Center, Beijing, 2003, Microsoft Technology Support Center, Shanghai 1998	USA	computer services and software	company website and Chen 2006
Motorola	Motorola China Research and Development Institute (MCRDI) Beijing, 1993, 19 R&D centers	USA	network and other communications equipment	company website, interviews
NEC	NEC Labs, Beijing 2003	Japan	Computers, office equipment	*company websites, China Daily* November 2006
Nokia	Nokia Research Center Beijing (1998), 6 R&D labs in China	Finland	network and other communications equipment	*Business Week*, Nov. 6, 2006, company website

Nortel	Beijing, 2003, Guangdong Nortel R&D Center, 1995	Canada	telecom	website; http://www.supplychain.cn/en/art/?1246
Novo Nordisk	Beijing, 2002	Denmark	pharmaceuticals	website, company presentation (Denmark, 2006)
Novozymes	Beijing, 1995	Denmark	pharmaceuticals	company website, company presentation, State Universiry of New York, August 2005)
Omron	Omron Institute of Sensing and Control Technology, Shanghai, 2006	Japan	automation	Omron Annual Report 2006 (company website)
Oracle	Beijing Development Center (does global/Asia Research), 2002, also centers in Shanghai (2007) and Shenzhen (2002?)	USA	computer services and software	Presentation by Kevin Walsh, OECD workshop (September 2006), company press release July 30, 2007, http://www.sinonews.biz/content/view/126/2/
Philips	Research East Asia, Shanghai (2000), Philips Innovation Campus (2005)	Netherlands	electronics, electrical equipment	company website http://www.isa.org/InTechTemplate.cfm?Section=Industry_News&template=/ContentManagement/ContentDisplay.cfm&ContentID=52672; http://www.hk.pg.com/career/career_r&d.htm
Procter & Gamble	Beijing Technology Center (BJTC) (1998)	USA	household and personal products	
Ricoh	Ricoh Software Research Center Beijing (2004)	Japan	computers, office equipment	company website; Telephone interview (January 2007)
Roche	Shanghai (2004)	Switzerland	pharmaceuticals	Tremblay 2006
Rohm and Haas	Shanghai (2006)	USA	specialty materials	Tremblay 2006
Samsung Electronics	Samsung Communication Technology Research Institute, Beijing (2000) 20 research centers	Korea	electronics, electrical equipment	company website, press clippings

(Continued)

Appendix 1 Continued

Company	Location, year of establishment	Home country	Sector	Source
SAP	Shanghai (2003)	Germany	computer services and software	company website; *Infoworld*, March 27, 2006
Siemens	Global Headquarters Voice-Centric Mobile Phones, Beijing, 2002; Shanghai, Nanjing, Hangzhou	Germany	electronics, electrical equipment	company website and *China Daily* October 2006
Sony	Design center, Shanghai (2005)	Japan	electronics, electrical equipment	company website, http://www.sony.net/Fun/design/profile/globallocations/shanghai.html
Sony Ericsson	Development Center Beijing (2002)	Japan/Sweden	network and other communications equipment	*China Daily Dec. 2006*
Sun Microsystems	Sun China Engineering and Research Center, Beijing (2001)	USA	computers, office equipment	company website, http://cn.sun.com/eri/English/index.html
Toray	Nantong, 2002, Shanghai, 2004	Japan	specialty materials/chemicals	Tremblay 2006
Unilever	Shanghai	UK	Specialty materials/chemicals/medicine	company website, http://www.unilever.com/ourvalues/sciandtech/How_where/On_the_map.asp

Note: The list is not comprehensive but rather constitutes an effort to create an inventory of global R&D centres by foreign MNCs in China. A number of discussions and interviews have also been conducted with representatives of around 20 companies in connection with the study. The companies varied in terms of size and industry. Representatives of chambers of commerce, employers' organizations, trade associations, universities and colleges, other government authorities, international organizations, professors, journalists and other experts were also interviewed. The aim of the interviews was to supplement the lack of reliable statistics on China, and to gain a realistic picture of the extent and drivers of foreign R&D operations in China. To gain an idea of how much strategic R&D is conducted by foreign companies in China, I have cross-referenced press clippings, e.g. on the establishment of a foreign R&D center in China, with companies' annual reports and websites, and interviewed experts and company representatives.

4
Extensive Growth and Innovation Challenges in Bangalore, India

Anthony P. D'Costa[1]

4.1 Introduction

Bangalore is a medium-sized Indian city in the southern state of Karnataka once known for its quaint houses with gingerbread trim, many parks, defence personnel, and pensioners. Today it is considered to be India's Silicon Plateau. Multinational information technology (IT) businesses are rushing in to outsource IT services from their subsidiaries and from Indian subcontractors. Some are also establishing R&D centres. Bangalore is also host to numerous universities, management and public policy institutes, engineering colleges, and several government research institutes in aerospace research, electronics, and communications. It also draws the cream of the country's technical talent. This clustering of high-tech economic and knowledge-based educational activities intuitively suggests synergy among local economic agents in a virtuous mode. If taken to its logical end it also suggests a thick institutional architecture between government, industry, and universities, institutional actors that are said to be critical for national innovation and competitiveness.

The issue is not *whether* India is doing well in the IT industry. It is, given the rapid growth of the IT industry and sizeable share of software exports. Several factors explain the rise of the Indian IT industry: the country's long-standing emphasis on technical education, fortuitous global demand in the 1980s and 1990s, the successes of local and expatriate Indian techno-entrepreneurs, dense social networks among professionals, Indian economic policy reforms, multinational investments, and state support for software exports (Arora et al., 2001; D'Costa, 2004a, 2003a; Kattuman and Iyer, 2001; Patibandla and Petersen, 2002; Heeks, 1996). Among these factors the role of human capital – embodied in universities and research and development – is given a prominent place. It is clear that universities are responding directly to the growing demand for software professionals by increasing enrolment capacity and adjusting curriculum based on market needs. This implies that a strong institutional relationship between university and industry must be in the making in Bangalore, if rising productivity (revenues per employee)

79

can be demonstrated (Okada, 2004: 291; Athreye, 2005). After all, value addition based on deepening skills suggests innovativeness and cutting-edge technologies, which presumably are consequences of collaborative efforts in knowledge-intensive output (see Tsai, 2005; Audirac, 2003; Doloreux, 2002). In fact, most high-tech industry clusters suggest a catalytic, if not critical role for universities and the government in sustaining productivity growth.

Contrary to such expectations, I argue that Bangalore's (and by extension India's) IT development is predicated on an Indian business model that does not encourage thick institutional linkages such as those encapsulated by either the national innovation system (NIS) or the triple helix model (THM) (Etzkowitz and Klofsten, 2005; Baber, 2001; Etzkowitz and Leydesdorff, 2000; Leydesdorff, 2000; Hayashi, 2003). The purpose of this chapter is to understand the institutional arrangements of THM in India with the intention of capturing India's innovative capability.[2] The purposive interactions among industry, government, and academia comprise THM, which presumably leads to a cross-fertilization of new ideas and new modes of organizing for competitiveness. In Bangalore, there is collective efficiency due to the spatial concentration of several hundred IT businesses in a milieu of numerous engineering and science colleges and high-end public sector research institutes. However, surveys as well as anecdotal evidence suggest that most Indian firms do not collaborate extensively with other Indian firms nor do they partner with academia or government institutions for well-identified research projects.[3] The supposed thick institutional architecture in Bangalore is in reality thin, with a significant gap between universities and industry (Dahlman and Utz, 2005: 91).

I argue that Bangalore's (and thus India's) dynamism stems from linear and extensive growth rather than non-linear and intensive growth (D'Costa, 2002a). This is a result of an export-oriented model based on offshore development of software services, targeted mainly at the US. Neither the domestic market nor non-US markets, such as East Asia, are pursued aggressively by Indian firms. Structural constraints and incentives work against market diversification, both in terms of geography and products and services, and as long as growth is possible the need for thick institutional linkages is redundant. Indian companies are well positioned to take on high-volume, but often low-value projects for foreign markets due to the availability of a relatively homogenized but capable IT workforce (Kambhampati, 2002: 27; Parthsarathi and Joseph, 2004: 100–4). Firms face low entry barriers and the IT industry is structurally locked into a model of mostly small, undifferentiated firms. The ensuing 'excessive' competition, though healthy for dynamic change, in the absence of deeper local institutional and intersectoral linkages discourages inter-firm cooperation, encourages high labour turnover, and contributes to a local wage-cost spiral and related distortions. Hence, the sustainability of the extensive model could be at stake if competitiveness shifts towards low-cost Europe and Asia. The drive towards intensive growth could also be constrained if Indian firms continue to specialize in services without significant diversification in skills profiles and geographic markets (D'Costa, 2006, 2002b).

To break out of the extensive growth trajectory, non-routine knowledge-intensive strategic endeavours are necessary (Looy et al., 2003: 225; OECD 2001: 7). THM will be an important institutional matrix whereby universities and public research institutions could create thick institutional links aimed at 'firm formation and regional development' (Etzkowitz and Klofsten 2005: 245). Currently, India has a rudimentary form of THM. The other option would be to continue with the extensive model, with few highly innovative firms with the rest as large and small firms engaged in relatively routine IT services. This option is dependent on a growing supply of IT workers and continuing demand for such services from India, neither of which could be simply assumed. A third option lies somewhere in between, forging some selective partnerships with THM institutions, diversifying product and geographical markets, and overhauling the university system in favour of research to take advantage of both imminent labour shortages in high-skill areas and new technological opportunities in the global economy (D'Costa, 2004b).

This chapter is divided as follows. Section 4.2 presents an outline of institutional linkages and their evolution in the Indian macro environment. Section 4.3 briefly discusses the sources of Bangalore's growth by reviewing the supply of technical talent and the characteristics of external market demand. Section 4.4 presents some of the shortcomings of the extensive model in operation by examining the structure of the industry, the impact of excessive competition, and learning constraints. I conclude that the incentives to set up thick institutional linkages do not exist at this time. However, section 4.5 outlines some intermediate steps to go beyond extensive growth. These steps include a greater focus on the domestic market, the diversification of export markets towards East Asia, anticipating new technological trajectories, and attracting and retaining expatriate talent.

4.2 The triple helix model

A cluster of firms is expected to emerge from the synergy of infrastructure, government policies, and educational institutions (Morosini, 2004). Of course, many variables, including historical contingency, contribute to the specific institutional and national differences among clusters. However, collectively firms share the high spatial concentration of economic activities with significant cooperation among institutions such as government, business, and academia to promote collective learning (Doloreux, 2002). They can be integrated horizontally or vertically and found in both labour-intensive manufacturing and high-technology industries and services. Examples include a surgical instruments cluster in Pakistan (Schmitz, 1999), textiles and garments in India (Cawthorne, 1995), automobiles in Japan and India (Smitka, 1991; D'Costa, 2005), ceramics in Italy (Best, 1990; Piore and Sabel, 1984), and computer-related manufacturing in Taiwan (Wong, 1995). Automobile manufacturing is a good example of vertical integration, whereby components suppliers are located in close proximity to buyers of semi-finished goods and various services (Odaka et al., 1988; D'Costa, 2003b, 2004c, 2005). Silicon Valley and Hsinchu Science Park in Taiwan are good examples of horizontal integration, where a variety of electronics-based manufacturing and design activities are carried out independently but within the parameters of the needs of the larger industry.

The physical clustering of economic activities suggests geographical proximity of firms, interdependence, and a local institutional milieu (Keeble and Wilkinson, 1999). Irrespective of the type of production units, interdependence involves cooperative buyer–supplier interaction and spin-offs. They are driven by one firm's output becoming the input of the other (Wever and Stam, 1999). The increasing returns from such externalities are a key feature of clusters.[4] The institutional environment for high-tech clusters supports a rapid rate of innovation, which is backed by a system of inter-sectoral networks of knowledge creation, growth, diversification, and mobility of professionals (Hayashi, 2003; Keeble and Wilkinson, 1999: 296). The state plays a role in education, training and skill enhancement, and research and development (Nelson, 2000; Acs, 2000). The role of universities and technical institutions assumes greater significance, especially in the formative stages of THM for latecomers (Etzkowitz and Leydesdorff, 2000) (Figure 4.1). A THM is 'nationally' anchored, relying on endogenous sources of innovation, which is different from endogenous growth theory (see Sharif 2006: 10), but draws on global knowledge frontier. However, for developing countries (latecomers) a THM tends to be weak due to thin national institutional links.[5]

In India in the 1950s the state assumed responsibility for the academic and investment component of high-tech industry (Figure 4.1a). With a weak private sector, state formed the core of an institutional architecture. Under state-led strategy, particular cities were often targeted for industrial investments. Academic institutions were created to meet the labour needs of industry, especially for technicians and engineers. However, state-dominated academic and industrial efforts were challenged by global competitive pressures, compelling more flexible arrangements under neoliberal policies. Consequently, the relationships between universities and the industry become stronger (Figure 4.1b). The institutional architecture is reconfigured among industry, government, and universities. This represents an incipient THM as the three institutional actors are theoretically more or less equidistant from each other (Figure 4.1b) with 'spin-offs, trilateral initiatives, and strategic alliances among firms, government laboratories, and academic research groups' (Etzkowitz and Leydesdorff, 2000: 112). However, under globalization Indian industries, especially IT, have forged strong links with the global

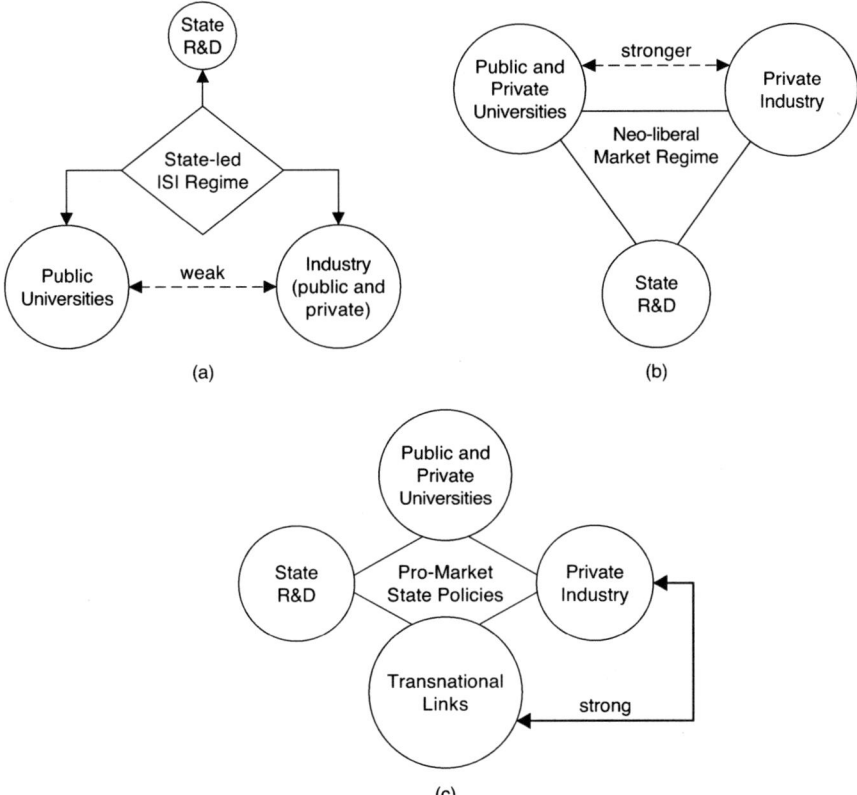

Figure 4.1 Co-evolution of the triple helix system in India

industry (Figure 4.1c), thereby crowding out alternative institutional linkages between local firms, the state, and public research institutions.

With 'value chain modularity' it is possible to disperse production horizontally across different locations (Sturgeon, 2003: 204) and create finely detailed global division of labour. High-technology cities such as Bangalore, while in the embrace of international networks, could be locked into particular form of specialization. There is considerable tension between external drivers and the endogenous 'local production system' (*à la* Lombardi, 2003) in the formation of an 'interlocking... collective order' (Scott, 1998 in Lombardi, 2003: 1444). If foreign economies, as clients, innovators, and markets influence what gets produced, how, and where it is sold, then the inter-institutional networks, such as inter-firm collaboration or industry–university linkages in a local production system are likely to be undermined. Or as Morosini (2004: 308) points out, 'industrial clusters can also experience high rates of employee turnover and non-cooperation between firms, which can jeopardize the entire cluster'.

'Jeopardize' is perhaps too strong a word. Latecomers can be hosts to *both* high- and low-technology segments in the broader category of high-tech sector. This is due to the massive wage and income differentials in the world economy, with considerable wage arbitrage in favour of developing countries, which also have good tertiary technical education systems. Hence, the desired evolutionary path based on intensive growth could be truncated by extensive growth as low wages continue to dictate production pattern. International offshoring firms confronting capital accumulation problems are likely to find this attractive. This is not to deny that that the production system cannot move towards intensive growth in select market niches. But in order to make the transition considerable institutional effort in the form of a THM is necessary. As of now, a THM, if there is one in India, is weak in part because the exogenous pull is much too powerful (D'Costa, 2002b, 2004d), leading to narrow band of IT service specialization based not on 'close-knit social communities' (Morosini, 2004: 309) that are so vital to cluster dynamics. Due to the structure of incentives facing IT firms in India, the Bangalore cluster exhibits weak institutional linkages among firms, universities, and the state.

4.3   Some sources of Bangalore's extensive growth

4.3.1   Human capital and the supply side

Bangalore's extensive growth has created an expanding cluster of IT firms, engineering and science colleges such as the Indian Institute of Science, and IT training centres. Many government high-technology entities, such as the Indian Air Force headquarters, the Indian Space Research organization, and Hindustan Aeronautics Ltd, predate this growth, while state-sponsored critical infrastructure such as satellite links in centrally organized software parks are more recent developments (Heitzman, 2004: 222–9, Taebe, n.d.) (Figure 4.2). However, the mere physical concentration of public sector academic and research institutions is not a sign of thick linkages since the earlier model of a top-down state-dominated academic and industrial agenda in a semi-closed, security-conscious economy has not been completely shed (Sridharan 2004, 1995). What has changed recently is the rise of active state promotion of the Indian IT industry.[6] For example, the Software Technology Parks of India (STPI), under the supervision of the Department of Electronics of the Ministry of IT and Communications, has provided critical infrastructure support for exports. In 2003–04, Bangalore exported $3.8 billion, 36 per cent of *all* STPI exports.[7] The state government of Karnataka, where Bangalore is located, established an Electronics City in 1985, a few kilometres away from Bangalore's city core. Electronics City houses numerous software firms, including one of India's largest and most successful firms, Infosys. Later, an international export-oriented high-technology park (ITPL)

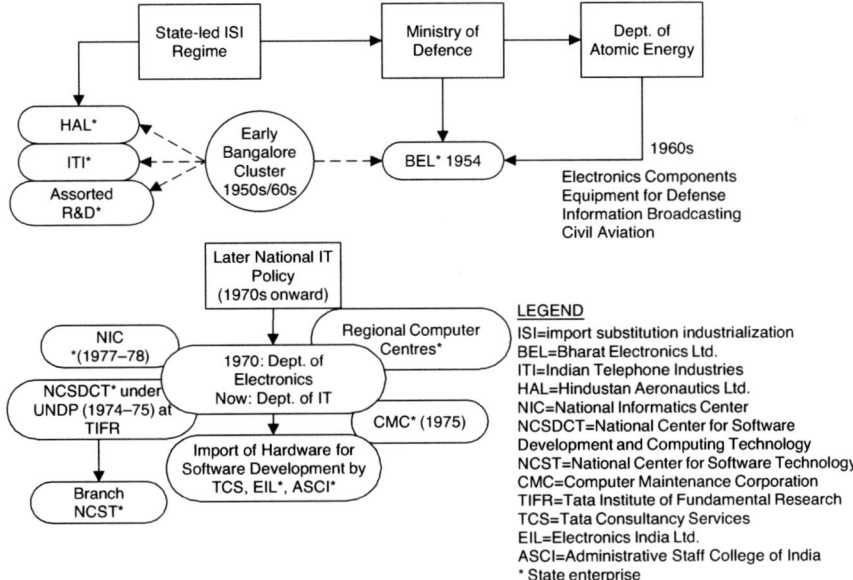

Figure 4.2 State, Department of Information Technology, and Bangalore Cluster
Source: Adapted from Naidu 2003: 119–122.

was established by a consortium of Singaporean companies led by Ascendas Land (International), Tata Industries (the investment arm of the Tata Group), and the government of Karnataka. ITPL is host to 107 foreign and domestic firms. Many clients of ITPL are global organizations in need of state-of-the-art information, communication, and physical infrastructural facilities.[8] ITPL presents a microcosm of a thick institutional system where local government and business, domestic and foreign, work collaboratively on a project. However, ITPL is more of a real estate project and remains enclave in nature.

Bangalore's extensive growth has been sustained by the supply of IT workers. Although the state of Karnataka has only 5 per cent of India's population it accounts for nearly 15 per cent of the nation's higher education enrollments.[9] In 2000–01 it had over 100 engineering colleges and nearly 12,000 students took IT-related courses (Table 4.1). Karnataka has 83 engineering colleges under Vishweshvaraiah Technology University offering Bachelors of Engineering degrees.[10] Of these, 25 colleges were located in Bangalore; 59 are in the Bangalore region. Most of these colleges graduate students proficient in basic engineering skills, mathematics, and programming. There are eight other non-engineering universities, two of which are in Bangalore. Bangalore University itself has over 50 colleges located within Bangalore.[11] Though not a source of engineers, these colleges contribute to

Table 4.1   The growing supply of IT professionals in Karnataka State

	Engineering colleges B-Tech			Diploma colleges		
	Engineering colleges	Intake in IT-related courses	% of total intake	Diploma colleges	Intake in IT-related courses	% of total intake
1980–81	41	–	–	44	–	–
1981–82	42	25	0.3	46	–	–
1986–87	50	1,240	8.4	150	n/a	n/a
1991–92	51	2,020	2.1	165	2,364	10.3
1996–97	53	2,758	14.1	178	4,379	15.2
1997–98	70	4,028	16.9	196	5,682	17.8
1998–99	71	4,190	17.0	184	5,585	17.1
1999–2000	106	5,802	22.0	186	6,120	18.0
2000–01	109	11,565	34.5	201	n/a	n/a

Note: n/a = not available.
Source: Government of Karnataka in Okada (2004: 299).

English-speaking science- and IT-proficient graduates. Karnataka has two of the nine national institutes of technical education including the recently created Indian Institute of Information Technology (IIIT) (now the International Institute of Information Technology due to partial privatization), the established Indian Institute of Science (IISc), and one of the 20 National Institutes of Technology (formerly Regional Engineering Colleges).[12] In all, Bangalore's home state has 12 per cent of the country's degree colleges under universities granting technical degrees and 15 per cent of the country's diploma-granting polytechnics (Okada, 2004: 298).

Notwithstanding the massive technical education infrastructure in Bangalore, only a few technical institutions can be considered world-class. On the whole, Indian tertiary education is plagued by shortages of high-quality staff, underinvestment in research facilities, and poor training (see Dahlman and Utz, 2005: 63–72). It also lacks a 'richer academic ecosystem' (in Cookson, 2005; NASSCOM, 2002a: 73–4). For IT training there is the added problem of the flight of instructors for the more lucrative software industry.[13] Compounding this problem is the unattractiveness of PhD-based research.[14] One industry professional pointed out the necessity of building up 'IIT's *graduate* student population, and [improving] links between universities and public research labs and industry' (in Cookson, 2005) rather than have them go abroad for post-graduate education. Despite the progress in technical education in India, the number of doctorates in engineering is low. In 1979 the figure was 506, which increased to a mere 546 in 1995 (Dahlman and Utz, 2005: 61). Even after a decade this figure has not changed much.[15] The

Table 4.2 Professional employment in the Indian IT and ITES sectors

	1999–2000	2000–01	2001–02	2002–03	2003–04E
Software Exports	110,000	162,000 (47%)	170,000 (5%)	205,000 (21%)	260,000 (27%)
Software Domestic Market	17,000	20,000 (18%)	22,000 (10%)	25,000 (14%)	28,000 (12%)
ITES-BPO	42,000	70,000 (67%)	106,000 (51%)	171,000 (61%)	245,500 (44%)
Total	284,000	430,000 (51%)	522,250 (21%)	661,000 (27%)	813,500 (23%)

Note: Figures in parentheses denote % change from the previous year.
Source: NASSCOM (2004: 186).

low number of engineering doctorates suggests un- or underdeveloped doctoral programmes and constrained R&D environment in India. NASSCOM fears the problem is going to get worse as faculty is pressured to raise non-public sources of revenues through consulting and teaching in non-degree programmes. It is clear that the raw talent needed by most Indian IT firms is substitutable and under the extensive growth model in-house training is adequate to serve the global customized services market at this time.

The growth in the IT-enabled services (ITES) and business process outsourcing (BPO) sector has outpaced both the domestic and export software sectors by a wide margin, which mostly represent call centres and back office operations (Table 4.2). This clearly represents extensive growth, a growth that is predicated on increasing absorption of labour and not necessarily on increasing productivity as measured by the level of revenues per employee. Based on NASSCOM's narrow definition of IT services (which excludes ITES, engineering and R&D services, software products, and hardware, all of which are higher value services and whose magnitudes can be assumed to be fairly low at this time) a simple calculation of productivity (revenues per employee) shows that it has remained virtually stagnant recently (in fact declined slightly) in terms of US dollars. For the fiscal year 2004 it was $33,953, for 2005 $33,670, and for 2006 (estimate) $33,165 (NASSCOM, 2006). Admittedly there are problems with such data but it is in the interest of the industry association to demonstrate rising productivity, which is not the case here. If this is a tacit acknowledgement of an extensive growth model then the incentive to forge links between the industry and the university appears muted.

Consistent with extensive growth of the Indian IT industry there has been an outflow of technical talent through temporary and permanent emigration of Indian science and engineering students and IT professionals to foreign markets (D'Costa, 2008). Roughly 44 per cent of the USA's H1-B visas have gone to Indians, allowing Indian professionals to work in US firms. Despite signs of some reverse flow, most Indian professionals do not return home (see Hira, 2004).[16] While this has given the Indian IT industry a foothold in export markets, generated a brand name, and established professional networks, it has also induced the local university system to be content with replenishing the outflows of students without much regard to enhancing postgraduate training in core engineering fields.

Weak R&D is also reflected in the small number of patents filed in the US (Dahlman and Utz, 2005: 81). Between 1991 and 2003 China was granted 2,038 patents compared to India's 1,555 (US Patent and Trademark Office, 2005: A2-1). In the same period Hong Kong and Taiwan were granted 4,191 and 45,127 patents respectively.[17] The bulk of research and innovation in India are undertaken by government institutions and a handful of universities (see Mani, 2004), whose links to other institutions are weak. In general, patents in software are difficult to secure and are not granted by the government of India. India's research strengths lie in the pharmaceutical and chemical industries, a reflection of India's larger output of science doctorates (almost six times the number of engineering doctorates). It also suggests a better R&D record in non-IT sectors such as pharmaceuticals, automobiles, and electronics (Dahlman and Utz, 2005: 84). More recent data suggest India's growing patent applications in product and technology within India – from under 4,000 in 2000 to 25,000 in 2006 (Siliconindia.com, 2007). But these numbers pale into insignificance when placed in a global context, with US and Japan filing more than 500,000 patents and UK and China about 200,000 each in their respective countries. Moreover, the bulk of patents filed in India are by multinational corporations (Siliconindia.com, 2007). Of the 14 software firms in India reporting patent applications, seven Indian and seven foreign, the share of Indian patent applications was only 24 per cent (McManus, Li, and Moitra, 2007: 54).

India is strong in high-science activities such as space research and high energy physics, but India's role in industrial innovations is negligible due to the enclave nature of government institutions (Sridharan, 2004).[18] Research by Indian faculty is weak as reflected in the low number of citations and the lack of original research. India declined in scientific publications from eighth position in 1973 to 15th in the world (Jayaraman, 2002a: 100). Between 1980 and 2000, India's science-related publications fell from 14,983 to 12,127, while China's rose from 924 to 22,061. China is also ahead of India within the IT industry despite its current export weakness in the software service area (Kshetri, 2005, see Table 4.3). India's selective innovative capability stems from the kind of training given to technical students. For example, NASSCOM identified lack of student exposure to projects, the inability to identify needed skill sets, and lack of curriculum standardization as sources of innovative weakness (NASSCOM, 2005). Without rectifying these shortcomings it will be an uphill task for the Indian education

Table 4.3 A Broad China-India Comparison in the IT Industry

	CHINA	INDIA
Export market	60% Japan	63% USA
High-tech exports	19% of manufacturing	4% of manufacturing
Yearly PC sales	11 million	3 million
Telephones per 1000 persons	167	40
Cell phones per 1000 persons	161	12
Internet users	87 million	18 million
CMM Level 5	2 companies	60 companies
Tertiary students in Science, mathematics, Engineering	53%	25%
Researchers in R&D	584 per million people	157 per million people
Average size of software firm	25 employees	174 employees

Source: Kshetri 2005.

system to obtain necessary international certifications and accreditations (NASSCOM, 2005).

4.3.2 Exports and the demand side

Bangalore's extensive growth is also reinforced by the particular business model adopted by the Indian IT industry. Its principal characteristic is outward orientation based on offshore development for exports with a focus on software services, and heavy dependence on the US market (nearly 65 per cent). There are nearly 100 multinational firms in Karnataka state, most of which are in Bangalore (http://www.bangaloreit.com 2005). Roughly 24 per cent of India's top software firms are located in Bangalore, of which over 63 per cent were multinationals (Okada, 2004: 286). In 2002, of 102 Bangalore-based NASSCOM members, close to 40 per cent of firms had 100 per cent exports, while 67 per cent of the firms had export ratios over 80 per cent (NASSCOM, 2002b). Over 80 per cent of the firms had export ratios exceeding 60 per cent (NASSCOM, 2002b). According to one estimate (Table 4.4), India's strength is in customized software, with about 25 per cent of the global market. However, this segment itself is estimated to be less than 4 per cent of the global IT market. Other segments, where India's prospects are considered to be high, do not require advanced skills except for network infrastructure management and packaged software.

The growth of the ITES sector is a welcome development for a labour-abundant Indian economy. But this is not a sector that can boast of being at

Table 4.4 Forecast of global IT services and India's opportunities

	Global Market (2001)		Global Market (2005)		India's Exports (2001, US $ b)	India's Global Share (2001, %)	Potential for Exports
	US $ b	% Share	US $ b	% Share			
Professional Services	142.9	32.5	238.7	34.1	5.3	3.7	
IT Consulting	21.3	4.8	31.5	4.5	0.1	0.3	Low-Medium
Systems Integration	81.1	18.4	142.1	20.3	0.1	0.1	Low-Medium
Custom Applications	19.3	4.4	25.3	3.6	4.5	23.1	High
Network Consulting & Integration	21.2	4.8	39.8	5.7	0.7	3.3	Low
Product Services	117.9	26.8	176.9	25.3	0.4	0.3	
IT Training & Education	25.5	5.8	40.9	5.8	–		High
H/W Support & Installation	44.4	10.1	49.4	7.1	–		Low
Packaged Software Support Services	48.0	10.9	86.6	12.4	0.4	0.7	Medium-High
Outsourcing Services	179.2	40.7	284.8	40.7	0.1	0.0	
Processing Services	78.4	17.8	103.8	14.8	–		High
IS Outsourcing	64.0	14.5	100.2	14.3	–		High
Application Outsourcing	13.4	3.0	39.0	5.6	–		Low-Medium
Network Infrastructure Management	23.4	5.3	41.8	6.0	0.1	0.3	Medium-High
Total	440.0	100.0	700.4	100.0	5.7	1.3	

Notes: IT = information technology; IS=information services; H/W = hardware
– not available or not applicable.
Source: Calculated from NASSCOM 2002a: 24, 46 in D'Costa 2004a.

the cutting edge of technology. The role of universities is limited for this sector since IT training institutes are likely to impart basic technical training to mostly English-speaking graduates. Firms remain responsible for job-specific training. Thus far, the low cost of labour relative to OECD norms makes ITES outsourcing to India attractive (Hoffman, 2003), especially when competition under recessionary conditions in OECD countries compels cost cutting. The wage arbitrage in this sector is powerful enough to reinforce extensive growth of the Indian IT industry.

4.4   The pitfalls of sustained extensive growth

One of the central sources for industrial development and innovative capability is the embeddedness of firms in the local production system (Parthasarathy, 2004; Hsu, 2004). According to UNCTAD (2004: 151), 'high-skill services', among other things, 'require advanced skills at high levels of specialization, often with strong educational institutions. They involve agglomeration economies, *with different skills, enterprises and institutions interacting with each other to share work, stimulate knowledge flows, and allow specialized skills to be fully utilized*' (author's emphasis). Bangalore enjoys some of the institutional dynamics of high-skill agglomeration economies. However, as will be shown below, the particular business model adopted by the India IT industry sustains extensive growth and thus discourages the kind of interactions necessary to be at the technological frontier. Related to this model is the significant influence of export markets. When foreign economies and firms are the principal clients and innovators, they influence what gets produced, how, and for which markets. This is not intrinsically unfavourable since diversified external market growth could translate into local technological learning. This is certainly evident in India. But institutionally, the tension between exogenous drivers such as export demand and technical talent and the endogenous 'local production system' could undermine an 'interlocking... collective order' (Scott, 1998 in Lombardi, 2003: 1444), resulting from the structural difficulties in establishing thick *local* institutional linkages such as those between universities, industries, and the government (see D'Costa, 2004d).

Extensive growth presumes continued demand for customized software services and an elastic supply of Indian IT professionals. Such growth results from maximizing revenue strategies by adding more employees. The consequences of this are a fragmented industry, institutional disconnectedness, and high labour mobility. Except for institutional incoherence, neither of the other two outcomes by definition are detrimental to intensive growth. If fragmentation means lots of small firms one could assume competition is rife and thus a basis for innovation. Furthermore, such an industrial structure is consistent with global patterns of export domination by large IT firms.

Similarly, international labour mobility suggests competition among firms, skill upgrading, knowledge transfer, rising wages, and thus innovation-led intensive growth. However, as it will be shown below transitioning to intensive growth in Indian IT without an explicit innovation policy incorporating a thick interactive institutional architecture is daunting.

4.4.1   Fragmented industry structure

Unlike hardware manufacturing, the entry barriers in software services are low and competition among small firms in India is intense. The proliferation of firms reinforces extensive growth based on low value-addition by an elastic supply of IT workers.[19] NASSCOM's membership, which represents 95 per cent of the industry's revenues, rose from 38 in 1988–89 to 402 in 1996–97 and 892 in 2004, with 24 per cent of all members being located in Bangalore (NASSCOM, 2005). The growth in India's software output is attributed largely to growth in the export market. This, at first glance, is a healthy sign of rising Indian entrepreneurship as global market opportunities are capitalized on by Indian and foreign firms (D'Costa, 2003b). A crude ratio of revenues per firm shows that the average Indian firm has increased its export revenues from US$7 million in 1995 to US$21 million in 2003. This is a threefold nominal increase, though in real terms it is likely to be less.

On closer inspection, the average revenue is misleading since most Indian software companies are small in terms of revenues and number of employees, indicating the easy phase of exports (Figure 4.3).[20] Most new firms end up clustering around low-end activities (see BusinessWeek online 2003).[21] The top 20 Indian software exporters still account for about 60 per cent of total exports, with more than 800 firms accounting for the remaining 40 per cent of the software market (Data Quest, various issues). The top ten firms based in Bangalore contribute more than half of Bangalore's exports (Okada, 2004: 286). In theory, size *per se* is not important since most software projects tend to be small and thus well suited for entrepreneurial initiatives. In practice, however, large firms are able to carry out multiple projects simultaneously and thus not only spread risks across projects and markets but also carry out large, complex projects by quickly mobilizing large numbers of professionals. Large enterprises, local or foreign, can exercise monopsonist clout relative to smaller firms, though they too could be subject to periodic shortages of specific skill sets. A shortage of over 1 million 'suitably qualified people' by 2010 has been predicted (Economist, 2005: 58). As we will see below, extensive growth itself is reproduced in a systemic way.

The fragmentation of the industry suggests that regardless of size most Indian companies pursue whatever projects they can secure and which can maximize the absolute difference between costs and revenues. This generalist and undifferentiated nature of Indian firms suggests competition based on price (Arora et al., 2001). Thus, most improvements in productivity

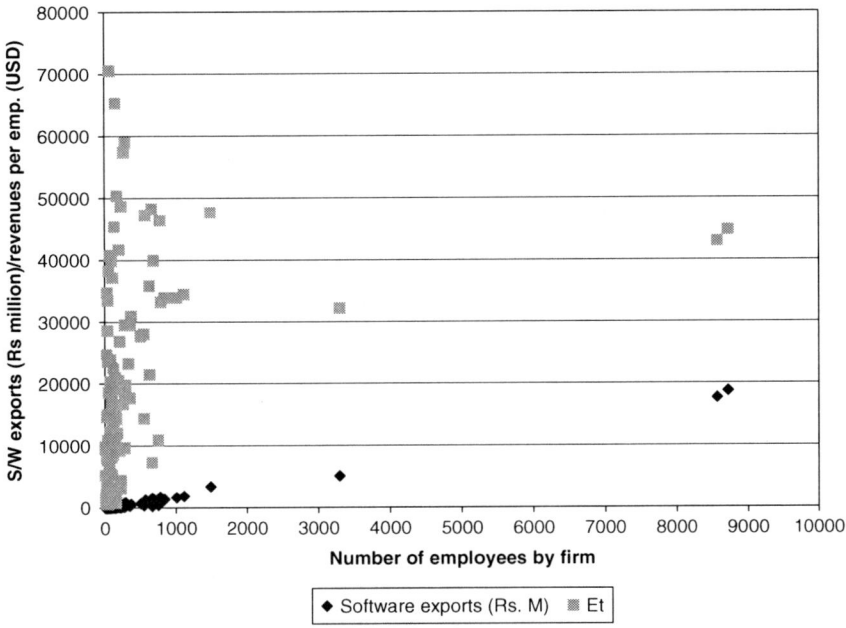

Figure 4.3   Bangalore's fragmented software cluster
Source: Author's compilation from NASSCOM 2002b.

are typically passed on to the foreign (read US) clients (Arora and Athreye, 2002: 255). The implication of this is that innovative capability is likely to be confined to a few large firms and a handful of medium-sized highly entrepreneurial firms, resulting from the founder's or team's particular technical strengths, professional networks, and firstcomer advantages. It also implies that the incentives to participate in an interactive set of institutions designed to foster innovations are not high at this stage of extensive growth.

4.4.2   Competition in undifferentiated software and software-enabled services

One important consequence of a fragmented industry is cut-throat competition. There are costs and benefits to such competition. An excessively competitive environment distorts compensation rates, induces a high labour turnover, a real estate bubble, and stress in the urban industrial and physical infrastructure. These would not be much of an issue over the medium term if supply responses by the public and private sectors were speedy and flexible. However, the political economy of Indian development suggests institutional impediments that slow down responses, as in the years of delay in constructing the new international airport in Bangalore. Such competition also undermines inter-firm collaboration, which most cluster experience shows contributes to industrial dynamism.[22] This can be also inferred from the number of alliances between Indian and US firms, which exceed the number of alliances among Indian companies (Basant, 2003). Intense competition is evident from the secrecy maintained by firms when discussing projects and clients (see Prabhu, 1999: 504). This has also been true for small electronic and engineering firms in Bangalore, where mutual suspicion among entrepreneurs has overshadowed cooperative ventures (Holmström, 1998: 225). Social trust is still weak in India and hence cooperation among IT firms remains limited.[23] A recent study observes that: 'The cluster is not characterized by cooperation alongside competition, but rather by competition alone... [A] central feature of the software cluster in Bangalore is that firms are not *to any significant degree* linked by input–output relations' (Lema and Hesbjerg, 2003: 142).

Lack of trust is reinforced by the business model that facilitates the compartmentalization of offshore development of software projects. This division of projects by the client structurally inhibits Indian inter-firm cooperation, constrains project capabilities, and restricts joint coordination of activities (Lema and Hesbjerg, 2003: 137–43). For example, a foreign client may outsource two components of the same project from two Indian firms, but the Indian firms operate independent of each other. They do not know what the component is for, how it might integrate with other software components, and do not have the technological understanding of the larger project to which they are contributing. Of course, this could be a client strategy to protect key technologies. However, the result is that the systems integrator, typically the client or an intermediate consultant, has the knowledge of diverse domains rather than the Indian software component suppliers. This can act as a systemic barrier to moving up the value chain. So even if Indian firms have mastered the production of components of complex software projects, they find moving up to the larger project architecture daunting. The lack of trust translates into a strategy of capturing whatever projects come by and a growth strategy based on human resource expansion and talent poaching (see Kumar and Joseph, 2005: 100).

The undifferentiated nature of most Indian software firms induces severe competition.[24] As a result there is high labour turnover (more than 20 per cent) (Athreye, 2005: 20–3) and a wage-cost spiral based on high IT compensation growth averaging 30 per cent per annum throughout the 1990s, poaching of talent by large firms, and curiously, labour shortages in certain areas.[25] Larger foreign and Indian firms are better placed to employ graduates from elite technology, science, and business schools, while smaller firms are compelled to rely on students from tier-two institutions. However, tight labour markets also compel productivity growth for the industry. Rising salaries lead to greater enrollments in technical education, subsequent investment in educational infrastructure, and further growth of the industry. At the same time, in the absence of institutional linkages, this virtuous cycle contributes to an extensive form of IT sector growth.

A recent survey shows that IT salary hikes in India exceeded 18 per cent over the 2004–05 period and the top three paymasters were all multinationals (Cadence Design, Sun Microsystems, and Philips), while the top 20 had five multinationals (Arora, 2005). On the surface, this is a good development as higher revenues are shared among a greater number of employers and employees. However, with international opportunities for students and software professionals, there is an outflow of technical talent and pressure on local wages.[26] There is also internal brain drain as engineers and other professionals exit non-IT sectors to join the more lucrative IT sector (Arora and Athreye, 2002: 266). But as costs rise in Bangalore, other lower-cost locations in eastern India such as Kolkata (formerly Calcutta) and Bhuvaneswar (the capital of Orissa state) become attractive. This is a welcome development if deconcentration of urban centres and the national diffusion of economic activities are intended. However, this is very much in line with extensive growth, meaning a repetition and a geographical dispersion of more or less similar activities. Such dispersion can also prematurely end the agglomeration economies, which Bangalore today enjoys.

Such wage pressures have been felt even in the IT-enabled services (ITES) segment, which consists mainly of call centres and back office processing (BPO). A recent report warns that India's advantage in ITES could be eroded by wage inflation, which is higher than in the US (Siliconindia.com, 2005a). Wages have risen from $114–136 a month to $159–204 a month – an increase of 40–50 per cent in four years (Siliconindia.com, 2005b). Rising costs could no doubt compel firms to pass some of the costs on to the client or force them to move up the value chain. The very rationale for foreign clients to reduce costs through outsourcing is undermined. Moving into technologically challenging markets, among other things, requires a fundamental realignment of the business model and an array of institutional linkages. The warning that India's cost advantage in ITES could be challenged by other lower-cost countries such as Vietnam, the Philippines, and those of Eastern Europe illustrates the predicaments of low-end service provision (Siliconindia.com, 2005c). The growth of the ITES sector in effect works against the anticipated inter-institutional architecture for high technology growth associated with an evolving triple helix system.

4.4.3 Other challenges to innovation

High turnover suggests inter-firm mobility of labour in an industrial cluster, leading to technology transfers and learning spillovers. But the compartmentalization of IT projects, with subcontractors responsible for a component or two of the entire project, and high labour attrition does not make knowledge transfer easy. An alternative interpretation suggests that high labour turnover could be detrimental to skill development and project completion, if there is a scarcity of particular skill sets. As long as there is a steady supply of raw talent, extensive growth can continue and hence high labour turnover can be accommodated by the industry. This is the likely scenario in the absence of an institutional network among the industry, the state, and universities. The problem arises when either the quality of engineers suffers due to unregulated growth in educational institutions or when external demand slows due to erosion of competitiveness. There are signs of both.

Extensive growth deepens domain expertise in a limited way as user feedback is constrained (D'Costa, 2004d; Parthasarathy, 2004). The modular type of production undertaken by Indian producers limits the understanding of 'kernel' technologies associated with high technologies and subsequently to an inability (real or perceived) to carry out systems integration.[27] Nor does the model provide the incentive to serve the domestic market, which in effect is priced out by foreign clients. Export revenues also discourage the local development of software products, which could be tested locally and further refined for subsequent export at higher returns. The current incentives not only encourage service exports but also discourage software product and hardware development (D'Costa, 2004a). This decoupling can be argued to constrain the technological learning of the Indian IT industry as a whole. The result is a form of disembeddedness in which local institutions operate as enclaves. Even multinational subsidiaries undertaking high-tech R&D in Bangalore operate as enclaves (Arogyaswamy, n.d.), though their links are articulated seamlessly to their parent firms. They have no local ties other than the professionals they hire, often poaching talent from financially less-endowed Indian firms. Confidentiality requirements discourage subcontracting to other local firms. They report directly to their parent R&D unit, and continue to remain captive markets for R&D output (D'Costa, 2002b; Parthasarathy, 2004). The insular security-conscious defence-related R&D public sector also operates as an enclave with respect to the commercial market.

4.5   Transitioning to intensive growth

Given the structural constraints of the business model adopted by the Indian IT industry, the question is how to break out of the extensive growth trajectory. The significance of innovation and the knowledge-driven economy suggests more robust institutional alliances between university and industry. There are a few such linkages in India (Basant, 2003). They are often *ad hoc* and confined to a handful of research universities and technical institutes and their collaborations tend to be mostly with foreign firms. Again, intrinsically this is not detrimental to building knowledge capabilities if there is sufficient spillover from such activities (see Parthasarathy, 2000). India's weak, albeit improving manufacturing capability for complex IT products such as semiconductors and a wide variety of third generation telecom products calls for deeper institutional linkages. New lines of IT hardware and embedded software comprise the future of the IT industry. Most Indian IT firms, for both market and technical reasons, are not seriously engaged with this segment.

Given the incentive structure of the current model of offshore-based software service exports to the US market, I suggest an intermediate set of approaches to support a potentially intensive form of growth. This involves reorienting the export business model by addressing the interrelated areas of domestic market needs, export market diversification, and expatriate talent. However, to meet these challenges adequately the role of universities cannot be ignored since technical education, improving the research environment, and anticipating new technologies are integral to innovative capabilities.

4.5.1 Developing the domestic market

India is weak in product development. Low-level IT diffusion constrains the development of local software products (Kambhampati, 2002). While costs for international marketing are prohibitive, software products for the home market can be used as a stepping-stone for the export market. Several India-made products are available in banking, finance, and software tools. There is good potential for software products in vernacular languages as evidenced by HCL Infosytems' unicode-compatible PCs to support seven Indian languages. Another area of software use is to provide critical government services explicitly for development purposes. Here again, India has a better record than most developing countries, but India needs to bolster software use even more for wider impact (Kaushik, 2006; Thomas, 2006). In addition, the recent efforts by Indian firms such as Encore Software and the Indian Institute of Science have led to the development of an indigenous, low-cost computer called the 'Simputer'. Relying on the open source Linux operating system, the Simputer promises to be a good alternative to relatively expensive foreign products (Jayaraman, 2002b: 359; Personal Interview, Encore, Bangalore, February 2005). Already the affordable Simputer is being marketed to other developing countries in Asia and Africa. Due to export controls, Indian research institutions have developed alternatives to supercomputers made by US firms. These are good examples of software and hardware application for low-income countries, complementing and diversifying export markets.

4.5.2 Diversifying export markets

The Indian IT industry is heavily dependent on the US. As a result, other technologically sophisticated markets are not at present widely served by Indian firms. Japan is a case in point. NASSCOM estimates that of the nearly US$10 billion software service exports by India in 2002–03 only 2 per cent went to Japan, whereas nearly two-thirds went to the US (D'Costa, 2004b: 17). Poorer regions of the world imported more software services from India than Japan, despite Japan being the second largest market in the world known for design and embedded software.[28] Among all the regions Japan had the lowest dependency ratio, which suggests India's penetration of the Japanese market is extremely low. While there are institutional and business reasons for India's limited participation, the opportunities are immense in the Japanese market.[29] The high-growth East Asian economies, including China, with their vast high-technology manufacturing base, offer new markets for the Indian IT industry.

It is in India's interest to tap the under-served products market by creating high-calibre engineering talent capable of design, development, and implementation of complex projects. There are many intermediate products, such as Bluetooth software applications and telecom-related hardware, that could be sold as intellectual property by Indian firms to foreign manufacturers. Such domain expertise requires advanced university technical training, project-specific learning, and market exposure. The first calls for a revamping of the engineering curriculum, especially in microelectronics; the second demands domain expertise (Basant, 2003). A few Indian firms, such as Mindtree Consulting, Sasken Communications, and Interra Systems, are engaged in such activities for Japanese clients, where the market for embedded systems and technology-related intellectual property (IP) is large. These are lucrative projects, even if the projects tend to be small (Field Research, February–March, India; May–June, Japan 2005). The global market for such services is growing rapidly, particularly in East Asia. Japan, Taiwan, and China are behind in software development and hence offer new opportunities for the Indian industry. But in order to serve these markets effectively the industry as a whole must be embedded in the larger institutional network involving both universities and research institutions in fostering a rapidly moving technological frontier.

4.5.3   Anticipating new technologies and markets

In the related area of market diversification, the direction of the Japanese IT market is instructive. In 2001, the Japanese Ministry of Economy, Trade, and Industry (METI) launched the Industrial Cluster Project, with 30 local governments proposing Knowledge Cluster Plans (Interview, Kitakyushu Science Research Park, June 2005). Later, ten clusters were finalized in 18 areas (Ministry of Education, Culture, Sports, Science and Technology (MEXT) 2004). These represent increasing specialization of knowledge, science intensity, and aggressive R&D efforts (MEXT, 2004: 43–4).[30] Hence, it is not surprising to see six educational institutions, including the Kyushu Institute of Technology, Waseda, and Cranfield Universities locating their engineering training programmes in Kitakyushu Park, one of Japan's high-tech clusters, along with semiconductor design and manufacturing firms.[31] The lesson for the Indian industry is to explore these technological options, given anticipated labour shortages in Japan and the increasing technological demands of future industries.

4.5.4 Attracting expatriate talent

To tap Indian talent from abroad, even on a temporary basis, both professional opportunities and high-quality urban amenities are necessary. India has to compete with rich countries in order to retain its own talent. Recently, the UK government announced plans to retain Indians in the UK (Siliconconindia.com, 2005d). While matching US salaries in India would be difficult, a high investment environment consistent with high macroeconomic growth would be the necessary first step. Bangalore is banking its growth and development on the far less capital- and R&D-intensive, service-driven IT export model based upon thin institutional arrangements. The select return of Indian expatriate talent is good for local production, especially if they encourage the inflows of venture capital (Dossani and Kenney, 2002). However, often enough they, too, subscribe to the off-shore development model, reinforcing extensive growth.

Some progress beyond extensive growth has been made with increased R&D by multinationals, the promotion of Bangalore as a biotechnology hub, and a shift towards the development of embedded systems (UNCTAD, 2001, 2005; NASSCOM, 2002a; Reddy, 1997). Bangalore now hosts about 70 firms engaged in embedded software (Personal Interview, STPI, Bangalore, February 2005). There are also reports of university initiatives to set up joint projects, fellowship programmes, and establishing endowed chairs. This evolution, if on a wider scale, is consistent with India's better patent record in chemicals and pharmaceuticals and the availability of more doctorates than in IT. Thus, the challenge for Bangalore and the Indian IT industry is to go beyond writing algorithms and move towards science and technology-based knowledge generation and diffusion.

It is too early to tell if recent manufacturing plans in India by Nokia, Samsung, and other major telecom and electronics manufacturers are the beginning of a major knowledge-intensive innovative thrust. To date, the Indian IT industry is decoupled from the hardware industry (D'Costa, 2004a; Kumar and Joseph, 2005). The Indian firm TCS intends to move from the design to the fabrication of semiconductor chips, which is a novel approach since it has been usually the other way around, with software following hardware development (Kash et al., 2004: 789, 795). It is a welcome development, given India's lack of chip fabrication plants. But it is imperative to forge appropriate institutional links with universities and public research institutions since semiconductor production is complex and capital-intensive as the Japanese efforts indicate. Such manufacturing activities could be a harbinger of high-end research, which could also retain and attract expatriate talent.

The Taiwanese experience offers some lessons. Like their Indian counterparts, Taiwanese students went to the US to study engineering and sciences. Most stayed, but the rate of return increased from 11 per cent in the 1970s to more than 25 per cent by the mid-1990s (Chang, 1999: 82–3). The government offered returning Taiwanese talent travel subsidies, job placement, opportunities for year-long research jobs before permanent employment, appointments for expatriate professionals in research and academic institutions at attractive salaries, and housing and schooling facilities (Chang, 1999: 70–5, Mathews and Cho, 2000: 191–2). Those who did not return nevertheless became integral to networks that linked Taiwan to Silicon Valley through the circular movement of Taiwanese techno-entrepreneurs (Saxenian, 2004: 171, 270; Dicken, 2003: 414; Breznitz, 2005: 167). Return migration has been facilitated by the systematic support of the state for private sector initiatives in the electronics and semiconductor industries. The state's establishment of the Industrial Technology Research Institute (ITRI) in the 1970s for semiconductor research and the Hsinchu Science-based Industry Park in the 1980s encouraged expatriate talent (Castells, 2003: 269). Today the government subsidizes private investments through equity participation in high-technology companies in the science park and more public–private partnerships for basic research on fundamental technologies, nanotechnology, and digital signal processing (Dicken, 2003: 186, Mathews and Cho, 2000: 194).

4.6   Conclusion

The story of Bangalore's evolution as a high-technology cluster is a mixed one. The success of Bangalore (and India) is readily recognized in terms of the growth of the software industry and the responsiveness of the state in IT infrastructure and educational institutions. However, several challenges remain to transition from extensive to intensive growth. The industry, as a whole, needs to be embedded in a thick institutional setting. The current incentive structure does not encourage creating such linkages. Nevertheless, the intermediate steps that leverage an elastic supply of IT workers suggest a diversification of India's IT markets by focusing on both the domestic and East Asian markets. Attracting expatriate talent is integral to this strategy, as they are likely to have advanced technical degrees from abroad and richer market and R&D exposure.[33] If successful, these measures are likely to incrementally induce intensive growth and possibly the evolution of a robust institutional context.

Bangalore is a good illustration of a developing country's success story pushed by its history and local and national institutions. It is also a warning of the structural challenges faced by poor countries in overcoming technological and market barriers in the world economy. Developing countries can learn a number of lessons from Bangalore's experience. First, in a knowledge-driven economy technical education in emerging industries is critical. Both public and private parties can be involved and collaboration between the industry and public educational institutions is important. Secondly, Bangalore's high rates of growth illustrate a cumulative outcome of history, changing business strategies, and global (mainly US) demand. It is difficult to recreate intervening variables. Thirdly, rapid growth does lead to increased educational enrollments but often results in a lower quality of applied technical education. Fourthly, some firms are pushed to upgrade their activities, but the challenge is how to place the sector on a wholly new technological trajectory. Fifthly, the retention and return of expatriate talent can assist in intensive growth. Small countries are likely to find this daunting unless offset by sustained high macroeconomic growth. India stands a good chance of creating the incentives for investments and the return of expatriates.

The final lesson is that Bangalore offers a cautionary tale about the rapid growth of a narrow sector requiring high skills, which might detract from the more fundamental needs of development such as basic education, health, and infrastructure. The economic and social polarization resulting from Bangalore's growth may not be politically sustainable. A strategy of domestic inclusive development, combined with a long-term national innovation policy in a global context for sectoral upgrading will go a long way towards ensuring greater returns to talent employed in Bangalore and India.

India's extensive growth trajectory means it is not as yet a threat to European countries. Should India transform its growth pattern to an intensive form by establishing the necessary institutional architecture, diversifying its export markets and forging regional ties to Asia's IT industry, India and the whole Asian region could be a formidable force. Future joint action by regional players such as China, South Korea, Japan, Taiwan, Singapore, Malaysia, and Australia could shift the axis of the IT industry to Asia. It could also mean expanded markets for US and European products and services to India. Under this scenario India is unlikely to remain as mere fodder for Silicon Valley's appetite for low-cost software services and could instead become a voracious user of innovation-based products and services.

Notes

1. Earlier versions of this Chapter have been presented at the '3rd International Convention of Asia Scholars', Singapore (2003), a workshop on 'Universities as Drivers of Urban Economies in Asia', Washington, DC (2005), and conference on 'New Asian Dynamics in Science, Technology, and Innovation', Gilleleje, Denmark (2006), and available as World Bank's Policy Research Working Paper; No. WPS 3887. I thank K.C. Ho, the World Bank and the Social Science Research Council, New York, and the Nordic Institute of Asian Studies, Copenhagen for their support. As an Abe Fellow I also gratefully acknowledge the financial support received from the Abe Foundation, which allowed me to collect Japan-related data. Janette Rawlings and Govindan Parayil provided substantial editorial support.
2. The institutional interactions between industry and government and other supporting institutions are part of NIS. While both NIS and THM are similar frameworks to understand innovation dynamics, I apply only the THM framework.

I admit that there are differences between NIS and THM, but for the purposes of this chapter, which is about institutional architecture, both approaches emphasize self-reinforcing links between the government, research institutes, industry, and universities. In THM, the role of universities is paramount. But the two will be used interchangeably with a clear preference for a THM since academia must be made the third pillar of the institutional architecture. The discussion on clusters is premised on industrial agglomeration (or districts) and flexible production in the classic Marshallian form (Malecki, 1995).

3. Field research conducted by the author in 1998, 1999, and 2005 in Bangalore, Kolkata, Chennai, Mumbai, Delhi, NOIDA, Gurgaon, India and in the greater Tokyo area and Nagoya in Japan.

4. Attempts by governments to consciously promote high-tech clusters have met with varied results. For example, the establishment of 26 'technopolises' in Japan designed to spawn commercial spin-offs (Malecki, 1995:300) really did not take off, while Taiwan's Hsinchu Science Park is considered to be a success.

5. There are shortcomings of THM framework. It captures innovation dynamics better in affluent societies. States in developing countries may still remain as a major player, albeit in a transformed role. In addition, THM assumes innovative activities are endogenous with 'networks of relations' (Etzkowitz and Leydesdorff, 2000: 112) but global influences of technology dynamics, as in the Indian case, can be strong.

6. In the 1970s the state set up the National Informatics Center, the Computer Maintenance Corporation (CMC), the National Center for Software Development and Computing Technology, and regional computer centres. There are other research organizations in astrophysics, defence, space, artificial intelligence, basic sciences, microwaves, power, biological sciences, and mathematical modelling and computer simulation.

7. The New Okhla Industrial Development Authority (NOIDA), an industrial agglomeration near Delhi, came a distant second.

8. www.intltechpark.com. Accessed 18 April 2005.

9. www.bangaloreit.com/html. Accessed on 15 April 2005.

10. Other sources put the number at 103. www.educationinfoindia.com/engg/karnatakaeng.htm. Accessed on 18 April 2005.

11. www.educationinfoindia.com/streamwisecolleges/others. Accessed on 18 April 2005.

12. A similar IIIT has been established in Hyderabad. Bangalore's IIIT is located in Electronics City to encourage close academic–business interaction with IT firms in the campus, including firm-specific training.

13. Recently, the Chancellor of Vellore Technology Institute lamented the fact that he could not increase enrollments despite the huge demand in part due to lack of good instructors. The scarcity was a direct result of talent flight resulting from the high salaries of the IT industry (Informal Presentation, Tokyo, 25 February 2006).

14. The ratios for salaries between a PhD doing research in a non-IT segment and a non-PhD working in the IT industry has increased from 1:2 to 1:4/5 (Personal Interview, Sasken, February 2005). NASSCOM fears the problem is going to get worse as faculty is pressured to raise non-public sources of revenues through consulting and teaching in non-degree programmes.

15. Arora and Athreye (2002: 263) report that the numbers fell from 675 in 1987 to 375 in 1995.

16. This dynamic is expected to lead to labour shortages in India itself (see NASSCOM, 2002a: 67).
17. At the beginning of January 2006, allegedly there was no Indian IT company with a patent although the top ten patent holders in the world were IT companies (Mishra, 2006).
18. Incidentally, Taiwanese firms are not known for new products, rather they innovate products developed elsewhere and in which markets are well-established (Breznitz 2005: 157), that is, they 'leverage' existing knowledge (Mathews and Cho 2000). Taiwan has an excellent hardware sector, which is effectively driving the weaker software sector. It does little research, but a lot of design (Lu and Liu 2004: 460). Between 1991 and 2002, the number of IC design companies increased from 57 to 225 (Breznitz 2005: 161). But even Taiwan seems to lack meaningful university–industry linkages.
19. Curiously, Khanna and Palepu (2004) do not consider that Indian IT companies may be facing monopsonistic situations compounded by the fact that the going international price is far above of what the local market can bear. Hence, there is no 'rent-seeking and entry-deterring behaviour' since India is still considered to be low-cost producer relative to the US. All of these are consistent with an externally-driven IT industry.
20. Software revenues per employee in 1995 were as follows: Israel $100,000, Ireland $142,000, India $9,000, and the US $126,000 (Arora and Athreye, 2002: 259).
21. The correlation coefficient between number of software employees and revenues per employee at the firm level for 2002 for Bangalore's 102 firms is not significant at 0.297.
22. This was verified by most of the 75 firms surveyed during 1998, 1999, and 2005 in India. Of the 30 firms surveyed in 2005, 17 of which were in Bangalore, this author found no evidence of inter-firm collaboration (survey carried out in February and March 2005).
23. For example, some large, successful Indian firms in the Japanese market went even to the extent of refusing to sell a successful product made by another Indian company in the Japanese market (Lema and Hesbjerg, 2003: 140). Similarly, research in Tokyo suggested that Indian firms in Japan are highly competitive and hence there is limited networking for professional support in contrast to Chinese entrepreneurs in Tokyo (see D'Costa and Kobayashi, this volume). Communication by Tomoko Kobayashi, Tokyo, March 2006).
24. This was confirmed through interviews of over 70 firms, carried out in several Indian cities in 1998, 1999, and 2005.
25. Large Indian firms are also responding to rising costs by investing in lower-cost countries such as China.
26. The issues surrounding brain drain and return migration are discussed in D'Costa (2008).
27. This has been the strategy of Taiwanese IT firms outsourcing R&D services from China (Lu and Liu, 2004: 460–2). This is no different from the asymmetrical relationship between Silicon Valley and India, which entails 'value chain modularity' (see Sturgeon, 2003: 204).
28. The relative dependency ratio, computed by taking the share of Indian exports to Japan (2 per cent) and divided by the region's share in world IT services spending ($34.9 billion/$349.1 billion), was 0.2.
29. The highly competitive Japanese hardware producers have always bundled their software, hence the development of an independent software industry in Japan

has been discouraged (Anchordoguy, 2000). However, this development also suggests that the Japanese are strong in hardware-intensive software development, which for technical reasons has its own entry barriers.

30. These included embedded systems, intelligent electronics (high precision controls, wireless networks), bioelectronics (multifunctional chip devices, nanotech materials), smart devices (nano carbon composites, organic nano materials), super visual imaging (medical, solid state), system LSI design (design methods, architectures, EDA technology), SoC technology (IP and design technology, sensor networks).

31. I am grateful to Shoichi Yamashita of International Centre for South East Asian Development for introducing me to Takao Kageyama, Project Director of Kitakyushu Park.

32. Of course, the physical quality of life in Bangalore must be improved drastically if it is to attract and retain talent. It took the author more than an hour travelling from the city centre to Electronics City, which was 12 kilometres away (Bangalore, February 2005). Urban congestion, unreliable electric supply, pollution, an ageing airport, and skyrocketing real estate prices are the norm (Fannin, 2004).

References

Acs, Z. J. (ed.) (2000) *Regional Innovation, Knowledge, and Global Change*, London: Pinter.

Anchordoguy, M. (2000) 'Japan's Software Industry: A Failure of Institutions?', *Research Policy*, 29(3): 391–408.

Arogyaswamy, B. (n.d.) 'Beyond Software: Is India on the Road to Technological Leadership', Syracuse, LeMoyne College (unpublished paper).

Arora, S. (2005) 'The DQ-IDC India: Salary Survey – 05, *Data Quest*, 12 September 2005, http://www.dqindia.com. Accessed 14 September 2005.

Arora, A. and S.S. Athreye (2002) 'The Software Industry and India's Economic Development', *Information Economics and Policy*, 14(2): 253–73.

Arora, A. et al. (2001) 'The Indian Software Service Industry', *Research Policy*, 30(8): 1267–87.

Athreye, S.S. (2005) 'The Indian Software Industry and Its Evolving Service Capability', *Industrial and Corporate Change*, 14(3): 393–418.

Audirac, I. (2003) 'Information-Age Landscapes Outside the Developed World: Bangalore, India, and Guadalajara, Mexico', *Journal of the American Planning Association*, 69(1): 16–32.

Baber, Z. (2001) 'Globalization and Scientific Research: The Emerging Triple Helix of State–Industry–University Relations in Japan and Singapore', *Bulletin of Science, Technology & Society*, 21(5): 401–8.

Basant, R. (2003) "US–India Technology Co-Operation and Capability Building: The Role of Inter-Firm Alliances in Knowledge Based Industries," Unpublished Paper, February.

Best, M.H. (1990) *The New Competition: Institutions of Industrial Restructuring*, Cambridge, MA: Harvard University Press.

Breznitz, D. (2005) 'Development, Flexibility and R&D Performance in the Taiwanese IT Industry: Capability Creation and the Effects of State-Industry Coevolution', *Industrial and Corporate Change*, 14(1): 153–87.

BusinessWeek Online (2003) 'The New Global Job Shift', 3 February, http://www.businessweek.com/print/magazine/content/03_05/b38180.

Castells, M. (2003) *End of Millennium* (The Information Age: Economy, Society and Culture, Volume III), Malden, MA: Blackwell Publishing.

Cawthorne, P.M. (1995) 'Of Networks and Markets: The Rise and Rise of a South Indian Town, the Example of Tirrupur's Cotton Knitwear Industry', *World Development*, 23(3): 485–502.

Chang, S.L. (1999) *Taiwan's Brain Drain and Its Reversal*, Taipei: Lucky Bookstore.

Cookson, C. (2005) 'Academia Seeks to Join Global Elite: Research and Development in Asia', *Financial Times*, 8 July.

Dahlman, C. and A. Utz (2005) *India and the Knowledge Economy: Leveraging Strengths and Opportunities*, Washington, DC: The World Bank.

Data Quest, Gurgaon, India, http://www.dqindia.com/, various issues.

D'Costa, A.P. (2002a) 'Technology Leap-frogging: The Software Challenge in India', in P. Conceição et al. (eds),*Knowledge for Inclusive Development*, New York: Quorum Books, pp. 183–99.

D'Costa, A.P. (2002b) 'Software Outsourcing and Policy Implications: An Indian Perspective', *International Journal of Technology Management*, 24(7/8): 705–23.

D'Costa, A.P. (2003a) 'Uneven and Combined Development: Understanding India's Software Exports', *World Development*, 31(1): 211–26.

D'Costa, A.P. (2003b) 'Capitalist Maturity and Corporate Responses to Liberalization: The Steel, Auto, and Software Sectors in India', in A. Mukherjee-Reed (ed.), *Corporate Capitalism in Contemporary South Asia: Conventional Wisdoms and South Asian Realities*, Basingtoke: Palgrave Macmillan, pp. 106–33.

D'Costa (2004a) 'The Indian Software Industry in the Global Division of Labour', in A.P. D'Costa and E. Sridharan (eds), *India in the Global Software Industry: Innovation, Firm Strategies and Development*, Basingstoke: Palgrave Macmillan, pp. 1–26.

D'Costa, A.P. (2004b) 'Globalization, Development, and the Mobility of Technical Talent: India and Japan in Comparative Perspectives', UN University, *World Institute of Development Economics Research*, Helsinki, Research Paper Series (WIDER RP2004/62). http://www.wider.unu.edu/publications/.

D'Costa, A.P. (2004c) 'Flexible Institutions for Mass Production Goals: Economic Governance in the Indian Automotive Industry', *Industrial and Corporate Change*, 13(2): 335–67.

D'Costa, A.P. (2004d) 'Export Growth and Path-Dependence: The Locking-in of Innovations in the Software Industry', in A.P. D'Costa and E. Sridharan (eds), *India in the Global Software Industry: Innovation, Firm Strategies and Development*, Basingstoke: Palgrave Macmillan, pp. 51–82.

D'Costa, A.P. (2005) *The Long March to Capitalism: Embourgeoisment, Internationalization and the Industrial Transformation in India*, Basingstoke: Palgrave Macmillan.

D'Costa, A.P. (2006) 'ICTs and Decoupled Development: Theories, Trajectories and Transitions', in G. Parayil (ed.), *Political Economy and Information Capitalism in India: Digital Divide, Development and Equity*, Basingstoke: Palgrave Macmillan, pp. 11–34.

D'Costa, A.P. (2008) 'The International Mobility of Technical Talent: Trends and Development Implications', in A. Solimano (ed.), *International Mobility of Talent and Development Impact*, Oxford: Oxford University Press, pp. 44–83.

Dicken, P. (2003) *Global Shift: Reshaping the Global Economic Map in the 21st Century*, fourth edition, New York: Guilford Press.

Doloreux, D. (2002) 'What We Should Know About Regional Systems of Innovation', *Technology in Society*, 24(3): 243–63.

Dossani, R. and Kenney, M. (2002) 'Creating and Environment for Venture Capital in India', *World Development*, 30(2): 227–53.

Economist (2005) 'The Next Wave: India's IT and Remote-Service Industries Just Keep on Growing', *The Economist*, 17 December 17, p. 57–8.

Etzkowitz, H. and M. Klofsten (2005) 'The Innovating Region: Toward a Theory of Knowledge-Based Regional Development', *R&D Management*, 35(3): 243–55.

Etzkowitz, H. and L. Leydesdorff (2000) 'The Dynamics of Innovation: From National Systems and "Mode 2" to a Triple Helix of University–Industry–Government Relations', *Research Policy*, 29(2): 109–23.

Fannin, R. (2004) 'India's Outsourcing Boom', *Chief Executive*, May: 28–32.

Hayashi, T. (2003) 'Effect of R&D Programmes on the Formation of University-Industry-Government Networks: Comparative Analysis of Japanese R&D Programmes', *Research Policy*, 32(8): 1421–42.

Heeks, R. (1996) *India's Software Industry: State Policy, Liberalization and Industrial Development*, New Delhi: Sage Publications.

Heitzman, J. (2004) *Network City: Planning the Information Society in Bangalore*, New Delhi: Oxford University Press.

Hira, R. (2004) 'U.S. Immigration Regulations and India's Information Technology Industry', *Technological Forecasting and Social Change*, 71(8): 837–54.

Hoffman, T. (2003) 'Gartner: One in 20 End-User IT Jobs to Move Offshore by Late 2004', *Computerworld*, http://www.computerworld.com/printhis/2003/0,4814,83568,00.html.

Holmström, M. (1998) 'Bangalore as an Industrial District: Flexible Specialization in a Labor Surplus Economy?', in P. Cadène and M. Holmström (eds), *Decentralized Production in India: Industrial Districts, Flexible Specialization, and Employment*, Pondicherry: French Institute of Pondicherry, pp. 169–229.

Hsu, J.-Y. (2004) 'The Evolving Institutional Embeddedness of a Late-Industrial District in Taiwan', *Tijdschrift voor Economische en Sociale Geografie*, 95(2): 218–32.

Jayaraman, K.S. (2002a) 'India's Scientists Agonize over Publication Rate', *Nature*, 419: 100.

Jayaraman, K.S. (2002b) 'India Online', *Nature*, 415: 358–9.

Kambhampati, U.S. (2002) 'The Software Industry and Development: The Case of India', *Progress in Development Studies*, (2)1: 23–45.

Kash, D.E., R.N. Augur and N. Li (2004) 'An Exceptional Development Pattern', *Technological Forecasting and Social Change*, 71(8): 777–97.

Kattuman, P. and K. Iyer (2001) 'Human Capital Development in the Move Up the Value Chain: The Case of the Indian Software and Services Industry', in M. Kagami and M. Tsuji (eds), *The IT Revolution and Developing Countries: Late-comer Advantage?*, Tokyo: Institute of Developing Economies and Japan External Trade Organization, pp. 208–27.

Kaushik, P.D. (2006) 'ICT Initiatives in India: Lessons for Broad-based Development', in A.P. D'Costa (ed.), *The New Economy in Development: ICT Challenges and Opportunities*, Basingstoke: Palgrave Macmillan, pp. 110–36.

Keeble, D. and Wilkinson, F. (1999) 'Collective Learning and Knowledge Development in the Evolution of Regional Clusters of High Technology SMEs in Europe', *Regional Studies*, 33(4): 295–303.

Khanna, T. and Palepu, K. (2004) 'The Evolution of Concentrated Ownership in India: Broad Patterns and a History of the Indian Software Industry', Cambridge, MA: National Bureau of Economic Research, Working Paper 10613.

Kshetri, N. (2005) 'Structural Shifts in the Chinese Software Industry', *IEEE Software*, July–August: 86–93.

Kumar, N. and K.J. Joseph (2005) 'Export of Software and Business Process Outsourcing from Developing Countries: Lessons from the Indian Experience', *Asia-Pacific Trade and Investment Review*, 1(1): 91–110.

Lema, R. and B. Hesbjerg (2003) *The Virtual Extension: A Search for Collective Efficiency in the Software Cluster in Bangalore*, Roskilde: University of Roskilde, Public administration and Public Economics & International Development Studies.

Leydesdorff, L. (2000) 'The Triple Helix: An Evolutionary Model of Innovations', *Research Policy*, 29(2): 243–55.

Lombardi, M. (2003) 'The Evolution of Local Production Systems: The Emergence of the "Invisible Mind" and the Evolutionary Pressures Towards More Visible "Minds"', *Research Policy*, 32(8): 1443–62.

Looy, B.V., K. Debackere and P. Andries (2003) 'Policies to Stimulate Regional Innovation Capabilities via University–Industry Collaboration: An Analysis and An Assessment', *R&D Management*, 33(2): 209–29.

Lu, L.Y.Y. and J.S. Liu. (2004) 'R&D in China: An Empirical Study of Taiwanese IT Companies', *R&D Management*, 34(4): 453–65.

Malecki, E.J. (1995) *Technology and Economic Development: The Dynamics of Local, Regional and National Change*, Harlow: Longman Group Ltd.

Mani, S. (2004) 'Institutional Support for Investment in Domestic Technologies: An Analysis of the Role of Government in India', *Technological Forecasting and Social Change*, 71(8): 855–63.

Mathews, J.A. and D.-S. Cho (2000) *Tiger Technology: The Creation of a Semiconductor Industry in East Asia*, Cambridge: Cambridge University Press.

McManus, J., M. Li and D. Moitra (2007) *China and India: Opportunities and Threats for the Global Software Industry*, Oxford: Chandos Publishing.

Ministry of Education, Culture, Sports, Science and Technology (MEXT) (2004) 'Cluster: Knowledge Cluster Initiative, 2004', Tokyo: Office of the Promotion of Regional R&D Activities, MEXT.

Mishra, G. (2006) 'India Lags in Number of IT Patents', www.rediff.com 1/05/2006. Accessed 23 March 2007.

Morosini, P. (2004) 'Industrial Clusters, Knowledge Integration and Performance', *World Development*, 32(2): 305–26.

Naidu, B.V. (2003) 'Tracing the History of IT in India', in M. Jussawalla and R.D. Taylor (eds), *Information Technology Parks of the Asia Pacific: Lessons for the Regional Digital Divide*, Armonk, NY: M.E. Sharpe, Inc., pp. 119–50.

NASSCOM (2002a) *The IT Industry in India: Strategic Review 2002*, New Delhi: NASSCOM.

NASSCOM (2002b) *Indian IT Software and Services Directory, 2002*, New Delhi: NASSCOM (CD ROM).

NASSCOM (2004) *The IT Industry in India: Strategic Review 2004*, New Delhi: NASSCOM.

NASSCOM (2005) www.nasscom.org. Accessed 18 April 2005.

NASSCOM (2006) www.nasscom.org. Accessed 6 September 2006.

Nelson, R. (2000) 'National Innovation Systems', in Z.J. Acs (ed.), *Regional Innovation, Knowledge, and Global Change*, London: Pinter, pp. 11–26.

OECD (2001) *Cities and Regions in the New Learning Economy*, Paris: OECD.

Odaka, K. et al. (1988) *The Automobile Industry in Japan: A Study of Ancillary Firm Development*, Tokyo: Kinokuniya Company Ltd.

Okada, A. (2004) 'Bangalore's Software Cluster: Building Competitiveness through the Local Labor Market Dynamics', in A. Kuchiki and M. Tsuji (eds), *Industrial Clusters in Asia: Analyses of their Competition and Cooperation*, Tokyo: Institute of Developing Economies and Japan External Trade Organization, pp. 276–314.

Parthsarathi, A. and K.J. Joseph (2004) 'Innovation Under Export Orientation', in A.P. D'Costa and E. Sridharan (eds), *India in the Global Software Industry: Innovation, Firm Strategies and Development*, Basingstoke: Palgrave Macmillan, pp. 83–111.

Parthasarathy, B. (2000) *Globalization and Agglomeration in Newly Industrializing Countries: The State and the Information Technology Industry in Bangalore, India*, Berkeley, CA: Department of Sociology, University of California, unpublished doctoral dissertation.

Parthasarathy, B. (2004) 'Political Economy of the Computer Software Industry in Bangalore, India'. Paper presented at the Conference on Asian Innovation System and Clusters, Bangkok, 1–2 April.

Patibandla, M. and B. Petersen (2002) 'Role of Transnational Corporations in the Evolution of a High-Tech Industry: The Case of India's Software Industry', *World Development*, 30(9): 1561–77.

Piore, M.J. and Sabel, C.F. (1984) *The Second Industrial Divide*, New York: Basic Books.

Prabhu, G.N. (1999) 'Implementing University–Industry Joint Product Innovation Projects', *Technovation*, 19(8): 495–505.

Reddy, P. (1997) 'New Trends in Globalization of Corporate R&D and Implications for Innovation Capability in Host Countries: A Survey from India', *World Development*, 25(11): 1821–37.

Saxenian, A. (2004) 'The Silicon Valley Connection: Transnational Networks and Regional Development in Taiwan, China, and India', in A.P. D'Costa and E. Sridharan (eds), *India in the Global Software Industry: Innovation, Firm Strategies and Development*, Houndmills: Palgrave Macmillan, pp. 164–92.

Schmitz, H. (1999) 'Collective Efficiency and Increasing Returns', *Cambridge Journal of Economics*, 23(4): 465–83.

Sharif, N. (2006) 'Emergence and Development of the National Innovation Systems Concept', *Research Policy*, 35(5): 745–66.

Siliconindia.com (2005a) 'India's Outsourcing Edge to Erode: Gartner', www.siliconinida.com. Accessed on 15 September 2005.

Siliconindia.com (2005b) 'Pay Hike Will Mar BPO Growth', www.siliconindia.com. Accessed on 20 April 2005.

Siliconindia.com (2005c) 'India Fast Losing BPO-Edge: Forbes', www.siliconindia.com. Accessed on 17 October 2005.

Siliconconindia.com (2005d) 'Efforts on to Prevent Talented Indians from Leaving UK', www.siliconindia.com. Accessed on 17 October 2005.

Siliconindia.com (2007) 'Indian Patents to Touch Global Levels', www.siliconindia.com. Accessed on 26 March 2007.

Smitka, M. (1991) *Competitive Ties: Subcontracting in the Japanese Automotive Industry*, New York: Columbia University Press.

Sridharan, E. (1995) 'Liberalization and Technology Policy: Redefining Self-Reliance', in T.V. Sathyamurthy (ed.), *Industry and Agriculture in India Since Independence*, New Delhi: Oxford University Press, pp. 150–88.

Sridharan, E. (2004) 'Evolving Towards Innovation? The Recent Evolution and Future Trajectory of the Indian Software Industry', in A.P. D'Costa and E. Sridharan (eds), *India in the Global Software Industry: Innovation, Firm Strategies and Development*, Basingstoke: Palgrave Macmillan, pp. 27–50.

Sturgeon, T.J. (2003) 'What Really Goes on in Silicon Valley: Spatial Clustering and Dispersal in Modular Production Networks', *Journal of Economic Geography*, 3(2): 199–225.

Taebe, F.A. (n.d.) 'Proximities and Innovation: Evidence from the Indian IT Industry in Bangalore', Danish Research Unit for Industrial Economics, Working Paper No. 04–10.

Thomas, J.J. (2006) 'Informational Development in Rural Areas: Some Evidence from Andhra Pradesh and Kerala', in G. Parayil (ed.), *Political Economy and Information Capitalism in India: Digital Divide, Development and Equity*, Basingstoke: Palgrave Macmillan, pp. 109–32.

Tsai, D.H.A. (2005) 'Knowledge Spillovers and High-Technology Clustering: Evidence from Taiwan's Hsinchu Science-Based Industrial Park', *Contemporary Economic Policy*, 23(1): 116–28.

UNCTAD (2001) *World Investment Report: Transnational Linkages*, New York: UNCTAD.

UNCTAD (2004) *World Investment Report: The Shift Toward Services*, New York: UNCTAD.

UNCTAD (2005) *World Investment Report: Transnational Corporations and the Internationalization of R&D*, New York: UNCTAD.

US Patent and Trademark Office (2005) 'PTMD Special Report: All Patents, All Types: January 1977–December 2004', Alexandria, VA: US Patent and Trademark Office.

Wever, E. and Stam, E. (1999) 'Clusters of High Technology SMEs: The Dutch Case', *Regional Studies*, 44(4): 391–400.

Wong, P.-K. (1995) 'Competing in the Global Electronics Industry: A Comparative Study of the Innovation Networks of Singapore and Taiwan', *Industry and Innovation*, 2(2): 35–61.

www.intltechpark.com. Accessed on 18 April 2005.

www.bangaloreit.com/html. Accessed on 15 April 2005.

www.educationinfoindia.com/engg/karnatakaeng.htm. Accessed on 18 April 2005.

www.educationinfoindia.com/streamwisecolleges/others. Accessed on 18 April 2005.

5
Innovation in India and China: Challenges and Prospects in Pharmaceuticals and Biotechnology

*Jayan Jose Thomas**

5.1 Introduction

India and China are important players in an evolving process of globalization of research and development (R&D). Focusing on pharmaceuticals and biotechnology industries, this chapter analyses the challenges and prospects facing the two countries in global innovation.

The *World Investment Report 2005* points out that there is now a fresh wave of R&D investments by multinational corporations (MNCs) in developing countries, particularly China and India. In a survey of the world's largest R&D-spending MNCs conducted by the United Nations Conference on Trade and Development (UNCTAD) in 2004–05, China was identified by the respondents as the most attractive location for future investments in R&D. India was the third most attractive location, behind the United States (USA) (UNCTAD, 2005: 22–6). The *Economist* described India and China as 'high-tech hopefuls' in a special report on technology in the two countries in its November 2007 issue.[1] Foreign direct investment (FDI), especially in technology-intensive industries, used to be circulated

* The author thanks the Institute of South Asian Studies, National University of Singapore for providing a research grant to carry out this study. Different versions of this study were presented at the Conference on *China, India, and the International Economic Order* organized by the Faculty of Law, National University of Singapore, 23–4 June 2006; a seminar at the Centre for Technology, Innovation and Culture, University of Oslo, 25 September 2006; Conference on *New Asian Dynamics in Science, Technology and Innovation*, Copenhagen, 27–9 September 2006; and the fourth annual conference of GLOBELICS (Global Network for Economics of Learning, Innovation, and Competence Building Systems) held at Thiruvananthapuram, India, 4–7 October 2006. The author is grateful to the participants at the different conferences/seminar, as well as to Anthony D'Costa and Govindan Parayil for their comments on this study. A special word of thanks to B.K. Keayla for granting permission to reproduce a part of a table from his work, and to Joe C Mathew for his generous help in the field study and collection of research material.

largely within developed countries. MNCs restricted their R&D activities in developing countries mostly to the adaptation of technologies for local markets (Kleinknecht and Wengel, 1998). Therefore, the recent interest shown by MNCs in shifting some of their core innovation activities to China and India marks the blossoming of a process of globalization of R&D.

There are many factors behind the growing prominence of India and China as R&D locations. The large supply of relatively low-cost skilled professionals in these countries is a key one. India and China have made significant public investments in science and technology over recent decades, and this provides a strong base for future growth. In addition, both countries have, in recent years, introduced rules ensuring greater protection to intellectual property rights (IPRs), in compliance with World Trade Organization (WTO)'s Agreement on Trade Related Aspects of Intellectual Property Rights (TRIPS). The assurance of IPR protection has been an important incentive for MNC investments. Furthermore, the expanding numbers of middle-class consumers in India and China promises a booming market for high-tech products, and this raises interest levels among multinational corporations.

At the same time, the challenges unfolding are many as the outsourcing of innovation by global corporations picks up speed. First, there are concerns on the future supply for the vast market for innovative products for the poor in developing countries. There exists demand for cheap medicines for poor patients, demand for biotechnological innovations that ensure food security in the third world, demand for novel telecommunication devices for rural areas, and so on. Leading corporations in the West have so far given a low priority to this market, focusing instead on the market for innovations for rich consumers. It appears that the rise of India and China as favoured destinations for the outsourcing of R&D is not likely to give any positive impetus to innovations targeting the poor. The IPR rules implemented by India and China encourage MNC investments, but also create new constraints for domestic firms in these countries. Many of the top domestic firms in China and India will benefit from the growing opportunities for contract research, but in the process they might concentrate their efforts to the innovation needs of the rich.

Secondly, as domestic firms in India and China progressively move away from their home markets and carry out R&D for export markets, questions are raised about their long-term growth prospects vis-à-vis western MNCs. Will they become capable of challenging western MNCs, or will they remain as junior partners in a global chain of innovation? Evidence from India's software industry indicates that its extreme reliance on export markets is likely to be a constraint on future growth. D'Costa (2002, 2004) argues that India's software industry is overly dependent on a single export market (the US market), locking the industry into a low innovation trajectory. According to Chandrasekhar (2006), the outsourcing and offshoring of business services to India aid the strategies of US corporations to maintain their high profit levels by tapping into the global supply of cheap labour.

This chapter discusses current and future trends in innovation in China and India in the context of the globalization of R&D, with a particular focus on pharmaceuticals and biotechnology industries. The next section discusses in greater detail the rise of India and China in the knowledge economy. Section 5.3 outlines some of the features of the demand for pharmaceutical and biotechnology innovations for the third world. Section 5.4 shows how India's pharmaceuticals industry could meet the demand for affordable medicines, and discusses the challenges facing the industry in the post-TRIPS phase. Section 5.5 is on pharmaceuticals and biotechnology industries in China, and section 5.6 concludes.

5.2   The rise of India and China in the knowledge economy

The most important factor that triggers the new wave of investments in R&D in India and China is the large supply of highly skilled professionals in these countries. Both China and India are now ahead of the United States with respect to enrolment in tertiary technical education. In 2000–01, the total numbers of students enrolled for tertiary education were approximately 12 million in China and 10 million in India (UNCTAD, 2005: 162). In China, in 2004, 13.3 million students were enrolled as undergraduates, while those enrolled for a Master's degree and Doctor's degree were, respectively, 654,286 and 165,610 (National Bureau of Statistics of China, 2005: 689–95). At the same time, the costs of employing skilled workers are relatively low in India and China. The annual cost of hiring a chip design engineer, in 2002, was found to be $28,000 in China (Shanghai province) and $30,000 in India compared to $300,000 in Silicon Valley in the United States (Ernst, 2005: 56).

Both India and China have a large population of emigrants working as skilled professionals in foreign countries. For example, in 1999 Indian professionals accounted for 47 per cent of all H-1 visas issued (to skilled workers) in the United States and professionals from China formed the second largest group, with a share of 5.0 per cent (cited in Chanda and Sreenivasan, 2006: 220). With regard to work permits issued to emigrants from different nationalities in the United Kingdom (UK), Indians topped the list with a share of 21.4 per cent of the total work permits issued in 2002 (Findlay, 2006: 78). The number of postgraduates studying abroad has steadily increased in the case of China: from 860 in 1978 to 20,381 in 1995, and 114,682 in 2004 (National Bureau of Statistics of China, 2005: 689–95). Today, India and China are encouraging return migration of their skilled professionals to energize high-technology entrepreneurship back home. The Chinese Academy of Sciences has introduced many attractive schemes to woo returnee researchers as part of a programme of 'reverse brain drain' currently promoted in that country (Zweig, 2006).

The state in post-independence India actively intervened to build a strong infrastructure for science and technology. R&D in India has been financed largely by the public sector. The combined share of the Central and State governments (including public sector units under their management) in the total national expenditure on R&D in India in 2002–03 (the latest year for which data were available) was 75.6 per cent. The share of the private sector was only 20.3 per cent while higher education accounted for the remaining 4.1 per cent (GOI, 2006: 3–8). In China, science and technology was a major plank in the 'four modernizations' the government embarked on after 1978 (Spence, 1999: 618–20). Today, the Chinese government promotes R&D through two major national initiatives: the national high-tech R&D Programme or the 863 programme and the national programme on key basic research or the 973 programme.[2] The 973 programme has identified life sciences, nanotechnology, information technology, and earth sciences as frontier areas for basic research. In 2004, of the total funding for science and technology enterprises in China, 22.8 per cent came directly from the government, 64 per cent of the funds were raised by enterprises themselves, and 6.1 per cent came through loans from financial institutions (National Bureau of Statistics of China, 2005: 714–17).

5.2.1 Challenges facing India and China in high-technology sectors

While India and China enjoy some advantages in science and technology, as noted above, both countries still have a long way to go. Table 5.1 shows that China and India lag clearly behind the United States in many national-level indicators such as R&D expenditure and researchers and patents granted per resident population. The evolving relationship between multinational

Table 5.1 Selected indicators of performance in research and development (R&D): India, China and the United States

	India	China	United States
R&D expenditure, billions of US dollars, 2002	3.7*	15.6	276.2
R&D expenditure as % of gross domestic product, 2000–2005	0.8	1.4	2.7
Researchers in R&D, per million people, 1990–2005	119	708	4605
High technology exports as % of manufactured exports, 2005	4.9	30.6	31.8
Patents granted to residents, per million people, 2000–05	1	16	244

Notes: *2001 data.
Sources: UNDP (2007), Tables 13 and 16; UNCTAD (2005), p. 105.

companies, on the one hand, and the state and domestic firms in India and China, on the other, will also be highly crucial. Dicken (1998) points out that the relationship between MNCs and states is conflictual as well as cooperative – with each trying to gain bargaining power over the other. In fact, according to Stopford and Strange (1991) (cited in Dicken, 1998), the 'balance of power' has moved over time from governments as a group to the multinationals. MNCs in the United States and Western Europe continue to reign supreme in high-technology industries, their R&D expenditures exceeding the national R&D expenditures in many developing countries. For instance, R&D spending by Pfizer of the United States in 2002 was US$4.8 billion, while the national R&D expenditure of India in 2001 was $3.7 billion (UNCTAD, 2005: 120). Given their dominant position, MNCs' investments in India, China and other developing countries will not be necessarily beneficial to the host country. Local R&D firms may be taken over by MNCs; local firms and universities may not receive fair compensation as they enter into partnerships with MNCs; and talented researchers in local firms may move into better paying jobs in MNCs (UNCTAD, 2005: 190–3).

Studies have pointed to several areas of weaknesses in the nature of growth of developing country firms, including China's high-technology firms. Lardy (2002) and Steinfeld (2004) argue that China's integration with the global economy has been a shallow one. In an ongoing process of modularization in global manufacturing – in which component manufacturing processes are spread out across locations and firms all over the globe – Chinese firms derive weak advantages on the basis of low costs and high volumes. China exports high-technology products but the country's role in these exports is limited largely to that of an assembler of high value-added components (Steinfeld, 2004). According to Branstetter and Lardy (2006), domestic value added accounts for only 15 per cent of the value of China's exports of electronic and information technology products. For instance, in the case of an Apple iPod, manufactured in China by a Taiwanese company and sold for $224 (in 2005), the value captured by China was just a few dollars, whereas Apple, an American company, claimed the largest share of profits (Linden et al., 2007). Based on an analysis of firms in the aerospace, oil and pharmaceuticals industries, Nolan and Zhang (2002) found that Chinese firms were weak vis-à-vis a few leading, mostly oligopolistic, global giants, which have emerged as core 'systems integrators' in their respective sectors. The relative weakness of Chinese firms was more marked in high-technology sectors (Nolan and Zhang, 2002). Despite such challenges, however, Naughton (2007) has expressed some amount of optimism on the future of technological development in Chinese firms.

One of the important challenges to the growth of high-technology firms in India is the limited degree of interconnection between institutional and industrial R&D in the country. With respect to total national R&D expenditure, government-owned R&D institutions account for the major chunk while the industrial sector's share is small (GOI, 2006). It is debatable whether public investments in science and technology have created strong 'national innovation systems' (NIS) in India.[3] D'Costa (2006) argues that the triple helix model, which refers to thick institutional linkages between industry, academia and government, has not put down deeper roots in India. Bangalore's software industry, for example, is characterized by a high degree of inter-firm competition – not cooperation – and limited degree of inter-action with academic and research institutions in the city (D'Costa, 2006). According to Dahlman and Utz (2005), there exists a deep gulf between the academic world and industry in India. However, there are indications of posi-tive changes occurring. A recent study by Basant and Chandra (2007) pointed to the gradual emergence of linkages between academia and industry in the Indian cities of Bangalore and Pune. The study also noted that these linkages are deepening: from interactions in the labour market to knowledge-based linkages (Basant and Chandra, 2007).

Finally, some of the provisions of the TRIPS Agreement are creating hurdles to and, as we shall see in this chapter, altering the nature of innovation in developing countries, especially India and China. Drahos with Braithwaite (2002) has shown that an alliance between a small group of United States (US) corporations and the US state was the driver of the sequence of events linking trade and intellectual property and culminating in the genesis of TRIPS. Representatives of US corporations, importantly of the pharmaceutical giant Pfizer, which stood to gain enormously from new rules on intellec-tual property, played an important role in setting the government agenda on intellectual property rights in the USA. The US state took forward this agenda and, employing a range of measures including the threat of trade sanctions through Section 301, succeeded in bringing other developed and develop-ing countries into compliance on TRIPS (Drahos with Braithwaite, 2002). The United States' advocacy of strict patent rules in developing countries in recent times must be seen against the fact that the IPR regimes in advanced countries including the USA during their periods of industrialization until the early twentieth century were characterized by laxity and frequent viola-tions (Chang, 2003). The rest of the chapter focuses on the changes in the nature of innovation in the pharmaceuticals and biotechnology industries in India and China, brought about to a large extent by the TRIPS agreement.

5.3   Demand from developing countries for innovations in pharmaceuticals and biotechnology

Extreme disparities exist between the developed and developing countries with respect to achievements in health and other human development indi-cators. Majority of the world's population living in developing countries suffer from food shortage and a lack of access to medical facilities. A person born in Sub-Saharan Africa in 2000–05 had a life expectancy of only 49 years,

Table 5.2   Selected indicators of achievements in health and human development, different regions of the world

	Population, millions (2005)	Life expectancy at birth, years (2000–05)	Population under-nourished, % (2002–04)	HIV prevalence, ages 15–49, % (2003)	TB cases, per 100,000 persons (2003)
LDCs	766	53	35	3.2	452
Developing Countries	5,215	66	17	1.3	289
Sub-Saharan Africa	723	49	32	7.3	487
South Asia	1,587	63	21	0.7	306
India	1,134	63	20	0.4–1.3	287
China	1,313	72	12	0.1	245
High Income OECD	932	79	–	0.4	18

Note: LDCs = Least developed countries; OECD = Organisation for Economic Cooperation and Development.
Source: UNDP (2005), Table 9; UNDP (2007), Tables, 5, 7 and 10.

whereas a person born in a high-income OECD country in the same years had a life expectancy of 79 years (see Table 5.2). As Table 5.2 shows, significantly large proportions of the population in both Sub-Saharan Africa and South Asia are undernourished. Tuberculosis is still highly prevalent in least-developed and developing countries (see Table 5.2). Malaria cases of more than 15 per 100 population were reported in the year 2000 in several African countries, including Botswana, Burundi, Zambia and Malawi, whereas none of the countries in Western Europe or North America reported incidence of Malaria in that year (UNDP, 2005).

Technological advances in pharmaceuticals and biotechnology open up tremendous opportunities for solving the problems of ill health and malnutrition in the developing world. However, while the majority of the world's population in need of medicines live in developing countries, the pharmaceuticals industry is effectively controlled by a small number of MNCs headquartered in the developed world. In 1999, the share of high-income countries (according to the World Bank's definition) in global pharmaceutical production was 92.9 per cent, middle- and low-income countries accounting for only the remaining 7.1 per cent. Research and development in pharmaceuticals is carried out largely in developed countries. Of the total global spending on health R&D, 42 per cent is privately funded, 47 per cent is funded by the public sector in high-income and transition countries, and only 3 per cent is financed by the public sector in low- and middle-income countries (WHO, 2004: 5–13). Not surprisingly, R&D activities are

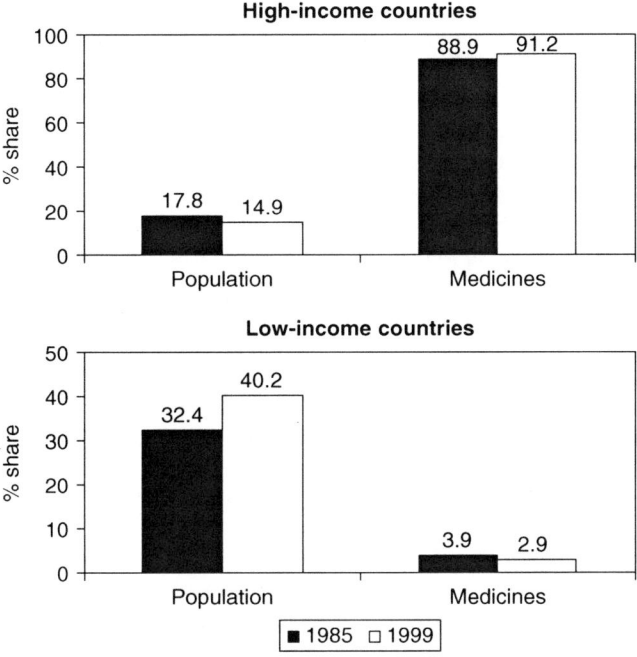

Figure 5.1 Shares of high-income and low-income countries in world population and global consumption (in value) of medicines, 1985 and 1999 (percentages)
Source: WHO (2004).

overwhelmingly directed towards the health needs of the rich in industrialized countries, towards lifestyle-related and convenience medicines. There are many 'tropical diseases' (also referred to as 'neglected diseases') such as dengue, diphtheria and malaria, which primarily affect people in poorer countries, but these diseases are given very low priority in pharmaceutical R&D (Lanjouw and MacLeod, 2005). It is pointed out that only 10 per cent of the worldwide spending on pharmaceutical R&D is directed toward 90 per cent of the global disease burden (WHO, 2004: 18–19).

Such unevenness in pharmaceuticals research and production capabilities is reflected in the statistics on the consumption of medicines. In 1999, low-income countries, accounting for 40.2 per cent of world population, had a share of just 2.9 per cent in the global consumption (in value terms) of medicines. Imbalances between high- and low-income countries in the consumption of medicines have been worsening, as shown through the data for the years 1985 and 1999 in Figure 5.1. It is reported that over one-third of world's population purchased less than one per cent of the pharmaceuticals sold worldwide. In 1999, 1,725 million people in the world, including

649 million in India, 267 million in Africa and 191 million in China, were without access to essential medicines (WHO, 2004: 31–3 and 62).

With the advent of biotechnology, healthcare and pharmaceutical industries are undergoing fundamental changes. The core scientific principles underlying pharmaceutical innovations are shifting 'from fine chemistry towards molecular biology' (Cooke, 2005: 333). Rather than waiting for 'chance discoveries', pharmaceutical innovation today is increasingly characterized by 'rational drug design' in which dedicated biotech firms play a prominent role (Cooke, 2005). However, dedicated biotech firms operate under the shadow of big pharmaceutical corporations. The therapeutic products they develop are licensed out to the big corporations. Biotech firms are located in clusters, and most of the leading clusters in biomedical sciences are in North America and Western Europe.

Advances in biotechnology also raise hopes for dramatic improvements in agricultural productivity, which are essential to meet the food supply requirements of a growing world population. However, research on agricultural applications of genetic engineering, including Genetically Modified (GM) crops, is carried out almost entirely by US-based MNCs. This is in contrast to the case of earlier innovations in agriculture, including those of non-GM hybrid crop varieties, which were born out of publicly funded research. The extreme dominance of US multinationals in GM research as well as concerns regarding biological safety and biopiracy largely explains the unpopularity of GM crops in Europe and in a majority of developing countries (Bernauer, 2003). GM research has so far covered only a limited number of crops, importantly, soybeans, maize and cotton, while it has almost neglected tropical subsistence crops such as cassava, millet and cowpeas grown by poor farmers in developing countries. Similarly, while GM research focuses almost exclusively on pest resistance and herbicide tolerance, some of the issues pertinent to developing country agriculture, such as drought resistance, have not been on its agenda (Paarlberg, 2001). Thus, the general picture in pharmaceuticals and biotechnology industries is one of neglect of developing country needs. It is against this general context that the case studies of pharmaceuticals and biotechnology industries in India and China presented in the following sections assumes relevance.

5.4   The pharmaceuticals industry in India

India's pharmaceuticals industry has been growing rapidly. India supplies 8 per cent of the world's output (in volume) of drugs, and 22 per cent of the world's output of generic drugs (Sampath, 2005: 15; Grace, 2004). As per the latest available statistics (for 2007), there were 75 manufacturing units in India approved by the United States Food and Drug Administration (FDA); India has the largest number of FDA-approved manufacturing facilities in any country other than the United States. Recent reports indicate that the Indian pharmaceutical industry consists of 300 large to moderate firms; the number of firms rises to 24,000 if small firms are also included.[4] In 2005–06, India exported drugs, pharmaceuticals and fine chemicals worth US$4.9 billion to a large number of countries including the United States, United Kingdom, Germany, Russia and China (CMIE, 2006).

5.4.1 State intervention and growth of the pharmaceuticals industry in India

State intervention has been an essential feature in the development and growth of the Indian pharmaceutical industry (Chaudhuri, 2005). A major component of this state intervention relates to the introduction of the Indian Patents Act of 1970 (which came into effect in 1972). Until 1970, the Patents and Design Act 1911 – a law framed during the British colonial period which guaranteed product patenting rights to drug companies – was a serious drag on the growth of domestic pharmaceutical firms in India. Production and distribution of medicines in India was almost entirely under the control of MNCs, and prices of medicines sold in India by the MNCs were reported to be one of the highest in the world.[5] The Indian Patents Act of 1970 introduced major changes. Section 5 of the 1970 Act disallowed product patenting in the case of drugs and food products; only the processes for manufacturing these products were eligible for patents, according to the Act. The period for which patents were granted was reduced from 16 years to five years (from the date of patent granting or seven years from the date of patent application). The 1970 Act made it mandatory for the patent holder to start domestic manufacturing using the patented process within three years from the date of sealing of the patent as well as to issue licenses to local manufacturers (for a royalty) after the three-year period (Lanjouw, 1997: 51; Chaudhuri, 2005: 36–8).

Pharmaceutical manufacturing and research organizations set up by the Indian government from the 1950s created a supportive environment for the growth of the domestic industry. Hyderabad's emergence as a centre for bulk drug manufacturing was aided by the early establishment of public sector units such as Indian Institute of Chemical Technology (IICT) and Indian Drugs and Pharmaceuticals Limited (IDPL) (inaugurated in 1956 and 1961 respectively) in the city. India's Council of Scientific and Industrial Research (CSIR) developed many technologies that were used by even the top pharmaceutical firms in India. At the same time, through a number of regulations in the 1970s, the government discouraged MNCs presence in low-technology areas, leaving these sectors for domestic firms (Chaudhuri, 2005). In addition, the government's Drug Price Control Order (DPCO) of 1970 took steps to check the unwarranted escalation of pharmaceutical prices.[6]

Under the protective cover of state support, domestic firms developed reverse engineering capabilities in chemicals-based processes for pharmaceutical production. Many of them grew to become leading producers of generic drugs and started supplying medicines for the Indian market. In 1970, of

Table 5.3   Selected indicators of growth of domestic firms in India's pharmaceuticals industry, 1970–2004

	1970	1998	2004
Share in % of domestic firms in India's pharmaceuticals market by sales	32	60	77
Number of domestic firms among top 20 firms by pharmaceutical sales in India	6*	12*	15

Notes: *Statistics refer to the years 1971 and 1996 respectively.
Sources: Chaudhuri (2005: 18) and Lanjouw (1997: 39). The statistics cited in Chaudhuri (2005) are based on the following: for 1970, Ministry of Petroleum and Chemicals (1971: 1); for 1998, Kalsekar (2003); for 2004, Sudip Chaudhuri's calculation using ORG-MARG (2004). The statistics relating to top 20 firms by pharmaceutical sales was cited in Lanjouw (1997) and was sourced from ORG, Mumbai.

the top ten pharmaceutical firms by retail sales in the Indian market, only two were Indian firms while the other eight were subsidiaries of multinational companies (Lanjouw, 1997: 3). The shares of domestic firms and MNCs in India's pharmaceuticals market by sales were 32 per cent and 68 per cent respectively in 1970. By 2004, these shares were altered upside down: the share of domestic firms rose to 77 per cent while the share of MNCs correspondingly declined to 33 per cent (see Table 5.3).

More importantly, domestic pharmaceutical companies were able to manufacture and sell generic versions of medicines at very low prices in India. Drug prices in India are much lower than the prices of similar drugs in several countries, including the United States and United Kingdom as well as Pakistan and Indonesia (see Table 5.4). India has been a major exporter of relatively cheap active pharmaceutical ingredients (APIs) and pharmaceutical formulations of several medicines, notably vaccines and anti-retrovirals (ARVs) (Grace, 2004: 13–15). The Indian pharmaceutical firm CIPLA supplies ARVs to over 250,000 HIV patients in poor countries.[7] When Ranbaxy, another leading Indian pharmaceutical firm, announced plans to launch the cholesterol drug atorvastatin in the USA and the UK, the media in the UK welcomed it as a move that would result in substantial financial savings to that country's National Health Service (Tomlinson, 2005).

5.4.2   The TRIPS Agreement and changes in India's patent laws

The WTO's Agreement on Trade Related Aspects of Intellectual Property Rights (TRIPS) came into effect on 1 January 1995. As WTO members, India and other developing countries were obliged to bring in legislation in line with the TRIPS provisions over a ten-year period. 'Mail box' facilities and exclusive marketing rights were to be introduced from 1 January 1995 itself; provisions regarding rights of patentee, term of patent protection,

Table 5.4 Prices of selected drugs in India and other countries, in Indian Rupees, 2002–2003

Drugs and dosage	India	Pakistan	Indonesia	UK	USA
Ciprofloxacin HCL, 500 mg	29	423.9	393.0	1185.7	2352.4
	(1.0)	(14.6)	(13.6)	(40.9)	(81.1)
Diclofenac Sodium, 50 mg	3.5	84.7	59.8	61.0	674.8
	(1.0)	(24.2)	(17.1)	(17.4)	(192.8)
Ranitidine, 150 mg	6.02	74.1	178.4	247.2	863.6
	(1.0)	(12.3)	(29.6)	(41.1)	(143.5)

Notes: Ciprofloxacin HCL is an Anti-infective. Diclofenac and Ranitidine are anti-ulcerants.
Figures in brackets show prices as indices with price in India = 1.
Drug prices refer to the following years: for India, 2003; for Pakistan 2002–03; for US, 2002; and for UK February 2004.
Retail prices in India and wholesale prices in other countries were considered. All prices were converted to Indian Rupees.
Source: Centre for Study of Global Trade System and Development (2004) and Keayla (2005).

compulsory licensing and reversal of burden of proof had to be legislated before 1 January 2000; and laws protecting product patents had to be legislated before the end of the ten-year transition period (therefore, before 1 January 2005).

The Indian Parliament debated the TRIPS provisions intensely, in the process delaying the introduction of obligatory legislations in several instances. The Parliament passed the Patents (Amendments) Act 1999, the Patents (Amendment) Act 2002, the Indian Patents Ordinance of 2004, and the Indian Patents (Amendment) Act of 2005. India's patent rules were made less liberal over the years with the introduction of these legislations. The Patents (Amendment) Act 2002 had made 64 changes to the Patents Act of 1970. There were criticisms that some of the new legislations had not even made full use of the flexibilities that the TRIPS regime had allowed for developing countries; there were also widespread concerns on the implications of these laws for public health. The Indian Patents Ordinance of 2004 had allowed patents on combinations and crystalline versions of known molecules; reduced the grounds on which a patent could be opposed during the pre-grant period; and required least developed countries (LDCs) to issue compulsory licenses for importing generic drugs from India. All these provisions generated strong criticisms as being unduly favourable to patent owners. The Indian Patents (Amendment) Act of 2005, which introduced product patent rules, rectified some of the drawbacks contained in the Ordinance of 2004 (Chaudhuri, 2005: 70–116; Grace, 2005).

The battle on patent rules is still going on in India. Multinational pharma corporations lobby for stricter patent laws and their stricter implementation. On the other side are a number of activists and international organizations raising concerns on public health. Currently, there is strong pressure particularly from the MNCs to bring in new provisions that allow data exclusivity in India. Data exclusivity specifies that the test data submitted by the patentee to the regulatory agencies will not be disclosed to the public. Generic drug firms, which need to prove bioequivalence of their generic versions of drugs, will be affected by this rule. Also data exclusivity allows the patent holders to extend their monopoly rights even after the expiry of the patent term (Chaudhuri, 2005: 80–3; Keayla, 2005).[8] At the same time, in August 2007, an Indian Court dismissed an appeal from the pharmaceuticals giant Novartis regarding patent cover for its drug Gilvec; the Court pointed out that incremental innovations will not be eligible for patent protection. This ruling has raised the hopes for health activists worldwide who argue that a strict patent regime is a deterrent to the supply of affordable medicines (Cookson and Yee, 2007).

5.4.3　TRIPS and strategies of Indian pharmaceutical firms

With the introduction of TRIPS-compliant product patent rules in 2005, the Indian pharmaceutical industry can no longer rely solely on its reverse engineering skills for future growth. India's domestic pharmaceutical firms have been growing in technological capabilities, and this enabled them to make two fundamental changes to their business models during the years of TRIPS implementation (1995–2005).[9] First, leading pharmaceutical firms in India have been making higher allocations for R&D spending and trying to acquire patents abroad. In the case of Dr. Reddy's Laboratories, an Indian firm, R&D charges as a proportion of sales revenue were 0.6 per cent and 2.8 per cent respectively in the three-year periods ending in 1987 and 1994–95. The proportion rose to 11.0 per cent in the three-year period ending in 2005–06.[10] Ranbaxy made 698 patent filings in the first nine months of 2005, compared to 428 patent filings in the first nine months of 2004.[11] Secondly, rising R&D intensity went hand in hand with export orientation, notably to the regulated markets of North America and Europe. For instance, in 2005, United States and Europe, together, accounted for 45.2 per cent of Ranbaxy's total global sales (of US$1,178 million).[12] The preference shown by Indian pharmaceutical firms to developed country markets is evident in Figure 5.2.

The Indian pharmaceutical industry's international ambitions have also received a boost from the changing nature of the global pharmaceutical business. The discovery of a new drug is an extremely lengthy and financially risky process. Bringing an experimental drug into the United States market takes an average of 12 years. According to some reports, of the 5,000 drug compounds that are evaluated at the preclinical stage, five compounds enter the phase of clinical trials, and only one compound ultimately gets the approval for marketing from the US Food and Drug Administration (FDA).[13] Reports in 2006 indicate that the cost of development and subsequent introduction into the market of a new drug compound range between US$0.8 and US$1.7 billion

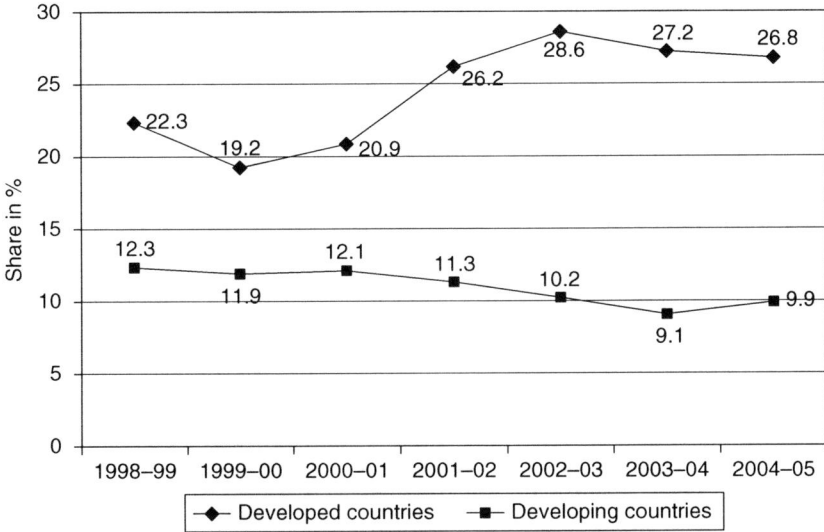

Figure 5.2 Exports of drugs, pharmaceuticals and fine chemicals by India to selected developed and developing countries, 1998–99 to 2004–05, shares in India's total exports (percentages)

Notes: Selected developed countries: United States, Germany, United Kingdom and Canada.
Selected developing countries: Nigeria, Viet Nam, Sri Lanka, Pakistan, Bangladesh and Nepal.
These ten countries have figured in the list of 21 leading destinations for India's exports of drugs, pharmaceuticals and fine chemicals throughout the period under study.
Source: Calculations based on CMIE (2005: 69).

(McKinnell, 2006). To reduce the cost of new drug discovery, pharmaceutical MNCs are entering into strategic alliances with smaller pharmaceutical firms, biotech companies and academic centres. Novartis, for instance, claims to have more than 400 collaborations in more than 20 countries.[14] India's cost advantages in pharmaceuticals R&D are encouraging global corporations to form research partnerships with Indian firms. Outsourcing of clinical trials to India has witnessed an especially fast growth; the number of ongoing clinical trials in India has risen to 270 in 2007 (Yee, 2007).

The strategies pursued by India's leading drug makers today consist of collaboration as well as, in some cases, competition with western pharmaceutical MNCs. Even the leading Indian drug firms are much smaller compared to pharmaceutical corporations, and they do not possess the skills or the resources to carry out the entire process of new drug discovery.[15] Therefore, Indian pharmaceutical companies conduct research and develop new molecules, but instead of proceeding further into the long and financially risky clinical trial and regulatory stages, they license out the molecule to bigger pharmaceutical MNCs (Chaudhury, 2005). At the same time, the high returns in the generic drugs market in North America and Western Europe is highly attractive to Indian drug firms. They have challenged big pharmaceutical corporations in the market for generic drugs in the West. Also, to consolidate their generic drugs business, many Indian drug firms have been pursuing an aggressive strategy of overseas acquisitions over the past several months.[16]

Success, however, is not guaranteed for India's leading drug makers in the generic business. Originator drug companies, some of which have launched their own branded generics, unleash long and expensive legal battles against their generic competitors (Rai, 2003; Jack, 2005). For originator drug companies, patent litigation to delay the entry of generic competitors even by a few months is a high return–zero risk strategy, whereas for generic drug firms, patent-related legal battles involve high returns as well as high risks (Chaudhuri, 2005: 205–6). Many of the top Indian drug firms are fighting IPR-related legal battles, incurring heavy costs. Ranbaxy has been engaged in a legal wrangle over its generic version of atorvastatin calcium, an anticholesterol drug, which Pfizer claims has violated its patent on Lipitor. Ranbaxy's fortunes have waxed and waned in this long-drawn-out litigation that is still being fought in the courts of several countries. Reports suggest that Ranbaxy spent US$30 million on legal expenses in the year 2005, while the company's R&D expenditure, for the year 2004, was US$75.1 million.[17] Similarly, Dr. Reddy's reportedly spent US$12 m on legal bills in 2004, which was equivalent to a quarter of the company's R&D budget (Economist, 2005).

5.4.4   The future of the industry and the supply of affordable medicines

It is doubtful whether the leading Indian pharmaceutical firms will ever grow to match the levels of western pharmaceutical MNCs, given the hurdles posed by the IPR regime and the disparities between the two in financial and research capabilities. No Indian firm has, so far, been able to fully develop an original drug, although some firms are close to achieving this feat. More crucially, a large number of relatively small Indian pharmaceutical firms are facing grim growth prospects in the post-TRIPS phase, and this is raising important questions for the future of the industry in India. It will be difficult for these small Indian firms to repeat the successes of their bigger counterparts in the export markets for generics: the regulatory barriers to entry and the stiff competition in the generic drugs markets in developed countries are important road blocks (Chaudhuri, 2005). In the domestic market, product patent legislations will eventually rein in the growth of small Indian firms. Manufacturing drugs for the domestic market using process innovations, the strategy that helped leading Indian drugs makers during their formative years, is no longer possible.[18] Furthermore, there is growingly intense competition in India's pharmaceutical markets, raising the costs of entry to smaller firms. In recent years, the Indian pharmaceuticals industry has witnessed a significant increase in mergers and acquisitions (M&As) and a consequent rise in concentration ratios.[19] Smaller Indian pharmaceutical firms are also affected by the tightening of regulatory restrictions in the Indian market as well as in the export markets of countries such as Brazil and Korea.[20]

A larger concern arising from the recent developments in the pharmaceuticals industry relates to their implications for the supply of medicines to poor patients. The advocates of a strict patent regime have argued that with the implementation of product patent rules, MNCs will step up investment in research on neglected diseases. However, this does not appear to have occurred in India. In a survey of 31 large pharmaceutical companies operating in India (which included companies under Indian ownership and MNC subsidiaries), Lanjouw and MacLeod (2005) found that only 10 per cent of the entire R&D investments by these companies in 2003–04 were targeted at developing country markets and tropical diseases. At the same time, multinational pharmaceutical corporations are sensing a major market opportunity for the supply of medicines for global diseases such as cancer and cardiovascular diseases prevalent among India's large middle-class population. It appears that MNCs' interests in India are limited to its growing pharmaceuticals market, and not so much in the country's potential as a pharmaceuticals manufacturer. MNC investments in India in the manufacture of bulk drugs have not recorded any appreciable increase in recent years (Chaudhury, 2005). Meanwhile, worries are surfacing about the outsourcing of clinical trials to India. Terming it as 'a new colonialism', Nundy and Gulhati (2005) have discussed some of the dangers of conducting clinical trials among poor and illiterate patients without putting in place a proper regulatory system.

The problems emerging as part of a post-TRIPS scenario are likely to become more severe in the future. Grace (2005), after examining previous studies, concluded that the share of patented drugs in the market value of medicines supplied in India in 2005 was in the range of 10 to 15 per cent. However, over time, as new medicines are invented, a greater proportion of the overall Indian market for medicines will come under the patent cover. New medicines are necessary in the treatment of most diseases including tuberculosis and malaria as older medicines turn ineffective with the setting in of drug resistance. In the case of combination drugs, even if only one drug in the combination is patent protected, that will escalate the cost of the entire therapy (Grace, 2005: 16–20). The pressures on the supply of affordable medicines brought about by changes in patent rules are reflected in the words of Dr. Y. K. Hamied of CIPLA, an Indian company that is an important supplier of medicines for tropical diseases:

...[India's product patent legislation implemented in 2005] will deprive the poor of India and also third world countries dependent on India, of the vital medicines they need to survive.... It will lead to a systematic denial of drugs to the three billion in the poorer nations, an act tantamount to selective genocide by the year 2015.[21]

5.5   Pharmaceuticals and biotechnology in China

The pharmaceuticals industry is expanding rapidly in China. According to Nolan and Yeung (2001), China's pharmaceuticals industry picked up in growth after the mid-1980s. This was a period when the government relaxed state controls over the industry and generated competition among the suppliers. The structure of ownership was also gradually liberalized, and raising funds on the stock market and mergers and acquisitions became instruments through which the pharma industry further consolidated its position in China (Nolan and Yeong, 2001). There were 4,296 pharmaceutical manufacturing facilities in China in 2003. Domestic industry supplies almost 70 per cent of the Chinese market for pharmaceutical products. Chinese firms have acquired great expertise in the manufacture of bulk drugs and active pharmaceutical ingredients (APIs). China is the world's second-largest producer of pharmaceutical ingredients, and the largest producer of many pharmaceutical products including penicillin (producing 60 per cent of world output), vitamin C (50 per cent of world output), terramycin (65 per cent of world output), doxycycline hydrochloride and cephalosporins (Grace, 2004: 13–14). The country is promoting innovative research in the area of traditional Chinese medicine; in April 2004, Chinese authorities approved the first HIV/AIDS treatment derived from traditional Chinese medicine (Grace, 2005: 10–11).

At the same time, however, the strength of China's domestic pharmaceuticals industry should not be overstated. Nolan (2002), while making a note of the impressive growth of the Chinese pharmaceutical firm Sanjiu during the 1990s, was quick to add that firms like Sanjiu would face massive hurdles as they compete with multinational pharmaceutical giants. Nolan (2002) points out that China's pharmaceuticals industry is 'relatively small and highly fragmented', and that the revenues and research capabilities of Chinese firms are only a tiny fraction of multinational companies (Nolan, 2002: 123).

5.5.1   Pharmaceuticals industry and the evolution of the IPR regime in China

Significant steps in the direction of establishing a patent regime began in China only after the late 1970s.[22] Chinese government's gradual implementation of an intellectual property rights (IPR) policy was shaped by two factors: a commitment to the development of domestic capabilities in science and technology, on the one hand, and, international pressure, particularly from the United States, pushing China to a strict patent regime, on the other. China joined the World Intellectual Property Organization (WIPO) in March 1980 and the Paris Convention for the Protection of Industrial Property in March 1985. The first Patents Law was introduced in China in 1984. This law, which came into effect on 1 April 1985, was rather narrow in its scope. It did not offer product patent protection to inventions in pharmaceuticals, chemicals, food, beverages and condiments (in much the same manner as India's Patents Act of 1970). The stipulations contained in the 1984 law ensured that foreign investments into China came along with technology transfer and contributed to the building of domestic innovative capabilities (Kong, 2005).

China introduced a stricter patent regime in 1992. In the early 1990s, China was integrating itself more closely with the world economy, and a commitment to IPR protection was important to attract foreign investments. In addition, from being an importer of technologies, China was gradually emerging as an exporter of technology-intensive products (Kong, 2005). Grace (2005) noted that the implementation of IPR policies in China was influenced, to a great extent, by the country's bilateral negotiations with the United States on trade and investment. Product patenting rules came into effect in China in 1993 – more than ten years before TRIPS would have required the country to implement them. An agreement between China and the United States in 1999 on China's WTO accession contained proposals for early introduction of TRIPS-compliant IPR rules. China joined the WTO in 2001, and the country introduced patent laws incorporating TRIPS provisions by the end of 2002. China was not given the transition period that was granted to other developing countries (Grace, 2005: 21–5; Kong, 2005). Chinese laws extend patent protection for twenty years and data exclusivity for six years (Grace, 2004).

Despite the implementation of product patent laws, China has been able to manufacture pharmaceutical ingredients that contribute to the supply of essential medicines for the developing world. China is an important producer of a wide variety of raw materials for second-line antiretrovirals (ARVs) (second-line ARVs are necessary for treatment once the patient develops resistance to first-line treatment).[23] One of the means through which China has bypassed the limitations set by patent laws is by manufacturing intermediates only up to the pre-API (active pharmaceutical ingredients) stage. Patent protection is usually applicable to APIs and finished products, and manufacturing a chemical that is one step away from formulation into an API will not be a patent violation. China then exports these intermediate pharmaceutical chemicals to other countries where it is processed into APIs and finished products (Grace, 2005: 23–5).

However, the pharmaceuticals industry in China is now facing challenges from the patent regime. The United States places continuous pressure on China to improve its record on IPR enforcement. In December 2006, on the occasion of the fifth anniversary of China's entry into the WTO, the US Trade Representative, Susan Schwab, slammed China's record in enforcement of IPR rules. In a one-hundred-page report submitted to the US Congress, the US Trade Representative claimed that piracy of software, videos, pharmaceuticals and other goods were rampant in China and that the Chinese government did very little to curb such practices (Weisman, 2006). Reports suggest that in 2004, 2,550 patent litigations were filed in Chinese courts and that the rulings in 80 per cent of these cases went in favour of the foreign patent holders (Seewald, 2006).

At the same time, western multinational pharmaceutical companies are waking up to the tremendous market opportunities offered by China's large (and expanding) middle-class population. Some estimates have suggested that by 2010 the Chinese pharmaceuticals market would be the fifth largest in the world.[24] It is noted that more than 20 of the world's top 25 pharmaceutical MNCs have already made their entry into China.[25] Pharmaceutical corporations are flocking in particular to China's eastern region comprising the Yangtze River Delta, which enjoys high levels of purchasing power.[26] With the introduction of product patent legislations, many multinational pharmaceutical companies are outsourcing pharmaceutical research and clinical trials to China. In 2007, there were a total of 510 clinical trials – including completed and ongoing – outsourced to China (Jack and Yee, 2007). Companies like Novo Nordisk and Novartis have set up R&D facilities in China in order to conduct research on global diseases by making use of China's cost advantage. They are equally keen on making inroads into the Chinese market for pharmaceuticals.[27]

5.5.2   Health and agricultural biotechnology in China

China has embarked on an ambitious programme in biotechnology. Since the late 1970s, life sciences has emerged as an important focus area in China, along with the state-wide promotion of science and technology in the country during this period. The National Centre for Biotechnology Development was established in 1983 under the State Science and Technology Commission (this centre later became part of the Ministry of Science and Technology) (Gross, 1995; Chervenak, 2005). By 1992, the government established 17 national biotechnology laboratories that were open to both domestic and foreign scientists; and by 1995, there were approximately 1,000 biotechnology projects in China, employing over 10,000 scientists in total. According to a report in 1995, almost one-third of the funds for biotechnology research in China came from the Central Government (Gross, 1995). Between 1996 and 2000, the Central Government invested over 1.5 billion Yuan (US$180 million) in biotechnology (Economist, 2002). Local governments as well as quasi-venture capital funds set up by the Central or local governments were active in the promotion of biotechnology in China. The government also encouraged Chinese firms to establish links with western biotechnology companies (Chervenak, 2005).

After 2000 the government enhanced funding for the biotechnology sector in China. As per estimates made in 2005, the Chinese government spends more than US$600 million per year on biotechnology R&D through its various funding programmes (Chervenak, 2005). This, however, must be compared to the investment in biotechnology R&D in the United States, which was US$15.7 billion in 2001 (Economist, 2002). In 2002, according to estimates published by China's Ministry of Science and Technology, 20,000 researchers were working in the field of life sciences in the country; in the same year, the biotechnology industry was reported to be employing 191,000 people in the United States. Approximately another 20,000 Chinese biotechnology researchers were working abroad in the year 2002, and their eventual return to the country was expected to give a further boost to the biotechnology industry in China (Economist, 2002).

The major centres of China's biotechnology industry include Shenzhen, Shanghai and Beijing. Beijing Genomics Institute (BGI), which was established as a state-sponsored research centre in 1999, took part in the Human Genome Project; China was the only developing country to participate in this project. Fudan University's Human Genome Laboratory in Shanghai is involved in the mapping and sequencing the human X chromosome (Gross, 1995; Chervenak, 2005).

China is also making rapid advances in the field of agricultural biotechnology. Agricultural biotechnology began to receive considerable policy attention in China from the late 1980s, as part of a response to the enormous challenges of feeding a large population and improving productivity in China's small farms. Reports suggest that the government under the former Prime Minister Zhu Rongji was highly concerned about the growing dominance of US biotechnology firms in Chinese agriculture.[28] That the seeds improved over decades by Chinese farmers could be appropriated by a few large US biotech corporations was a worrying prospect to policy makers in China (Chen, 1999).[29] Under these circumstances, the government stepped up funding for research on GM crops that are highly suited to local growing conditions. In 1999, government expenditure on agricultural biotechnology research in China was US$112 million. This figure was nearly ten times the agricultural biotechnology research budgets of India and Brazil in the same year, although it was still considerably smaller than the US$1–2 billion that the United States spent in 1999 on plant biotechnology research (Karplus, 2003).

Public investment in biotechnology research in China has produced impressive results. As per reports in 2002, Chinese research institutes developed 141 types of GM crops, of which 65 were undergoing field trials. Today, China is carrying out research on genetically modified tomatoes that take longer to rot (which helps in their transportation, processing and storage) and vitamin A-enriched rice that will help improve nutrition in many parts of the developing world. China has recorded great success in research on *Bt* cotton. Chinese research laboratories introduced 18 varieties of pest-resistant *Bt* cotton by 2002 (Karplus, 2003). Area under *Bt* cotton cultivation in China increased from 1.5 million hectares in 2001 to 3.3 million hectares in 2005, and over 4 million small-scale farmers were involved in *Bt* cotton cultivation in China in the year 2001 (Karplus, 2003).

5.5.3   India and China: interactions in pharmaceuticals and biotechnology

Over the past few years, there has been a growing engagement between India and China in the economic sphere, and this has also extended to the pharmaceuticals and biotechnology sectors. While several major Indian pharmaceutical firms have set up joint ventures and production facilities in China, China has emerged as a very important supplier of APIs and bulk drugs for the pharmaceuticals industry in India. As a source of India's imports of medicinal and pharmaceutical products, China's share is the highest, at 34.6 per cent in 2005–06, having risen from 6.2 per cent in 1993–94. Correspondingly, as a destination for India's exports of drugs, pharmaceuticals and fine chemicals, China's share increased from 0.4 per cent in 1993–94 to 3.5 percent in 2005–06 (CMIE, 1997, 2005, 2006).

It may be noted that the import of bulk drugs and APIs from China is viewed as a threat by many small-scale bulk drug manufacturers in India. Small-scale drug firms in India have been encountering several growth constraints (some of which have been discussed earlier), which are aggravated by competition from Chinese bulk drug makers. Chinese firms are large-scale producers of many of the intermediates for bulk drug manufacturing, enjoying clear advantages in their specific areas of production. At the same time, industry observers say that India possesses superior technology and skills in pharmaceutical formulations. Representatives of small-scale pharmaceutical industry in India have expressed fears about Indian drug companies transferring technological skills to their Chinese counterparts.[30]

While some of the genuine growth concerns of small-scale drug makers in India need to be addressed, these should not be a reason to stall greater positive interaction between pharmaceuticals industries in India and China. Rather than competing with each other, pharmaceutical firms in the two countries must seek avenues for cooperation at the higher plane of innovation. China's biotechnology sector and India's pharmaceuticals industry should feed into each other's expertise. Together, they can strive for pharmaceutical innovations that address the health needs of poor patients in the developing world.

India and China can also enhance their levels of engagement in agricultural biotechnology. India has given approval for commercial sale of 12 varieties of *Bt* cotton hybrids, all of which carry the gene developed by the multinational giant Monsanto and marketed in India by joint ventures (Mahyco-Monsanto) or sub-licensees of Monsanto (Chaturvedi, 2005). However, reports from Andhra Pradesh indicate that genetically modified (GM) cotton crops sold by Mahyco-Monsanto were a failure in all of the three years after the crop's introduction. Many farmers who took loans to buy GM seeds fell into huge debt-traps, but Mahyco-Monsanto refused to compensate the farmers for their losses (Venkateshwarlu, 2006). Given such experiences, public and private sector seed companies in India should strive to develop new GM technologies that are suitable for India's agro-climactic and socio-economic environment. Collaboration with China, which has a very successful programme in agricultural biotechnology, will be highly useful in this regard. It is indeed a positive sign that, in March 2006, Agriculture Ministers of India and China identified a number of areas for cooperation, including agriculture biotechnology and exchange of plant and animal germplasm.[31]

5.6 Conclusions

Globalization picks its winners and losers. According to Hoogvelt (2001), populations and countries on the periphery of the global economy lose out progressively while those who are part of the core strengthen their positions. This chapter shows that the geography and nature of innovation in the world economy are being altered fundamentally, in part due to global rules on intellectual property rights enshrined in the TRIPS and imposed on developing countries. India and China are emerging as important destinations for global R&D on account of their large supplies of highly skilled professionals and well-established science and technology infrastructures. However, even as the globalization of R&D gathers steam, the poor in India, China and other developing countries are likely to be left out of the new innovations, and big corporations in the West are likely to consolidate their strengths.

The case studies of pharmaceuticals industries in India and China presented in this chapter gives some credence to such apprehensions. India has been a major supplier of generic drugs at affordable prices within the country and outside, while China has been an important producer of raw materials and active pharmaceutical ingredients for the manufacture of several essential drugs, including antiretrovirals for the treatment of HIV/AIDS. Product patent rules in compliance with the TRIPS were implemented fully in China by 2002 and in India by 2005 as part of the WTO obligations of these countries. The leading Indian pharmaceutical firms have responded well to the challenge of a strict IPR regime by increasing their R&D spending and, simultaneously, targeting their sales to the generic drugs markets in North America and Europe. However, even as India's top drug firms have been growing in technological capabilities, they have also been shifting their focus away from the market for medicines for poor patients. Smaller pharmaceutical firms in India are facing several growth challenges, especially from the new IPR regime, which hinder their potential to supply drugs for the domestic market. The United States puts continuous pressure on China to improve its record on the enforcement of IPR rules. MNCs have increased their presence in India and China, conducting contract research and clinical trials on global diseases, eyeing the market of rich patients in these countries and outside. Contrary to expectations, the implementation of TRIPS has not led to any marked increase in MNCs' R&D spending on neglected diseases. In fact, there are growing uncertainties today on the future supply of affordable medicines in the developing world.

In the light of such experiences, developing countries, importantly India and China, need to initiate strong policy measures to counter the negative effects of globalization of R&D. First, developing countries should strengthen their national programmes in science and technology to ensure that they are not overshadowed by global corporations. In this regard, China's attempt to reinvigorate the biotechnology sector is commendable. Secondly, developing countries need to cooperate in the area of innovation. Innovative firms in India and China should explore areas for complementary growth, rather than competing with each other to obtain a slice of the market for R&D outsourcing. Blending India's expertise in pharmaceutical formulations and China's growing capabilities in biotechnology could result in new drugs for neglected diseases. China and India have the responsibility for and, indeed, the capabilities to lead developing countries in innovations that could solve world's problems of ill health and deprivation.

Notes

1. See 'High-tech Hopefuls: A Special Report on Technology in India and China', *The Economist*, 10 November 2007.
2. See the Ministry of Science and Technology of the People's Republic of China, at www.most.gov.cn/eng/programmes/programmes1.htm. Accessed on 18 January 2006.
3. For a discussion on 'national system of innovation', see Freeman (1995).
4. See the report 'Pharma Industry Aims High: Headed for a Place in the Global Top 10: Country Focus: India', *Chemical Week*, 21 November 2007, p. 38. See also Grace (2005: 8).
5. According to Kefauver Committee of the United States, cited in Keayla (2005).
6. The DPCO, which underwent several modifications, was finally replaced by the National Pharmaceuticals Policy of 2002.
7. See www.cipla.com. Accessed on 15 January 2006.
8. See also Dhar and Gopakumar (2006).
9. See Lanjouw (1997), Sampath (2005), and Chaturvedi et al. (2007) (which provides a detailed update on the strategies of Indian pharmaceutical firms before and after TRIPS). See also Ramanna (2005) on the emergence of a strong pro-patent lobby in the country prior to 2005.
10. See Dr. Reddy's Laboratories Ltd. Annual Reports, various years.
11. See www.ranbaxy.com. Accessed on 15 January 2006.
12. See *Ranbaxy Annual Report 2005*, downloaded from http://www.ranbaxy.com. Accessed on 17 January 2007.
13. According to information given at the Website of Alliance Pharmaceutical Corporation. See www.allp.com/drug_dev.htm. Accessed 15 July 2006.
14. See www.allp.com/drug_dev.htm and www.nibr.novartis.com/OurScience/drug_development.html. Accessed on 16 September 2006.

15. For example, in 2005, the sales revenue of the Indian company Ranbaxy was US$1.17 billion and that of Pfizer was US$51.3 billion. See Knowledge@Wharton (2006).
16. Information obtained from various issues of *Indian Industry: A Monthly Review* for the year 2007, published by Centre for Monitoring Indian Economy, Mumbai.
17. See Mahapatra (2006) and Ranbaxy's website www.ranbaxy.com. Accessed on 14 December 2005.
18. Interview with Mr. B. K. Keayla, 11 December 2006.
19. Concentration ratios of the largest four and largest eight firms in Indian pharmaceutical industry increased after 1995–96. See Chadha (2006).
20. Interview with Mr Lalit Kumar Jain, an entrepreneur having a long association with the small-scale pharmaceutical industry in India, New Delhi, 10 December, 2006.
21. Address by Dr. Y. K. Hamied, Chairman and Managing Director, CIPLA, Sixty-Ninth Annual General Meeting, 6 September 2005, at http://www.cipla.com/corporateprofile/financial/cm69.htm. Accessed on 14 December 2005.
22. See http://www.china.org.cn/e-white/20050421/index.htm. Accessed on 15 January 2006.
23. See 'India, China or Brazil – Who will Produce the Second Line ARVs?', *Health and Development Networks*, 12 July 2005, at http://www.aidsmap.com/en/news/24B33 FA6-89CB-42BA-880F-18D774FF85D6.asp. Accessed on 17 September 2005.
24. See 'The Chinese Pharmaceuticals Market is Forecast to Become the World's Fifth Largest by 2010', *Biotech Business Week*, 9 July 2007.
25. Information from *Biotech Business Week*, 14 May 2007.
26. See 'Pall Magnifies Focus on Asia; Brings Top Talent, More Resources to the Region to Address Growing Biopharmaceutical Market', *Business Wire Inc*, 21 March 2006.
27. See the report 'A Novel Prescription', *Economist*, 11 November 2006; see also Kjersem and Gammeltoft (2006).
28. In the late 1990s, the US biotechnology companies were in a dominant position in Shijiazhuang, Hebei and Langfang areas. Chinese biotechnology firms had the upper hand in Henan and Anhui provinces. See Chen (1999).
29. These are the views expressed by Chen Zhangliang, Vice Chancellor and Professor of Beijing University, in an interview he gave in 1999. See Chen (1999). According to Chen Zhangliang, the Chinese Premier expressed his concerns regarding US multinational corporations' dominance in Chinese agriculture after a visit to the north-eastern province of Jilin.
30. Interview with Mr. Lalit Kumar Jain, New Delhi, 10 December 2006. See also Chaudhuri (2005).
31. See the report 'India, China Sign Agriculture Cooperation Pact', *Financial Times*, 30 March 2006.

References

Basant, Rakesh and Chandra, Pankaj (2007) 'Role of Educational and R&D Institutions in City Clusters: An Explanatory Study of Bangalore and Pune Regions in India', *World Development*, 35(6): 1037–55.
Bernauer, Thomas (2003) *Genes, Trade, and Regulation: The Seeds of Conflict in Food Biotechnology*, Princeton and Oxford: Princeton University Press.

Branstetter, Lee and Lardy, Nicholas (2006) 'China's Embrace of Globalization', National Bureau of Economic Research, Working Paper 12373, Cambridge MA, July, at http://www.nber.org/papers/w12373. Accessed 17 December 2007.

Centre for Monitoring Indian Economy (CMIE) (1997) *Foreign Trade*, Mumbai: CMIE.

Centre for Monitoring Indian Economy (CMIE) (2005) *Foreign Trade and Balance of Payments*, Mumbai: CMIE.

Centre for Monitoring Indian Economy (CMIE) (2006) *Foreign Trade and Balance of Payments*, Mumbai: CMIE.

Centre for Study of Global Trade System and Development (2004) 'Report of the Fourth Peoples' Commission on Review of Legislations Amending Patents Act 1970', October 2004, Commissioned by National Working Group on Patent Laws and Public Interest Legal Support and Research Centre, New Delhi.

Chadha, Alka (2006) 'Destination India for the Pharmaceutical Industry', *Delhi Business Review*, 7(1): 1–8.

Chanda, Rupa and Sreenivasan, Niranjana (2006) 'India's Experience with Skilled Migration', in Kuptsch and Pang (eds) (2006), pp. 215–56.

Chandrasekhar, C.P. (2006) 'The Political Economy of IT-driven Outsourcing', in G. Parayil (ed.), *Political Economy and Information Capitalism in India: Digital Divide, Development and Equity*, Basingstoke: Palgrave Macmillan.

Chang, H.-J. (2003) *Kicking Away the Ladder: Development Strategy in Historical Perspective*, London: Anthem Press.

Chaturvedi, Kalpana, Chataway, Joanna, and Wield, David (2007) 'Policy, Markets and Knowledge: Strategic Synergies in Indian Pharmaceutical Firms', *Technology Analysis and Strategic Management*, 19(5): 565–88.

Chaturvedi, Sachin (2005) 'Dynamics of Biotechnology Research and Industry in India: Statistics, Perspectives and Key Policy Issues', Directorate for Science, Technology and Industry, Organisation for Economic Cooperation and Development (OECD), at www.oecd.org/dataoecd/43/35/34947073.pdf. Accessed 17 December 2005.

Chaudhuri, Sudip (2005) *The WTO and India's Pharmaceuticals Industry: Patent Protection, TRIPS, and Developing Countries*, New Delhi: Oxford University Press.

Chen, Zhangliang (1999) 'Unlimited Prospects for Biotechnology', Interview with Chen Zhangliang, *Knowledge Economy* [*Zhishi Jingji*], December, pp. 22–8, at www.usembassy-china.org.cn/sandt/biotechch.html. Accessed on 15 March 2006.

Chervenak, Mathew (2005) 'An Emerging Biotech Giant?', *China Business Review*, May–June, at www.chinabusinessreview.com/public/0505/chervenak.html. Accessed on 15 March 2006.

Cooke, Philip (2005) 'Rational Drug Design, the Knowledge Value Chain and Bioscience Megacentres', *Cambridge Journal of Economics*, 29(3): 325–41.

Cookson, Clive and Yee, Amy (2007) 'Novartis Loses Challenge to India Patent Law', *Financial Times*, 7 August.

D'Costa, Anthony P. (2002) 'Software Outsourcing and Development Policy Implications: An Indian Perspective', *International Journal of Technology Management*, 24(7–8): 705–23.

D'Costa, Anthony P. (2004) 'Export Growth and Path-Dependence: the Locking-in of Innovation in the Software Industry', in A.P. D'Costa and E. Sridharan (eds), *India in the Global Software Industry: Innovation, Firm Strategies and Development*, Basingstoke: Palgrave Macmillan.

D'Costa, Anthony P. (2006) 'Exports, Institutional Architecture, and Innovation Challenges in Bangalore's (and India's) IT Industry'. Paper presented at *New*

Asian Dynamics in Science, Technology and Innovation, Gilleleje, Denmark, 27–9 September.

Dahlman, Carl and Anuja Utz (2005) *India and the Knowledge Economy: Leveraging Strengths and Opportunities*, Washington DC: The World Bank.

Dhar, Biswajit and K.M. Gopakumar (2006) 'Data Exclusivity in Pharmaceuticals: Little Basis, False Claims', *Economic and Political Weekly*, 9 December.

Dicken, Peter (1998) *Global Shift: Transforming the World Economy*, 3rd edition, London: Paul Chapman Publishing Ltd.

Drahos, Peter with Braithwaite, John (2002) *Information Feudalism: Who Owns the Knowledge Economy?*, New York: The New Press.

Economist (2002) 'Biotech's Yin and Yang', *Economist*, 12 December, at www.economist.com/science/displayStory.cfm?story_id=1491569. Accessed on 17 December 2005.

Economist (2005) 'Prescription for Change: A Survey of Pharmaceuticals', *Economist*, 18–24 June.

Ernst, Dieter (2005) 'Complexity and Internationalisation of Innovation – Why is Chip Design Moving to Asia?', *International Journal of Innovation Management*, 9(1): 47–73.

Findlay, Allan M. (2006) 'Brain Strain and Other Social Challenges Arising from the UK's Policy of Attracting Global Talent', in Kuptsch and Pang (eds) (2006), pp. 65–86.

Freeman, Chris (1995) 'The "National System of Innovation" in Historical Perspective', *Cambridge Journal of Economics*, 19(1): 5–24.

Government of India (GOI) (2006) *Research and Development Statistics 2004–05*, New Delhi: Ministry of Science and Technology, Department of Science and Technology.

Grace, Cheri (2004) 'The Effect of Changing Intellectual Property on Pharmaceutical Industry Prospects in India and China: Consideration for Access to Medicines', Department for International Development (DFID) Health Systems Resource Centre, London, June, at http://www.dfidhealthrc.org/shared/publications/Issues_papers/ATM/Grace2.pdf. Accessed on 28 December 2005.

Grace, Cheri (2005) 'Update on China and India and Access to Medicines', DFID Health Resource Centre, London, November, at www.dfidhealthrc.org/what_new/Final%20India%20China%20Report.pdf. Accessed on 28 December 2005.

Gross, Ames (1995) 'Opportunities in the China Biotechnology Market', *Pacific Bridge Medical*, January–February, at www.pacificbridgemedical.com/publications/html/ChinaJanFeb1995.html. Accessed on 28 December 2005.

Hoogvelt, Ankie (2001) *Globalization and the Postcolonial World: The New Political Economy of Development*, 2nd edition, Basingstoke: Palgrave Macmillan.

Jack, Andrew (2005) 'Patently Unfair? Makers of Branded Drugs Struggle to Counter the Generic Onslaught', *Financial Times*, 22 November.

Jack, Andrew and Yee, Amy (2007) 'China Overtakes India in Drug Testing', *Financial Times*, 28 August.

Kalsekar, Mahesh (2003) 'The Indian Pharmaceuticals Market – An Overview', Mumbai: ORG-MARG, AC Nielson.

Karplus, Valerie (2003) 'Global Anti-GM Sentiment Slows China's Biotech Agenda', *YaleGlobal*, 26 September, at http://yaleglobal.yale.edu/display.article?id=2526. Accessed on 15 October 2006.

Keayla, B.K. (2005) 'Amended Patents Act 1970: A Critique', Centre for Study of Global Trade System and Development, New Delhi. Also available at www.combatlaw.org.

Kjersem, Julie Marie and Gammeltoft, Peter (2006) 'Investing in High-tech in China: the Cases of Novo Nordisk, GN Resound and BenQ Siemens Mobile', Paper presented at *New Asian Dynamics in Science, Technology and Innovation*, Gilleleje, Denmark, 27–9 September 2006.

Kleinknecht, Alfred and Wengel, Jan ter (1998) 'The Myth of Economic Globalization', *Cambridge Journal of Economics*, 22(4): 637–47.

Knowledge@Wharton (2006) 'Where will Indian Drug Companies be in Five Years? Everywhere – If They Innovate', Report Prepared by Knowledge@Wharton in collaboration with Bain & Company, at www.bain.com/bainweb/pdfs/cms/marketing/bain%20India%20Pharma%20FINAL%203-21-06.pdf. Accessed on 16 September 2006.

Kong, Qingjiang (2005) *WTO, Internationalization and the Intellectual Property Rights Regime in China*, Singapore: Marshall Cavendish Academics.

Kuptsch, Christiane and Pang, Eng Fong (eds) (2006) *Competing for Global Talent*, Geneva: International Institute for Labour Studies, International Labour Organization.

Lanjouw, Jean O. (1997) 'The Introduction of Pharmaceutical Product Patents in India: "Heartless Exploitation of the Poor and Suffering"?', Yale University, Economic Growth Centre, Discussion Paper No. 775.

Lanjouw, Jean O. and Margaret MacLeod (2005) 'Pharmaceutical R&D for Low Income Countries: Global Trends and Participation by Indian Firms', *Economic and Political Weekly*, 24 September.

Lardy, Nicholas R. (2002) *Integrating China into the World Economy*, Washington, DC: Brookings Institution Press.

Linden, Greg, Jason Dedrick and Kenneth Kraemer (2007) 'Who Captures Value in a Global Innovation System: The Case of Apple's iPod', Personal Computing Industry Centre (PCIC), California, June, at <http://pcic.merage.uci.edu/papers/2007/AppleiPod.pdf>, accessed 17 December 2007.

Mahapatra, Rajesh (2006) 'India's Ranbaxy Labs Blames Pricing Pressure in US Market for Sharp Drop in 2005 Profit', Associated Press Financial Wire, 18 January.

McKinnell, Henry (2006) 'With Patent Reform, India can be a Medical Innovator', *Financial Times*, 18 May.

Ministry of Petroleum and Chemicals (1971) 'Drugs Prices Control: Aims & Achievements', New Delhi: Government of India.

National Bureau of Statistics of China (2005) *China Statistical Year Book 2005*, Beijing: China Statistics Press.

Naughton, Barry (2007) *The Chinese Economy: Transitions and Growth*, Cambridge, MA: The MIT Press.

Nolan, Peter (2002) 'China and the Global Business Revolution', *Cambridge Journal of Economics*, 26(1): 119–37.

Nolan, Peter and Godfrey Yeung (2001) 'Big Business with Chinese Characteristics: Two Paths to Growth of the Firm in China under Reform', *Cambridge Journal of Economics*, 25(4): 443–65.

Nolan, Peter and Zhang, Jin (2002) 'The Challenge of Globalization for Large Chinese Firms', *World Development*, 30(12): 2089–107.

Nundy, Samiran and Chandra M. Gulhati (2004) 'A New Colonialism: Conducting Clinical Trials in India', *New England Journal of Medicine*, 352(16): 1633–6.

ORG-MARG (2004) 'ORG IMS Retail Store Audit for Pharmaceutical Products in India', ORG-MARG Private Limited, AC Nielson, Baroda, December.

Paarlberg, Robert L. (2001) *The Politics of Precaution: Genetically Modified Crops in Developing Countries*, Baltimore and London: International Food Policy Research Institute, Johns Hopkins University Press.

Rai, Saritha (2003) 'Generic Drugs From India Prompting Turf Battles', *New York Times*, 26 December.

Ramanna, Anitha (2005) 'India's Patent Policy and Political Economy of Development' in Kirit S. Parikh and R. Radhakrishna (eds), *India Development Report 2004–05*, New Delhi: Oxford University Press, pp. 112–25.

Sampath, Padmashree Gehl (2005) 'Economic Aspects of Access to Medicines after 2005: Product Patent Protection and Emerging Firm Strategies in the Indian Pharmaceutical Industry', United Nations University-Institute for New Technologies, at www.who.int/entity/intellectualproperty/ studies/PadmashreeSampathFinal.pdf. Accessed on 17 December 2005.

Seewald, Nancy (2006) 'Chinese Producers Pin Hopes on Innovation', *Chemical Week*, 19 July.

Spence, Jonathan D. (1999) *The Search for Modern China*, 2nd edition, New York and London: W. W. Norton & Company.

Steinfeld, Edward S. (2004) 'China's Shallow Integration: Networked Production and the New Challenges for Late Industrialization', *World Development*, 32(11): 1971–87.

Stropford, J.M. and Strange, S. (1991) *Rival States, Rival Firms: Competition for World Market Shares*, Cambridge: Cambridge University Press.

Tomlinson, Heather (2005) '$12bn Battle that Pits an Indian Upstart against the Mighty Pfizer', *The Guardian*, 29 August, at www.guardian.co.uk/business/story/0,3604, 1558296,00.html. Accessed 20 January 2006.

UNCTAD (2005) *World Investment Report 2005: Transnational Corporations and the Internationalization of R&D*, New York: United Nations Conference of Trade and Development. Downloadable from www.unctad.org.

UNDP (2005) *Human Development Report 2005: International Cooperation at a Crossroads: Aid, Trade and Security in an Unequal World*, New York: United Nations Development Programme. Downloadable from http://hdr.undp.org.

UNDP (2007) *Human Development Report 2007/2008: Fighting Climate Change: Human Solidarity in a Divided World*, New York: United Nations Development Programme. Downloadable from http://hdr.undp.org.

Venkateshwarlu, K. (2006) 'A Saga of Failures', *Frontline*, 23(1) January: 14–27.

Weisman, Steven R. (2006) 'Before Visit to China, A Rebuke', *New York Times*, 12 December.

World Health Organization (2004) *The World Medicines Situation*, World Health Organization, at http://w3.whosea.org/LinkFiles/Reports_World_Medicines_Situation.pdf. Accessed on 7 January 2005.

Yee, Amy (2007) 'Rule Changes Ease Pharma Pain', *Financial Times*, 5 September.

Zweig, David (2006) 'Learning to Compete: China's Efforts to Encourage a "Reverse Brain Drain"' in Kuptsch and Pang (eds), pp. 187–214.

6
The Indian Pharmaceutical Industry: Firm Strategy and Policy Interactions

Kalpana Chaturvedi and Joanna Chataway

6.1 Introduction

Since independence, the pharmaceutical industry has been one of India's most successful industrial sectors. In 1947, the industry was of very modest size with a market of about $28.5 million (Ahmad, 1988). Until the 1970s the domestic market was dominated by foreign-held patents and companies. In 1970, eight of the top 10 firms were foreign multinationals. In the last three decades, however, the dynamics has changed. In 2003, six out of the top 10 firms were of Indian origin (OPPI, 2003) and these are acknowledged as knowledge-driven and globally competitive firms. The process patent regime in the 1970s, together with price ceilings under the Drug Price Control Order (DPCO), first helped the nascent industry to establish itself in the domestic market. Economic liberalization that followed in the late 1980s then broadened the horizons for Indian firms to grow beyond national boundaries. The emergence of the TRIPS regime in 1995 and the adoption of the New Patent Act by India in 1999 added further dynamism and research orientation to this sunrise industry. Following this hot policy trail are Schedule M[1] enforcement in 2003 and Schedule Y[2] amendment in 2005. Given this context, how Indian firms are adapting to IP changes along with other policy amendments and changing market dynamics is an interesting question to investigate. This chapter investigates the evolution of research and marketing strategies in six leading Indian pharmaceutical firms transitioning from process to product innovation.

We argue that although interventions influenced by public policy have shaped technological trajectories and policy support has been crucial in creating an 'enabling environment', firms have contributed equally by making the policy work for the growth and evolution of this sunrise industry. At the macro level, the case of growth and evolution of the Indian pharmaceutical industry traced in this chapter provides a highly instructive example of the complexity of building domestic technical capability and expanding market (global) reach. At the micro level, the case studies highlight the strategic initiatives taken by industry leaders from time to time and demonstrate that firms who responded strategically to market and public policy changes attained leadership while others who resisted the change declined gradually. The rise and fall of the public sector firms in this context provides some key insights in this study.

Finally, stretching our thinking beyond the realms of policy, firms and innovation linkages the analysis in this chapter opens new lines of enquiry. Are firms capable of changing the environment such that new strategies become necessary? Our research is not necessarily focused on this question, but while investigating our main research question stated earlier and further thinking on these issues indicates the increasingly important role that firms are playing in shaping the policy and strategy for domestic and global stakeholders alike. For example, Schedule Y amendments have been instrumental in facilitating the return of MNCs to India and clinical research collaborations between foreign and domestic firms. Our company interviews indicate that these amendments premised on Indian firms being capable of carrying out clinical research and custom manufacturing to international standards and they have devised strategies to exploit new opportunities opened up in the 'R&D-led services' market. Thus firm-level strategies are critical not only in 'making the policy work' but, more importantly, in the 'making of the policy itself'. Market-pull and policy-push both appear to be at work at present. This new dynamics triggers new opportunities, new challenges and, therefore, new strategies to tackle them. This chapter examines how these opportunities and challenges are being met by leading Indian pharmaceutical firms.

In section 6.2 we briefly review the literature on resource-based views of strategy creation and dynamic capability framework which guides this research. This section also describes the methodology of the study and rationale behind using such a research design. Section 6.3 documents the events and evidence related to the three phases of development of the Indian pharmaceutical industry. Understanding of the capacities and capabilities created within the corresponding development phase, serves as a foundation for technological choices and strategic initiatives of Indian firms at present. Section 6.4 investigates research and market strategies of the six leading Indian firms in detail and analyses their novel strategies as they move from process engineering to drug discovery in the new policy environment. Section 6.5 presents the analysis of key strategies and trends and we draw some conclusions in section 6.6.

6.2 Theoretical framework and methodology

6.2.1 Interplay of dynamic capability and resource-based framework

The theory of how firms grow links firm-specific capabilities and assets with their strategies, often referred to as the resource-based perspective (Penrose,

1959). Elaborating on the 'stick to the knitting' concept associated with the resource-based perspective, Nelson and Winter (1982) argued that the technological and organizational knowledge and behaviour of a firm is conditioned by its past leanings. This implies that firms have different strategies that suit their capability resources. A rich literature on strategic management is built on this perspective and is often referred to while analysing the future direction for strategies at the firm level (Rumelt, 1984; Porter, 1991). More recently, the dynamic capability framework proposed by Teece et al. (1997) has further emphasized the function of firm's resources and capabilities as foundational to firm strategy. This framework proposes that 'competitive advantages is not just a function of how one plays the game but also a function of the assets that one has to play with and how these assets can be deployed and re-deployed in changing markets' (Teece, 1998: 72). An important, yet less explored dimension in this framework is the changes prompted by exogenous events beyond the control of firms. The present chapter shows that the equation between endogenous resources and exogenous factors is much more complex than envisioned in the theory. Changes in the external environment such as policy, market structure and the emergence of new knowledge fields such as biotechnology and bioinformatics destabilize even the most established patterns in an industry or firm. Exogenous events prompt firms to think new ways of not only adding value to the rents of existing capabilities but also to take on board high-risk strategies such as acquiring completely new resources/assets from external sources and exploring un-treaded paths.

Modern literature provides enough evidence to show that the capability creation process has become extremely dynamic in nature and that, over the years, it has become progressively more intangible in nature (Leonard-Barton, 1995; Malerba and Orsenigo, 2001). This literature identifies knowledge as a critical resource and research and innovation as firms' core competencies. Identifying new opportunities and organizing effectively and efficiently to embrace them are more fundamental to knowledge and wealth creation in the present dynamic environment. Firms are increasingly focusing on the intangible resources of diverse knowledge bases that differentiate them from the rest of the pack and lead to superior performance. The evolution of pharmaceutical sector-specific regulatory frameworks and their impact on firm-level strategies explored in this chapter confirms that at the level of the firms the bases of wealth creation have changed over time. The Indian pharmaceutical industry provides an interesting case study for analysing the shift from a purely resource-based static approach to a knowledge-based dynamic approach while simultaneously highlighting the role of a firm's resources and capabilities as the foundation for emerging strategies. We will show that strategic diversity is necessary to gain competitive advantages in a given context and necessary for dynamic capabilities to emerge. The chapter reflects on the increasing complexity and maturity in which Indian firms are tackling the management of new policy, products and markets in the current rapidly changing environment.

6.2.2 Research methodology

The central focus of this chapter is on investigating firm-level innovation strategies alongside changes in the national regulatory environment, knowledge levels and market dynamics. However, the theoretical discussions in the previous section highlight that firms devise strategies that suit their capability and resources. In this context, first knowing the history of technological developments in Indian firms during previous policy regimes and market environments is essential. Existing literature has been used to understand the capacity and capability development in this industry in the past. This literature is sourced from the annual reports of the Indian Ministry of Science and Technology, the Ministry of Commerce and Industry, Parliamentary reports and reports published by business and consulting firms (KPMG, 2006; OPPI, 2003).

Multiple case-designs are used to analyse how Indian firms are dealing with the changing policy, knowledge and market requirements in the post-TRIPS regime. A heterogeneous mix is selected for in-depth case studies with a number of criteria in mind such as: (i) the R&D base of the firm; (ii) the size of the firm in terms of its annual turnover and market share; (iii) collaborations with domestic and foreign organizations for research, manufacturing and marketing purpose; and (iv) the technical, operational and management capabilities of the firm. Two public sector companies are selected for their historical role in knowledge and technology diffusion in the formative years of the Indian pharmaceutical industry. Although currently both these units are declared as 'sick' or financially distressed companies by the Board for Industrial and Financial Reconstruction (BIFR) and are practically non-existent, the cases provide key insights into why some firms fail under changing circumstances while others succeed.

6.3 Growth and evolution of the Indian pharmaceutical industry

The Indian pharmaceutical industry has experienced three different policy thrusts, each corresponding to a different patent regime and each with a different aim and outcome. The three main components of the Indian pharmaceutical industry are the public sector, the private sector, and foreign multinationals. The industrial policy shifts in the pharmaceutical sector had a major impact on market dynamics and the rise and fall of these three main components. This section maps the progressive development of the Indian pharmaceutical industry vis-à-vis three key element – policy, markets and research – clustered under three major policy time frames.

6.3.1   The Patents and Designs Act, 1911 (1950–70)

The Patents and Designs Act, 1911 introduced by the British colonial administration allowed foreign inventors to patent their products in India and gave them exclusive privileges for 16 years to operate in an open market. This act was inherited by the newly independent nation in 1947. In addition, India's Industrial Policy Act of 1948 adopted a 'no discrimination policy', assuring a level playing field for foreign multinationals. Although these national policies were aimed at attracting foreign capital, technological resources and industrial know-how in the industries within the domestic sphere was limited or non-existent. The policy environment in the 1950s provided a free market to foreign investors and paved the entry of several more multinational corporations (MNCs) in India. The MNCs themselves were not keen on local manufacturing or any R&D activities in India and prevented Indian firms from doing so by using their patent rights. The benefit of technological spillovers and industrial know-how to the local industry that generally accompanies an open market dynamics was denied. Research was non-existent in India and manufacturing by MNCs meant mere refining and bottling of imported bulk drugs from their home countries.

6.3.2   The rise and fall of the Indian public sector

The government's vision to 'reduce dependency on MNCs for life saving drugs' and 'access to medicines for all' was the starting point of building local production facilities in the pharmaceutical sector. In 1954, the Indian government established Hindustan Antibiotics Limited (HAL), followed by Indian Drugs and Pharmaceutical Ltd. (IDPL) in 1961. The government democratized access to drugs by establishing several state-owned pharmaceutical firms. Public enterprises, particularly HAL and IDPL, played an important role not only in starting domestic production of key bulk drugs but also in diffusing substantial spillovers in terms of technical know-how, technology transfer, technology innovation process/system, and, more importantly, in generating entrepreneurs such as the founder of the immensely successful Dr. Reddy's Laboratories (DRL). Another important spin-off from government initiatives was the establishment of educational institutes, pharmacy colleges, and small and medium enterprises for up and downstream businesses (Felker, 1997). All this acted as the mainstay for the growth of the bulk drug sector in later years.

HAL started penicillin production with technical assistance from the WHO and UNICEF in 1954. HAL was the first to manufacture Penicillin, Streptomycin, and Gentamycin in the country with the help of imported technologies. However, indigenous R&D efforts to improve upon the acquired technology from the western leaders such as Merck and Glaxo could not match international standards. The problems encountered during technology sourcing, transfer and diffusion from the foreign companies had a negative impact on the overall performance of the firm. Building the indigenous technological base is a crucial consequence of technology transfer in order to deploy new technology effectively in an economic or operational sense (Cusumano and Elenkov, 1994). Not only did HAL implicitly limit the technological (knowledge) content of the transfer, but also it did not use existing capacities to push forward the technical and market performance of its products and processes.

The establishment of IDPL in 1961 by the government with Soviet technical know-how mainly for the manufacture of bulk drugs is regarded as a great initiative in the history of the Indian pharmaceutical industry. IDPL became a role model for many enterprises and provided tremendous boost to indigenous efforts. However, the firm succumbed to the regulatory and market pressures in the 1990s. In the absence of subsequent technology, knowledge and skill development, in-house R&D commercialization efforts, and overall management, the growth and performance of HAL, IDPL and other state-owned firms declined gradually. Nonetheless, the contribution from these firms and the government's early investments were instrumental in laying the basic foundation of the sector. Spin-offs of HAL, IDPL and other state-owned companies that went to private hands reached high quality standards and established modern managing practices.

In addition to establishing public sector firms, the government invested in industrial research through ministerial departments, especially through the Council for Scientific and Industrial Research (CSIR). The CSIR, consisting of 38 industrial R&D laboratories, was established solely to provide a research base for Indian industry. However, due to the lack of linkages between national laboratories and business firms, the laboratory research was hardly ever commercialized. For a long period, support systems and finances were made available only to the public sector firms and research organizations. The domestic private sector had neither the support of the government nor the technology spillover benefits from the multinationals. The combination of these provisions had a negative impact on the industry. Their urgent need for a system that encouraged technology acquisition, transfer, development, diffusion and incremental innovation was obvious. Patent laws were used as a tool to establish this system in India.

6.3.3 The Patents Act 1970 (1970–1990)

The Patents and Design Act of 1911 from the colonial era was replaced by the Patents Act 1970 that reduced the scope of patentability in food, chemicals and pharmaceuticals to only processes. The Indian Patents Act 1970 totally changed the complexion of the pharmaceutical industry in India and subsequently introduced complex laws and policies to regulate and promote the industry. Product patents were not allowed or recognised. In addition, the term of process patents was reduced to seven years from the date of patent

Table 6.1   Market share of national sector and MNCs

Sector	Bulk drugs		Formulations	
	1976–77	1983–84	1976–1977	1983–84
National Sector (including public sector)	85	290	40.6	930
MNCs (both FERA or ex-FERA)	63	65	292	615

Source: Indian Drug Statistics 1984–85, Ministry of Chemicals and Fertilisers, New Delhi.

grant or five years from the date of sealing, whichever was shorter, in food, drugs and chemicals, and to 14 years for other products. The provision of compulsory licensing was also introduced which provided for the opening of the patented drug for generic replication if the drug was not available to the general public in India at a reasonable price.

Subsequent policy changes to promote the development of local R&D facilities involved the elaborate use of an industrial licensing system. Whilst public firms and national laboratories were given incentives and promoted in an attempt to strengthen design and engineering component, new obligations were imposed on foreign firms. Foreign firms with an annual turnover of more than 50 million Rupees were obliged to have local R&D facilities with a capital investment of at least 20 per cent of their net block. These firms were also required to spend at least 4 per cent of their sales turnover as recurring expenditure on R&D facilities. The Drug Prices Control Order (DPCO) was first issued in 1970 with a ceiling on overall profit on bulk drugs. This acted as an added disincentive for the multinationals. The extremely short patent terms, compulsory licenses, profit ceilings and licenses of right ensured that MNCs could not use patents to exploit the Indian market any more. The overall policy environment in the 1970s had a major impact on the structural composition and production patterns of the pharmaceutical industry in India as depicted in Table 6.1.

Local production capacities were built through a combination of access to international technology and expertise through imports, establishment of education and health research systems and the right policy environment providing opportunities for learning by doing. Technology required for facilitating imitative process R&D that was first acquired by the public sector form foreign sources were passed on to the private sector (Srinivas, 2004). The private sector, a large part of which developed through a spin-off entrepreneurship of public firms, honed the reverse engineering skills and developed extensive capacity in synthetic chemistry which constitutes the core (competency) of the Indian private sector until today.

Table 6.2 Growth of pharmaceutical production in India by ownership groups as % of total production

Year	Public sector	MNC affiliates	Organized Indian sector	Small-scale Indian sector
1975	6.25	50.75	43.00	N.A.
1976	6.25	53.57	40.18	N.A.
1977	6.71	41.71	34.43	17.14
1979	5.71	N.A.	76.19	18.10
1980	6.26	N.A.	67.65	26.09
1983	6.25	40.00	27.69	26.06
1984	6.25	40.00	27.67	26.08
1985	6.24	40.01	27.64	26.11
1986	6.22	40.00	27.66	26.12
1975	36.7	37.8	25.6	N.A.
1976	33.1	40.0	19.2	7.7
1977	32.0	42.0	19.3	6.7
1979	24.5	28.0	37.5	10.0
1980	26.1	23.5	39.8	10.6
1981	25.8	23.3	39.6	11.3
1982	23.1	25.2	41.4	10.3
1983	20.6	22.2	37.2	20.0
1984	17.2	18.3	43.7	20.8
1985	17.0	18.0	44.0	21.0
1986	17.1	18.0	44.0	20.9

Source: Kumar and Pradhan, 2003.

6.3.4 Emergence of the domestic private sector

The protected nature of the India economy acted as a pull mechanism for small, medium-sized and large enterprises. There was an instant impact on the structural composition of the industry. Local investment rose sharply and so did production as shown in Table 6.2. The favourable environment not only encouraged new firms to enter this field such as Dr. Reddy's Laboratory, Sun Pharmaceuticals, Nicholas Piramal, Torrent and Lupin but also encouraged the older firms like Ranbaxy, Cipla, Unichem, Cadila and Alembic to transform and expand their production base and marketing networks. In the 1980s, the industry grew rapidly at the rate of 11 per cent per annum, which further accelerated to 17 per cent per annum during the 1990s (Kumar and Pradhan, 2003).

The legal and policy reforms relating to pharmaceuticals from 1970 onwards were both wide-ranging and effective. Technologically this started the era of reverse engineering where firms developed new products by simply changing a few steps in their production processes at a fraction of the cost otherwise involved in new drug development. Research and development work expanded from formulations and process technology to the discovery of new therapeutic substances. Several MNCs decided to exit India due to fierce competition and price wars with the local firms. Gradual expansion without market pressures provided time and space to Indian firms for enhancing their technical expertise, infrastructure, range of drugs, production capacities and market reach within the country. Once the domestic firms had established themselves in the local market, the top firms started to look beyond the national boundaries. The passing of the Drug Price Competition and Patent Term Restoration Act of 1984, commonly known as the Hatch–Waxman Act in the United States[3] provided this opportunity for Indian firms and expanded business horizons beyond the national boundaries. It enabled chemical drug manufactures to sell lower cost generic drugs immediately upon expiration of the product patent. Backed by policy initiatives at the international level and emerging international markets, Indian firms exported their products to semi-regulated markets initially. Export competitiveness was driven by low prices due to low R&D and production costs. Besides, there were no price controls, excise duty or sales tax on exports and profits were tax free as well. The lucrative export opportunities were taken on board immediately by the leaders. The fat profit margins for the bulk drugs in western markets gave an instant boost (economic performance) to first movers like Ranbaxy, Cipla, Lupin and Sun. The export-led business model was feverishly followed by others and continues to dominate the corporate strategy of major Indian firms until today.

6.4   Economic reforms and TRIPS Agreement (since the 1990s)

6.4.1   The new innovation dynamics

The protection of domestic production and local technological efforts in the 1970s and 1980s enabled India to build up a diverse and fairly sophisticated base in the pharmaceutical sector. However, technological backwardness and lack of genuinely innovative products was evident (Lall, 1987). The industry suffered setbacks in terms of R&D investments from both national and foreign sectors. MNCs did not introduce newer technologies in India fearing reverse-engineering by Indian firms. Also local firms did not feel the need for research and innovation since there was no reward for innovation and there was no penalty for imitation. Besides, there was a sizeable captive market available for generic drugs within India and later in several foreign countries (initially Russia, several countries in Africa and Asia). Research, innovation and quality were not major concern for Indian firms during this period.

The liberalization programmes initiated by the Indian government in 1991 became an important policy tool to initiate innovation-based competition. Steps towards serious research and development were taken during this period by Indian firms. Several private firms enhanced their focus on R&D and did their own value additions to the products. They gradually advanced to creative imitation stage (chiral synthesis) and produced good quality generics.

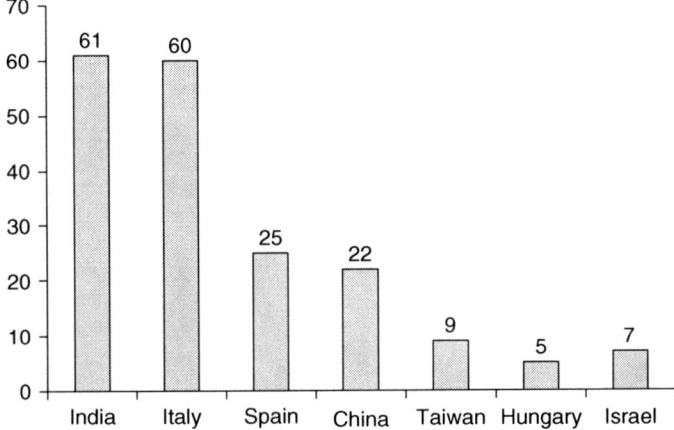

Figure 6.1   Number of USFDA-approved plants outside the USA
Source: IBEF, 2003.

Much new technology was still imported, but firms like Ranbaxy, DRL, Sun, Torrent, and Wockhardt started investing more in in-house R&D as a move to build a proprietary technological base. World-class production facilities were built by Indian firms in record times during this period. As a result, today India has the largest number of US FDA approved manufacturing facilities outside the USA (Figure 6.1). The capacities created during this time period ushered 'contract manufacturing' dynamics in the sector utilized by many Indian firms, details of which to follow.

Technological changes during this era were driven by competition facilitated by the government's liberalization policies. However, the changes were brought about by other factors. Ambitious and visionary firms such as Ranbaxy, Lupin, Torrent and Sun took their first few steps towards technological maturity even before the liberalization era. Nonetheless all the firms interviewed for this chapter agreed that liberalization added further momentum and pace to their efforts. Driven by the large global generic markets in the developed world, Indian firms gradually created capabilities for generics R&D and started moving towards creating non-infringing process innovations. Liberalisation also greatly steered their strategic directions for R&D and marketing.

6.4.2   Growth indicators

The amount of investment which stood at 1,400 million Rupees in 1965–66 rose to 5,000 million Rupees by 1982, an almost threefold increase over a period of nine years. It further increased to 18,400 million Rupees and 45,000 million Rupees in 1997–98 and 2002–03 respectively (Table 6.3). The

Table 6.3   Indian pharmaceutical industry: growth indicators in million Rupees (in US$ million calculated with the base rate of 40.8 Rupees to one US$)

Particulars	1965–66	1980–81	1997–98	1999–00	2000–01	2002–03
Capital	1,400	5000	18,400	25,000	29,000	45,000
Investment	(34)	(122.5)	(450.9)	(612.7)	(710.7)	(1122)
Production	1,680	14,400	146,910	197,370	228,870	392,547
	(41)	(352.9)	(3600.7)	(4837.5)	(5609.5)	(9621)
Trade						
Export	30.5	464	53,530	72,300	87,340	128,260
	(0.75)	(11.37)	(1312)	(1772)	(2140.6)	(3143.6)
Import	82	1125	28,680	16,160	29,800	28,650
	(2.0)	(27.6)	(702.9)	(396)	(730)	(702)
R&D	30	147.5	2,200	3,200	3,700	6,600
expenditure	(0.73)	(3.6)	(53.9)	(78)	(90.7)	(161.8)

Source: Nauriyal 2006.

bulk drug sector that was non-existent in the 1970s grew exponentially from the 1980s. The sector manufactures more than 400 key bulk drugs and contributes to 6 per cent of the international bulk drug market at present (Mani, 2006).

Technological development in India is reflected in the wide range of bulk drugs being produced from basic stages, through complex multi-stage synthesis and intricate fermentation and extraction techniques. R&D investment in Indian pharmaceuticals has been very low and it started picking up only in the early 1990s (see Figure 6.2). Until the mid-1980s, R&D investment by the public sector was much higher than by the private sector, but it started to change by the early 1990s (Bowonder and Richardson, 2000). Ranbaxy and DRL enhanced their focus and R&D investments in new chemical entities (NCEs) and novel drug delivery systems (NDDS). As early as 1993 DRL initiated research related to NCEs. Similarly, Ranbaxy began investing more in R&D even before liberalization. The investments in R&D, particularly for new drug discovery research (NDDR), were further enhanced in the wake of the intellectual property rights requirements of the Uruguay Round. Early success achieved by DRL and Ranbaxy due to higher value additions in their products propelled other private firms to invest and undertake R&D-related activities. Several private firms, including Cipla, Lupin, Wockhardt, Sun, Torrent, and Dabur, are today engaged in R&D activities.

Thus even without strong patent protection, Indian pharmaceutical firms and industry matured during the 1990s. In particular, large firms moved away gradually from reverse engineering and developed capabilities for NDDS. Evidence indicate that the combined effect of national and international policy changes and market shifts facilitated a 'research tradition' in the Indian pharmaceutical firms to a very large extent during this period.

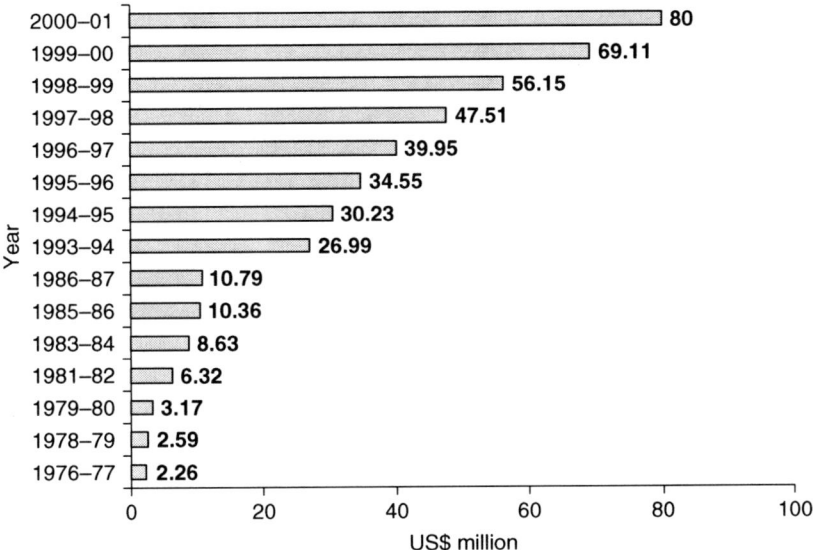

Figure 6.2 R&D Investment in the Indian Pharmaceutical Industry
Source: OPPI, 2001, 2003.

6.4.3 Policy experiments in the mid-1990s and progress thereafter

In the mid-1990s the Indian pharmaceutical industry experienced a number of policy levers, all focused on inducing research and innovation-based competition and trade expansion. For example, further dilution of DPCO in 1995 removing 50 per cent of drugs from price controls and then again in 2004 removing another 24 per cent of drugs from the list threw a large number of products for open competition (Figure 6.3). Liberalization of national and international financial markets followed in quick succession in 1995. Following this hot trail was India's decision to sign the TRIPS Agreement in 1995, the New Patent Act in 1999 and Patent (Amendment) Act of 2002 and 2005. Besides providing for product patent protection, the Patent (Amendment) Act 2005 has significantly extended the life term of pharmaceutical patents from seven years to twenty in one giant leap, thus wiping out the scope and benefits of reverse engineering of the patented products completely. Thus, TRIPS has totally reversed the 1970s policy initiatives.

The general perception in the studies concerning the Indian pharmaceutical industry is that the economic opportunities (generics in domestic and western markets) offered by liberalization and the Hatch–Waxman Act in the 1990s would cease to exist after 2005. Contrary to this belief, the fact that generics worth over US$55–65 billion are going off patent in 2007 in the US alone presents immense economic opportunities for Indian generics

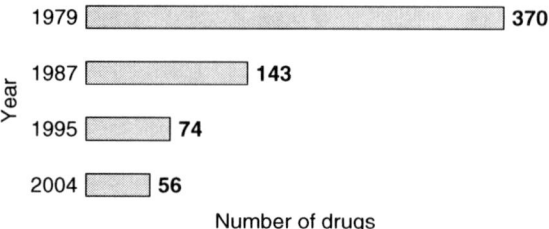

Figure 6.3   Decreasing number of drugs under price control
Source: Plotted from the data collected from various drug policy texts.

manufacturers. This is further triggered by opening up of the generics markets in EU/Western Europe (IBEF, 2003). Leading Indian firms have already made forays in these markets and have been doing extremely well in the US, Germany and Europe. Notable among these firms are Ranbaxy, DRL, Wockhardt, Cipla, and Lupin.

On the research front, it is obvious that the reverse engineering of patented molecules will no longer be possible but leading Indian firms are turning the prospect of increased patent protection to their advantage by spearheading new drug discovery programmes, albeit with different strategy strokes and varying time-frames (short-medium-long). Sceptics assert that Indian firms are not large enough to discover, develop and launch their own drugs. Firms have responded to these speculations strategically and have devised a variety of strategies to overcome financial, technology and commercialization constraints associated with NDDR. Depending on their core competencies, firms have positioned themselves as partners of choice for joint manufacturing, basic research, advanced research, and/or marketing and distribution for western multinationals. Indian firms are strategizing to leverage opportunities and appropriate values existing in formulations, bulk drugs, generics, and NDDS, before transitioning to NCE and biotechnology research.

The case studies of six well-established Indian firms, Ranbaxy, Dr. Reddy's, Cipla, Nicholas Piramal, Sun and Lupin, and macro-level data sourced from secondary sources provide ample evidence that the grace period allowed under the TRIPS agreement for harmonizing India's product patent law with 'international norms' (ten years of transition from 1995 to 2005) has been utilized well by the firms. The interim strategies whereby they could reap the reward of production of generics for developed markets while simultaneously building their drug-discovery pipeline for the post-2005 era has so far worked in their favour. The average growth rate of the industry in the last few years has been around 12 per cent (Ganguli, 2003). Due to extraordinary growth of the domestic Indian firms and their stronghold on the domestic market, the market share of MNCs has dwindled from 75 per cent in the 1970s to around 27 per cent in 2003. The industry had an overall production value of an estimated US$7 billion in 2003 (FICCI, 2005). The extraordinary efforts made by Indian pharmaceutical leaders and their strategies during the transition period are discussed below.

6.5 Strategic integration of policy, knowledge and markets

The transition from process to product patent regime, the advent of new knowledge fields such as biotechnology and bioinformatics, further liberalization of the economy and foreign trade and globalization of the world economy all at once have pressurized Indian firms and policy makers to make strategic changes to their business decisions. At the national level, these challenges require policy inputs that are directed more towards promoting research in the pharmaceutical industry and towards making it more internationally competitive. The Drug Policy Act 2002 was framed against this backdrop, which could not be fully implemented due to litigations against it. The overall policy framework governing the industry until then had been the Drug Policy Act of 1994. However, the Draft National Pharmaceutical Policy, 2006[4] has been imposed as a result of several rapid developments that have taken place during the course of last few years. In a major initiative to promote research in the country, the Indian government has established a corpus of 1,500 million Rupees, which is called the Pharmaceutical Research and Development Support Fund (PRDSF) for funding R&D projects in the country.[5] We discuss here only those policy aspects that explicitly affect the research and innovation strategies of selected cases for this study although other amendments are also touched upon while discussing the overall scenario in the sector.

Besides becoming fully TRIPS compliant in 2005 with the grant of product patent protection, India has made amendments in Schedule M and Schedule Y of Drugs as we mentioned earlier. While the Schedule M amendment that necessitates introduction of good manufacturing practices (GMPs) applicable uniformly across the sector puts tremendous pressure on the unorganized sector, the Schedule Y amendments allowing parallel phase clinical trials for new compounds provide tremendous opportunities for clinical research. The cumulative impact of these amendments on the pharmaceutical market dynamics is hotly debated. Industry experts claim that the Act will help India tap into the huge US clinical research field. Although the outcome of these initiatives is yet to be seen, the evidence so far confirm reconfiguration in the research, innovation and marketing strategies of the Indian firms and MNCs alike. First, it has encouraged the return of MNCs in India not only for manufacturing and marketing purposes, but also for clinical research. Secondly, clinical research opportunities have facilitated the emergence of several Contract Research Organisations (CROs) and small and medium-sized biotech firms. Large firms are also establishing their own CROs and collaborating with hospitals to seize clinical research opportunities. The analysis of case-studies indicates that firms are strategizing and pursuing three options.

These are: new drug discovery research, generics and R&D-led business services. Within these three broad categories, differentiation is achieved through interesting strategy-mix as discussed below.

6.5.1 Ranbaxy

6.5.1.1 Research focus

The largest pharmaceutical firm in the country, Ranbaxy was incorporated in 1961 and went public in 1973. Ranbaxy initiated new drug discovery research from its facilities in Okhla, New Delhi in 1994 and established its first state of the art multidisciplinary R&D facilities at Gurgaon (near New Delhi) in 1995. With the commissioning of its new R&D centre in August 2005, Ranbaxy now has in place, a total of three modern state-of-the-art multidisciplinary research facilities with skilled scientists, management and human resources. While two centres focus on the development of generics and Novel Drug Delivery Systems (NDDS) research, the third centre is dedicated to New Drug Discovery Research (NDDR). The R&D expenditure has gradually risen from 2 per cent in 1995 to 7 per cent in 2006. In the past few years the research team has developed five new drug molecules (company interview by authors and annual reports).

The firm has followed a fast-paced incremental strategy (two staged) to move from generics to NDDS first and then to NDDR. The robust R&D infrastructure created in the last one decade with equal focus on both new drug development and generics exhibits a hybrid strategy to sustain market position through branded generics while developing capabilities for new drug discovery. The generics R&D strategy first proved its mettle in 1991 with Ranbaxy's pioneering work for developing a non-infringing process for the production of Cefaclor, a molecule owned and heavily protected by Eli Lilly since 1979.[6] Ranbaxy's foray into Novel Drug Delivery Systems (NDDS) in the mid 1990s has paid rich dividends in the form of unique proprietary control release platform technologies. The first significant international success using the NDDS technology platform came in 1999 when Ranbaxy licensed its once-a-day Ciprofloxacin formulation, on a worldwide basis to Bayer AG, originator of this molecule. The focus on expanding its platform technologies and developing newer prescription products based on them has since been sharpened. Ranbaxy's current NDDS focus is mainly on the development of NDA/ANDAs of oral controlled- release products for the regulated markets.

The NDDR activity has made significant strides in areas like anti-infectives, urology, respiratory/inflammatory and metabolic diseases and has a number of pre-clinical leads in these segments. Ranbaxy's anti-malarial molecule has successfully completed proof of concept Phase II studies in 2005 (Table 6.3). Presently, the company has ten programmes in the area of NDDR, including one NCE in Phase-II clinical trials (Table 6.4). In addition, Ranbaxy also

Table 6.4 Ranbaxy's NCE pipeline

Molecule	Therapeutic area	Status
RBX11160	Malaria	Undergoing Phase IIb dose range finding
RBX9841	Urinary incontinence	Phase II clinical trials
RBX10558	Dyslipidemia	IND application filed

Source: Annual reports and company interviews.

has a global alliance in the area of drug discovery and development with GlaxoSmithKline Plc. While Ranbaxy leverages its early product development strengths, GlaxoSmithKline imparts its expertise at late development stage to complete the development process.

6.5.1.2 *Generics market strengthening*

Ranbaxy has classified its markets into developed, emerging and developing categories, based on each category's potential pharmaceutical market. In the Indian domestic market Ranbaxy has gained leadership owing to its strong brand building and marketing strategy. Ranbaxy has aggressively increased its product offerings in order to make up for the domestic market share losses due to its new policy of no imitation or reverse engineering. In order to gain access to new brands, Ranbaxy has acquired small firms in product segments in which it has been traditionally weak, like dermatology and anti-inflammation. To further expand its product portfolio, Ranbaxy has entered into licensing agreements with MNCs to manufacture and sell their products in India.

In foreign markets, Ranbaxy has established its presence in generics through its wholly owned subsidiaries or joint ventures in several countries while sourcing bulk actives from India. Ranbaxy has an expanding international portfolio of affiliates, joint ventures and representative offices across the globe with a presence in 23 top pharma markets of the world. With robust operations already in place in the USA, the UK, France, Germany, Russia, India, Romania and South Africa, Ranbaxy is strengthening its business in Japan, Italy, Spain and several other markets in the Asia Pacific. Ranbaxy's key corporate strategy is a balanced geographical presence and strong product flow from a wide therapeutic range including emerging CVS, anti-diabetic, antiretrovirals and pain management segments of the industry.

6.5.2 Dr. Reddy's Laboratories (DRL)

6.5.2.1 *Research focus*

Dr. Reddy's, India's third largest pharmaceutical firm, was founded by Dr. Anjii Reddy in 1984, as spin-off enterprise from IDPL. DRL opted for a

Table 6.5   Dr. Reddy's R&D pipeline

Compound	Therapeutic area	Development status
RUS 3108	Atherosclerosis	Phase I
DRL 16805	Atherosclerosis	Preclinical
DRF 2593	Diabetes	Late phase II (co-development with Rheoscience Denmark)
DRL 16536	Diabetes	Preclinical
DRF 10945	Dyslipidemia	Early Phase II
DRL 12424	Dyslipidemia	Preclinical
DRF 1042	Solid Tumour	Late Phase II (co-development with ClinTec International, UK
DRL 11605	Obesity	Phase I
DRL 15725	Rheumatoid Arthritis	Preclinical

Source: Annual reports and company interviews.

much riskier path, new drug discovery since its inception. Driven by Dr Anji Reddy's pioneering vision of taking molecules from lab discovery to the markets of the world, Dr. Reddy's Research Foundation (DRF) dedicated to new drug discovery research was established in 1993. The drug discovery efforts at DRF focus on studies of newly synthesized compounds for the treatment of cancer, diabetes, dyslipidemia, cardiovascular and metabolic disorders. DRL is planning to launch its first original drug for diabetes in 2010–2011 (Financial Times, 2006).

In 1998 DRL became the first Indian company to out-license molecules, when it agreed to license DRF 2725 and DRF 2593 in diabetes and DRF 4158 in cancer to Novo Nordisk for advanced phase development. Its key strategy was to move up the value chain incrementally and offset technical and commercial risks involved in NCE development. Although the molecules did not succeed and had to be abandoned, the integrated discovery strategy facilitated diffusion of research skills and market management talent related to NCE commercialization into DRL. According to a senior DRL executive, 'The path followed well ahead of its competitors has definitely provided an edge to DRL' (Author interview with Vice President-Formulations). Currently, DRF has nine NCEs in various stages of development (Table 6.5). A strategy mix of in-house development and out-licensing is being adopted for further advancement of these molecules.

6.5.2.2   Generics market strengthening

Like other Indian firms, DRL has also used innovation in process development and finished dosage to build a profitable and strong business. In 1985 DRL was the first Indian firm to export bulk actives (APIs) to Europe. In the domestic market, DRL launched its first branded product Norilet in 1987 followed by Omez. In 1992, DRL started building up its base for the US generics business and filed its first ANDA[7] in 1997 for Ranitidine followed by a Para IV filing for Omez (Omeprazole) in 1999. The firm initiated its generic foray in the western markets with the launch of Fluoxentine, a generic version of Eli Lilly's Prozac, in 2001. DRL was the first Indian firm to receive a 180-day exclusivity for a generic drug in the USA. The marketing success of Flouxentine was followed by the launch of Ibuprofen in the USA under its own brand name in 2003. The combined success of Fluoxentine and Ibuprofen provided a solid footing for DRL's generics base in the USA. 'The core business strategy is to sell company's product line (APIs, branded formulations and generics) in new markets and the challenge is to create global brands' said the Vice President-formulations (author interview). These businesses provide a cushion against risks involved in NDDR by building critical mass for the organization. Over the next few years, the focus will be on driving growth in these businesses to strengthen cash flows, build the US front-end and pursue DRL's ultimate dream that is new drug discovery.

6.5.3 Chemical Industrial and Pharmaceutical Laboratories (Cipla)

6.5.3.1 Research focus

Cipla, India's second-biggest generic drug maker, is one of the oldest pharmaceutical firms in the country. Founded in 1935, Cipla is better known for its reverse engineering skills than research. Cipla's innovation expertise spans the full spectrum of product development from new molecules to modern drug delivery systems. Cipla's focus is on analogue research with emphasis on NDDS and chiral chemistry. Development of new drug delivery systems includes sustained and modified release dosage forms; innovative transdermal delivery systems; innovative delivery systems based on inhaled forms; CFC-free metered dose inhalers; and ready-to-use formulations. Cipla has already developed its first chirally resolved molecule, salbutamol, which is an anti-asthma drug. Research on new chemical entities focuses in the area of antifungals, antihistamines and anti-AIDS compounds. According to the Cipla's Director Dr Hamied, 'While our skills are among the best in the world, what makes us different is our multi-disciplinary approach to research. Our research capabilities are extensive, from chemical synthesis, delivery systems and medical devices to process engineering, neutraceuticals and biotechnology' (author interview).

Given the public knowledge that Cipla is working on improved chemical entities and drug delivery systems, the research pipeline of Cipla appears to be moving on a similar trajectory as Ranbaxy's NDDS partnership with Bayer. Discussions with the company's director suggested that it is possible for Cipla to get into joint development partnerships with present patent holders of blockbuster molecules to develop improved versions by using chiral chemistry.

6.5.3.2   Generics market strengthening

In spite of the risk of being termed a 'patent-buster', Cipla opted for an aggressive product development (reverse engineering) strategy that catapulted it to the top league of generic makers in the world. Much of Cipla's progress is attributed to its strategy and focus on 'from inception-to-R&D-to-commercialization' and its thrust in making available the very latest in modern drugs and advanced delivery systems in a wide range of areas. The quest for making latest drugs available at a very low cost in the Indian market gained momentum with manufacturing ampicillin for the first time in the country in 1968. Cipla's achievements are notable such as medicinal aerosols for asthma in 1976, anti-cancer drugs vinblastine and vincristine (in collaboration with the National Chemical Laboratory, Pune in 1984), and etoposide, a breakthrough in cancer chemotherapy, in association with Indian Institute of Chemical Technology in 1991. The strategy has been immensely successful in the price-sensitive domestic market.

Cipla gained global recognition in 2001 when it offered to export anti-AIDS drugs to the poor in the developing world for less than US$1 a day per patient while the price charged by Western MNCs was many times higher and totally out of reach of patients in poor countries. Presently, Cipla manufactures and markets bulk drugs and formulations in over 40 therapeutic categories including anti-asthmatics, anticancer, anti-AIDS, anti-infectives and cardiovasculars. The company has twin-fold strategies for the international markets. First, in the sophisticated and highly regulated markets such as the USA and Europe, Cipla continues to register both active bulk drugs and their formulations. The firm has entered into strategic alliances with leading local generic companies both in the USA and Europe in order to enhance its reach and penetration into these markets. Secondly, Cipla continues to make fresh inroads into selected markets in Eastern Europe, Africa, Australia, the Middle East, Latin America and South East Asia. The emphasis is more on external markets and the firm aims to have equal domestic sales to exports ratio in the coming years. With a business model based on exploiting opportunities in the emerging generics markets, it has been a very profitable journey for Cipla in the last two-three years.

6.5.4   Nicholas Piramal India Limited (NPIL)

6.5.4.1   Research focus

NPIL was created in 1988 when the Piramal Group (initially involved in the textile business) acquired Nicholas Labs from the US company Sara Lee. After its first acquisition in 1988, numerous pharmaceutical acquisitions followed in the last 15 years. NPIL has become a core component of the Piramal group's business strategy. NPIL strategically acquired advanced technological and manufacturing bases created and nurtured by multinationals to get ahead of its competitors and to gain its leadership position in the domestic market.

NPIL operates in three key areas. The first key area is branded pharmaceuticals including prescription drugs for treating cardiovascular disease, psychiatric disease, diabetes, antibiotics, analgesics, antihistamines and anti-infectives. NPIL accumulated its product portfolio mostly by acquiring the Indian operations of foreign firms or licensing the Indian rights to their products. NPIL's second core business is custom manufacturing and research, an area where its existing or prospective customers tend to be the same foreign firms from which it has bought businesses or licensed product rights. NPIL's third and newest business area is inventing and testing new drugs to be sold under patent throughout the world. 'NPIL aims to be an integrated pharmaceutical company with a commitment to discovery, development, manufacture and marketing of indigenous pharmaceutical products' informed Dr. Sikka, Senior President, NPIL (author interview).

NPIL's R&D programme is divided into four strategic business areas: basic research, natural products, clinical research, and genomics research. R&D investment is quite low compared to other domestic pharmaceutical firms at 2 per cent of sales. However, recognizing the importance of basic research in this knowledge-based industry, NPIL has made strategic R&D investments. For example, in 1998 NPIL acquired the research centre of Hoechst Marion Russel located in Mumbai which since its inception in 1972 was focused on new drug discovery research and herbal research. Capex, a new R&D facility, has also been created in Mumbai to pursue NCE research in four key therapeutic areas – diabetes, oncology, anti-infective and rheumatology. The screening activities so far have produced some novel lead molecules. In oncology, the lead molecule P276 is currently in Phase I/II clinical trials in Canada and India; in inflammation, a lead molecule is undergoing further optimization and preclinical development and a lead candidate is being evaluated for safety and efficacy in two Phase II clinical trials in India; and in the field of infectious diseases, an anti-fungal herbal product is currently in Phase II clinical trials in India. A highly potent antibiotic drug candidate is also in late preclinical development (Piramal, 2007).

6.5.4.2 *Generics market strengthening*

Despite potentially huge profit opportunities in generics, NPIL gives priority to its custom manufacturing business. Nicholas Piramal is best known outside India for its custom manufacturing work, an area where both its capabilities and customer base are expanding rapidly. NPIL has several plants throughout India, all of which were bought from other companies and upgraded afterward to international standards. The strategy is to exploit upcoming outsourcing (by MNCs) opportunities for the production of finished formulations like it does for APIs. Near Hyderabad, the company's huge active pharmaceutical ingredient production complex consists of two adjacent plants that it bought as recently as 2003 and upgraded to comply with current Good Manufacturing Practices (cGMP). With the acquisition of

Rhodia's inhalation anesthetics business in December 2004 and the acquisition of Avecia's custom manufacturing business in December 2005, NPIL has access to over 90 countries for exports. It has state-of-the-art manufacturing facilities in the United Kingdom and India. Continuing its aggressive entry into Europe's custom manufacturing arena, Nicholas Piramal has acquired Pfizer's manufacturing facility in Northumberland in 2006. Thus NPIL, unlike Ranbaxy and DRL is not targeting US markets for generics. Instead it is strategizing to generate financial resources through alliances with western MNCs for custom synthesis and contract manufacturing.

6.5.5   Sun Pharmaceuticals Limited

6.5.5.1   *Research focus*

Sun began operations in 1983 with just five products to treat psychiatry ailments. Following the same path as NPIL, growth through mergers and acquisitions, Sun Pharmaceuticals has taken major strides. The firm manufactures and markets specialty medicines and APIs for chronic therapy areas such as cardiology, psychiatry, neurology, and gastroenterology. Recognising research as a critical growth driver, Sun established its first research facility, Sun Pharma Advanced Research Centre (SPARC) in 1993 which created the base for strong product and process development skills. According to one of its senior managers, Sun was one of the earliest among Indian companies to invest in research. Serious resources were committed to research in 1993 when it was a much smaller company. Ever since, 4 per cent of an increasing turnover has been invested every year on time-bound projects at the company's research centre (SPARC), that works on process synthesis, dosage forms, NCEs and NDDS. Sun has adopted a three staged research strategy. In the first phase, which has been accomplished, the emphasis was on reverse engineering and on high yield processes for specialty bulk drugs. The processes developed for specialty bulk actives at SPARC enabled the company to commercialize more than 60 specialty bulk actives in just seven years (1993–2000), several of which are based on novel and non-infringing processes. The second phase focused on innovative drug delivery systems and alternate patentable routes. Another research centre was set up in Mumbai in 1997 dedicated to developing innovative dosage forms and generics for the developed markets in the US and Europe. The output of the first and second phase coupled with increasing investments in the infrastructure and resources (financial as well as technical) equips the company for drug design in the long term, that is the third on-going phase.

In 2004 a new research campus was established in Baroda which conducts research both for complex generics and also for new-to-the-world drug discovery. According to company sources, about 35–40 per cent of the current research budget (2005) is allocated to innovation-based projects – this number is expected to exceed 50 per cent over the next three years. In the first phase of innovation, a drug discovery initiative is taking shape in three specific therapy areas and across four drug delivery system platforms. The intellectual property rights from these projects are envisaged to earn revenue streams globally in the future.

6.5.5.2 Generics market strengthening

With a strong market focus on specialty products, Sun taps the domestic niche markets as well as the vast international opportunity for specialty APIs. As a part of its generics market strategy, the first step to break into the important US healthcare market was taken in 1996 with its purchase of a controlling share of Detroit-based Caraco Pharmaceutical Laboratories. Caraco gave Sun Pharma its first USFDA-approved production plant. Sun also acquired an API plant from the multinational Knoll Pharmaceuticals in 1996 and acquired Gujarat Lyka Organics in 1999 which, in addition to adding its production of cephalexin bulk active, brought Sun a USFDA-approved manufacturing facility. Continuing its acquisition drive, Sun bought a stake in MJ Pharmaceuticals Ltd., a UK MCA-approved plant, based in Halol. MJ Pharma brought Sun its strong insulin production and facilitated its entry into the European market.

High-value brands in therapy areas like cardiovascular, neurology, respiratory, ophthalmology and orthopaedics appear to be the secret of Sun Pharma's higher than industry growth rate. Six large brands feature among India's largest-selling prescription brands (author interview). The company aggressively introduces new products every year in the domestic market. Over the past decade, Sun has expanded its presence with ethical brands in 26 markets across Asia Pacific, China, the erstwhile CIS countries, Africa and the Middle East. The key strategy is to focus on three large market opportunities that are branded products, specialty bulk actives, and generics. Branded products, that enjoy physician's confidence in India, have grown to become the mainstay of company's therapy focus, and are also marketed internationally. In several other markets where entry barriers are high the strategy has been to compete with speciality bulk actives. In the high potential markets of North America and Europe, Sun Pharma is gearing itself to compete as a value-added generic player by gradually moving up from generics to differentiated generics and finally to branded generics.

6.5.6 Lupin

6.5.6.1 Research focus

Lupin was established in 1968. The company's R&D Park, in Pune near Mumbai, conducts leading-edge research in generics, new chemical entities (NCEs), novel drug delivery systems (NDDS), oral-controlled release systems (OCRS) and phytomedicines. Other principle therapeutic areas are cardiovasculars and non-steroidal anti-inflammatory drugs (NSAIDs). The research park has four technology centres dedicated to formulations, value-additions (complex generics), NDDS and NCEs (natural products). While the Formulation Centre is classified as business driven, the rest of the three centres are classified as innovation driven. In an interview, Dr Sen, President of Pharma Research and Regulatory Affairs for Lupin, informed, 'A cross-functional team taking the existing processes and systems into account has designed a roadmap for building a knowledge-centric organization with globally benchmarked information and process infrastructure' (author interview).

Lupin has devised twin strategies for both the short and the long term. Its short-term goals are driven by market opportunities rather than research. The long-term research focus is on developing NDDS, herbal products and new chemical entities. Research on new chemical entities uses herbal-based leads in addition to the customary chemical synthesis route. NCE research is focused on four therapy areas: anti-migraine, anti-TB, anti-psoriasis and anti-inflammatory. Lupin developed a molecule for intra-nasal administration and filed an investigational new drug (IND) application in India in September 2002. In the anti-TB segment, it has identified three compounds that have demonstrated significant in vitro and in vivo (pre-clinical studies) activity against sensitive and resistant strains of M. tuberculosis. In other two segments also new molecules are undergoing pre-clinical studies. Lupin is leveraging its expertise in complex finished products and development and manufacturing to exploit market opportunities for NDDS. A proof of Lupin's NDDS capability is the world's first once-a-day cephalexin tablet launched in May 2003 and the world's first once-a-day cefadroxil tablet (Odoxil OD) launched in November 2003. Lupin has also filed seven patent applications for NDDS platforms in major areas.

6.5.6.2   Generics market strengthening

A global leader in the anti-TB segment, Lupin is the world's largest manufacturer of Ethambutol and Rifampicin. Its Rifampicin plant is one of the only three plants in the world approved by the United States Food and Drug Administration (USFDA). Building upon its existing R&D competencies in the anti-TB sector, Lupin is augmenting its product portfolio and broadening its market and distribution network. The strategy is to identify and implement profitable operations without new drug discovery in the short term. As part of its global strategy, Lupin has increased its participation in the high margin/high value advanced markets of the USA and EU. To date, Lupin has introduced its first generic-Cefuroxime Axetil tablets and its first branded Suprax dry suspension in the US market. Global alliances play a pivotal role in the company's international strategy. Examples are with Merck Generics (UK), the UK subsidiary of E. Merck for the marketing of Injectable Cephalosporins; with Wyeth Lederle for supplying advanced intermediate Ethambutol; with American Pharmaceuticals Partners, USA for supplying injectable Cephalosporin bulk-actives; and with Quatromed, South Africa for marketing anti-TB products. The firm has entered into alliances with Watson Pharmaceutical Inc. and with Baxter Healthcare Corporation for marketing and distribution of its products.

6.6 Key strategies and trends

As discussed earlier, a variety of strategies such as licensing product patents from patent holders, acquiring own patents through indigenous R&D, and integrating new knowledge and resources from external sources are being adopted by technology-based firms to address their post-patent needs. Pharmaceutical firms have collaborated aggressively with premium research institutions and international drug companies with dual objectives of changing their market image (to be known as research-driven) and for integrating new knowledge and skills from external sources. In this section, we summarize the key adaptive strategies as observed through the cases studied for this research.

6.6.1 R&D investments and patent filings

With the increasing realization that copying will no longer be permissible after signing up to the TRIPS regime in 2005, R&D investments have achieved serious momentum. From about 2 per cent of total sales at the start of the decade, the average R&D investment of the leading research-based domestic firms rose to around 5–6 per cent in 2004–2005. Of these companies Ranbaxy, Dr. Reddy's, Wockhardt, Torrent and Sun, are among the most prominent. Dr. Reddy's R&D expenditure increased from 7 per cent in 2002–03 to 12 per cent in 2003-04 (Table 6.6). Other firms, such as Cipla, Lupin, Zydus Cadila and NPIL, have invested more in the 'D' content for enhancing process efficiencies, economies of scale and large product baskets rather than in the pure 'R' part of the R&D system. Nonetheless, R&D is considered as the 'survival kit' by the firms analysed for this chapter.

Besides increasing their R&D investments since the mid-1990s, Indian firms have aggressively pursued multiple approaches to create intellectual property through such means as filing for Indian and international, Abbreviated New Drug Applications (ANDAs) claims, and Drug Masters Files (DMFs). All of the firms studied rated intellectual property creation as an important factor. Firms like Dr. Reddy's and Ranbaxy have made good use of first-to-file (Para III)[8] and Para-IV[9] filings in the ANDA category while other firms like Cipla, Lupin, and Sun have opted for the less demanding DMFs[10] route. The strategy to register their products gives them access to the high-growth international generic market. The dual focus on research and product registration indicates that firms are not focusing solely on NCEs but safeguarding their business by adopting hybrid models including generics research as well.

Table 6.6   R&D investments in selected Indian pharmaceutical firms

Company	R&D investments as % of sales					
	1999–2000	2000–01	2001–02	2002–03	2003–04	2004–05*
Ranbaxy	3.6	4.2	4.0	5.0–5.5	6.3	7.0–8.0
Dr. Reddy's	2.7	3.5	4.0–4.5	6.8	9.9	12.9
Cipla	3.5	3.5	4.0–4.5	4.5–4.8	4.8–5.0	5.0–5.5
NPIL	2.0	2.0	2.0	2.0	4.0	5.0
Sun	4.0	4.0	4.0	7.0	10.0	10.0–12.0
Lupin	1.7	2.0	2.0	3.0	3.0–3.5	8.0

Note: *represents projected investments.
Sources: Compiled by the author from various sources (Company interviews, Annual Reports, Journal Articles and Press releases.

6.6.2   Research collaborations

Evidence from our case studies suggests that collaborative alliances with national laboratories and universities have grown in the last five years. The availability of drug discovery research skills and other resources with public sector CSIR laboratories such as the Central Drug Research Institute (CDRI), the Indian Institute of Chemical Technology (IICT), the National Chemical Laboratory (NCL), and the Centre for Cellular and Molecular Biology (CCMB) are cited as the main reasons for these collaborations. Of the small number of new drugs that were developed indigenously, a lion's share came form Central Drug Research Institute (CDRI), one of the CSIR laboratories (Mani 2006). The market-pull and government-push to commercialize research from these laboratories has added further momentum to industry–Institutes collaborations. Firms are collaborating with academic institutions for lead identification. Ranbaxy is currently working with Anna University, Chennai, on herbal natural products, the National Institute for Pharmaceutical Education & Research, Punjab, on anti-inflammatory leads, and with the National Chemical Laboratory Pune, on anti-infective leads. DRL has research collaborations with the Indian Institute of Chemical Technology (IICT), Hyderabad, the National Institute of Nutrition (NIN), Hyderabad, the Centre for Cellular and Molecular Biology (CCMB), Hyderabad, and Nizam's Institute of Medical Sciences, Hyderabad.

India's Department of Science & Technology (DST) has also introduced funding schemes to encourage links between government and university research laboratories and biotech firms. The DST intends to duplicate the recent success of partnership between Lupin and CSIR laboratories, from which has emerged a drug for psoriasis treatment now in Phase II testing and a tuberculosis drug now in Phase I testing (Mashelkar 2007). Other Indian drug firms, including Nicholas Piramal, Cipla and Dr. Reddy's, have similar collaborations with the CSIR laboratories and universities (Table 6.7).

Table 6.7 Industry collaborations with CSIR labs

CDRI (Lucknow)	IICT (Hyderabad)	CCMB (Hyderabad)	NCL (Pune)
Cipla	Dr. Reddy's	Dr. Reddy's	Lupin
Lupin	Lupin		Cipla
Nicholas Piramal	Sun Pharmaceuticals		Torrent
Ranbaxy	Cipla		Ranbaxy
Wockhardt			Emcure,

Source: Adopted from FICCI 2005.

6.6.3 Marketing alliances, mergers and acquisitions

Acquiring firms with existing innovative product lines or knowledge base is increasingly favoured by Indian leaders. NPIL, Sun and Ranbaxy are pursuing this strategy as discussed earlier. Ranbaxy has concluded 15 acquisitions since 2004, including eight in 2006 (Table 6.8). An important observation is that earlier Indian firms acquired and merged solely to attain critical mass and elevate their market position (size, brands and so on) but now the M&As are increasingly driven by technology strengthening pursuit and market penetration. The shift in the motives behind strategic alliances pursued by the top Indian firms is evident.

Penetrating the international generics market through marketing alliances with MNCs is a common strategy adopted by Indian firms. Nicholas Piramal has entered into a five-year strategic alliance with US-based Biogen Idec to market in India and Nepal the latter's branded drug Avonex, a therapy for multiple sclerosis. Similarly, Lupin has entered into generics markets in US and Europe through strategic marketing alliances.

6.6.4 Technology licensing (licensing in and out)

Ranbaxy and DRL has set the trend of licensing-out molecules to multinationals in order to gain advantages related to speed to market, early launch and cost savings. The strategy of licensing out molecules to MNCs makes perfect business sense for Indian firms as it mitigates the risk involved in NDDR besides addressing skill constraints for taking molecules from lab to the marketplace. Ciprofloxacin to Bayer AG has been a breakthrough success for Ranbaxy. Other firms like Cipla and NPIL have been involved in intensive technology buying and selling worldwide. Cipla has licensing agreements with Canadian generics manufacturer Novopharm, MCPC of Saudi Arabia for formulations, Cipharm in Ivory Coast for formulations, Geneva Pharma in the USA (which is now a part of Novartis), and a host of manufacturers of antibiotics, anti-cancer and other life saving drugs in China. Many Indian companies have seized upon in-licensing of products to address current market needs. Recent licensing deals (Table 6.9) have ranged from products aimed at the Indian domestic market to products meant for global markets.

Table 6.8 Cross-border acquisitions in the Indian pharmaceutical sector

Indian firm	Acquired firm	Year	Purpose of acquisition
Ranbaxy	Ohm Laboratories (USA)	1995	FDA approved manufacturing facility in the US
	Bayer's (Germany) Generic business	2000	Entry into European generics market
	RPG Aventis (France)	2004	Entry into European generics market
	Generic products of Efarmes S.A. (Spain)	2005	Entry into European generic market
	Veratide from Procter & Gamble (Germany)	2005	Expansion into European generics market
	Unbranded generic business of GSK in Spain & Italy	2006	Expansion into European generics market
	Terapia (Romania)	2006	Expansion into European generics market
	Mundogen; a GSK subsidiary in Spain	2006	Expansion into European generics market
	Belgian Company Ethimed NV	2006	Expansion into European generics market
DRL	Benzex Laboratories (India)	1988	To expand Bulk Active Business
	American Remedies Ltd (India)	1999	Expansion into Indian Domestic Market
	BMS Laboratories and Meridian Labs (Germany)	2002	Entry into UK generics market
	Tregenesis (US)	2004	US generics, specialty products, dermatology segment strengthening
	Roche's Generic Business (Mexico)		Expansion into US generics market
	Betapharm (Germany)	2006	Entry into European Generics market
NPIL	Roche Products (I) Ltd (India)	1993	Entry into Indian domestic market
	Boehringer Mannheim (I) Ltd (India)	1996	Expansion into Indian domestic market
	Hoechst Marrion Russel (I) Ltd	1998	R&D
	Rhone-Poulenc (I) Ltd (India)	2000	Expansion into Indian domestic market
	ICI (I) Ltd (India)	2002	Expansion into Indian domestic market
	Rhodia's International Business (UK)	2004	Entry into European generics market
	Avecia Pharma (UK)	2005	Entry into European generics market
	Biosyntech (Canada)	2005	R&D Capability

(Continued.)

Table 6.8 (Continued)

Indian firm	Acquired firm	Year	Purpose of acquisition
Sun	Caraco Pharma (US)	1997	USFDA approved facility for solid oral dosage forms.
	Two facilities of Valeant Pharma (Hungary, US)	2005	APIs, Branded generics
	Able Laboratories (US)	2005	APIs Branded generics

Sources: KPMG 2006 and other publicly available sources.

Table 6.9 Some of the recent in-licensing deals by Indian pharma companies

Indian firm	Foreign firm	Product	Therapeutic fegment
Lupin Lab	ItalFarmaco, Italy	Enoxaparin Sodium	Cardiovascular
Ranbaxy	Ethypharm, France	1 NDDS	Cardiovascular
Ranbaxy	Ethypharm, France	NDDS- Tramadol	Analgesic
Ranbaxy	Eurodrug, Netherlands	Xanthine bronchodilator	Asthma
Dr. Reddy's	Foamix Ltd., Israel	Emollient Foam	Dermatology

Sources: Compiled by Chaturvedi from company interviews, annual reports and news reports.

6.6.5 Subcontracting of research

As firms move up the value chain and compete to launch new drugs, sub-contracting for research, clinical trials, custom synthesis, marketing and sales support is gaining popularity. It has been estimated that the pharmaceutical industry spends $800 million to bring a new molecule to the market (CSDD Report, 2003). Around a third of this total goes towards clinical trials, particularly Phase III trials that use large numbers of human subjects. Overall clinical development costs in India are estimated to be 40–60 per cent lower than those in the West (OPPI, 2003). Western multinationals are vigorously scouting for clinical development services and the painstaking chemical synthesis work for early drug development in India. They are also shopping for promising new treatments that may emerge from India's own drug discovery efforts. Backed by the recent government notification amending Schedule Y, multinationals like Pfizer, Eli Lilly, GlaxoSmithKline and Aventis have started simultaneous and stand-alone clinical trials in various therapeutic segments. Eli Lilly has more than 17 large and small clinical research projects running in 40 hospitals across India, while GSK has started seven simultaneous clinical trials of its vaccines and drugs. Others on the clinical trial trail include Sanofi-Aventis and Roche. To seize this opportunity, many big and small CROs have emerged the last five years. Organizations such as

Wellquest – a subsidiary of Nicholas Piramal, Aurigene – a sister concern of Dr. Reddy's and Clingene International – a group company of Biocon India, are offering contract/clinical research all along the discovery chain.

6.7 Discussion and conclusion

This chapter described how Indian pharmaceutical firms have responded to various patent regimes and other policy changes in recent years. While national policies can reinforce or change the behaviour of local actors with regard to innovation, it is also important to recognize that the institutional set-up at the global level has become a powerful force that shapes the parameters within which local actors make critical decisions with respect to innovation. The Hatch–Waxman Act of the United States enacted in 1984 and signing of the TRIPS agreement in 1995 heralded major changes at the firm level. Micro-level analysis instructs dynamic changes at strategic thinking and at the operational level in Indian firms. More recently, firm-level strategies are resulting in 'new lines of thinking and action' for policy, research and markets. Interestingly, these changes are not restricted to national boundaries alone, but also influence international policies, markets, firms and global industry landscaped. Thus, firms are playing an increasingly important role not only in making the policy work but also in the making of policy itself.

The evidence presented in this chapter amply demonstrates that the competitive environment for Indian firms has intensified due to the globalization of innovation and the introduction of a product patent regime and firms are pursuing some combination of international expansion to their generics business, contract manufacturing, and investment in new drug discovery programmes. Indian firms are following a combination of strategies that make use of their existing capabilities while creating new ones. Within this approach 'cooperate' rather than 'compete' has become the mantra for success. Foreign firms are not opting for aggressive competition with domestic firms or with the established international players in the near future, though some firms in the top league with strong financial muscle and better infrastructure may use a combination of cooperation and competition strategies.

Notes

1. The Ministry of Health and Family Welfare, Government of India had set 31 December 2003 as the deadline for the companies to evolve good manufacturing practices (GMP) and fulfill all the requirements of premises, plant and equipment for pharmaceutical products.
2. Schedule Y of the Drugs and Cosmetics Act 1940 stipulated that permission for clinical trials for new compounds in India be given for one phase behind the development status in the rest of the world i.e. if the compound was going phase III in its source country, permission would be granted for only phase II trials in

India. This phase lag has now been removed and India allows parallel phase clinical trials for new compounds.

3. The Hatch–Waxman Act granted chemical drug manufacturers the option of filing an Abbreviated New Drug Application (ANDA). The rationale for the Hatch–Waxman Act was to enable manufacturers to sell lower cost drugs in the US immediately upon expiration of a pioneer manufacturer's patent.

4. Department of Chemicals and Petrochemicals, http://chemicals.nic.in/npp_circulation_latest.pdf, accessed on 15 October 2006.

5. For detailed discussions on the overall policy framework, see Mani 2006.

6. Eli Lilly had filed more than 70 patents for subsequent process improvements on its original Cefaclor molecule to protect the drug form generic competition.

7. An application for a license to market a generic (or a duplicate) version of a patented drug that has already been granted an approval under a full NDA (that is, the drug has already met the statutory standards for safety and effectiveness).

8. A Para III filing is made when the ANDA applicant does not have any plans to sell the generic drug until the original drug is off patent.

9. A Para IV filling is made when the ANDA applicant believes its product or use of the product does not infringe on the innovator's patent listed in the Orange Book or where the applicant believes such patents are not valid or enforceable. If successful the generic drug company gets an exclusive marketing right (EMR) for 180 days to sell the drug.

10. A drug master file is a document prepared and submitted to the FDA for examination and provides confidential detailed information about facilities, processes, or articles used in the manufacturing, processing, packaging, and storing of one or more human drugs. There is no requirement by law or FDA regulation to present a DMF.

References

Ahmad, H. (1988) *Technological Development in Drug and Pharmaceutical Industry in India*, New Delhi: Navrang Publications.

Bowonder, B. and P.K. Richardson (2000) 'Liberalization and the Growth of Business led R&D: the Case of India', *R&D Management*, 30(4): 279–88.

Chaturvedi, K. and J. Chataway (2006) 'Strategic Integration of Knowledge in Indian Pharmaceutical Firms: Creating Competencies for Innovation', *International Journal of Business Innovation and Research*, 1(1/2): 27–50.

CII (2003) 'Opportunity India: Drugs and Pharmaceuticals', New Delhi, Retrieved on 23 June 2005 from the website:http://www.indopaktrade.com/sectoral_info/India/drugs-cii.pdf.

Company Interviews and Annual Reports 2002–2005.

Cusumano, M.A. and D. Elenkov (1994) 'Linking International Technology Transfer with Strategy and Management: a Literature Commentary', *Research Policy*, 23: 195–215.

Desai, A.V. (1980) 'The Origin and Direction of Industrial R&D in India', *Research Policy*, 9(1): 74–96.

Felker, G. (1997) 'Introduction: The Pharmaceutical Industry in India and Hungary', World Bank Technical Paper No. 392.

FICCI (2005) 'Competitiveness of the Indian Pharmaceutical Industry in the New Product Patent Regime', report for National Manufacturing Competitiveness Council, New Delhi, India: FICCI.

Financial Times (2006) 'Dr. Reddy's Charts New Path for Indian Drugmakers', 19 December.

Ganguli, P. (2003) 'The Pharmaceutical Industry in India: a Report', Business briefing PHARMATECH 2003. Retrieved on 15 June 2005 from the website: http://www.bbriefings/pdf/17/pt031_r_15_ganguli.pdf.

Indian Drug Statistics (1984–85). Ministry of Chemicals and Fertilisers, Government of India, New Delhi.

IBEF (2003) *India Fastest Growing Economy: Pharmaceuticals*, India Brand Equity Foundation, CII Gurgaon, India.

KPMG (2006) *The Indian Pharmaceutical Industry: Collaboration for Growth*, Mumbai: KPMG Consulting Private Limited.

Kumar, N. and J.P. Pradhan (2003) *Economic Reforms, WTO and Indian Drugs and Pharmaceutical Industry: Implications of Emerging Trends*. CMDR Monograph Series No. 42, The Centre for Multidisciplinary Development Research, India.

Lall, S. (1987) *Learning to Industrialize: The Acquisition of Technological Capability by India*, London: Macmillan Press.

Leonard-Barton, D. (1995) *Wellsprings of Knowledge: Building and Sustaining the Sources of Innovation*, Boston, MA: Harvard Business School Press.

Malerba, F. and L. Orsenigo (2001) 'Innovation and Market Structure in The Dynamics of the Pharmaceutical Industry and Biotechnology: Towards a History Friendly Model'. Paper presented at the DRUID Nelson and Winter Conference, Aalborg, 12–15 June.

Mani, S. (2006) *The Sectoral System of Innovation of the Indian Pharmaceutical Industry*, Working Paper No. 382, Kerala, India: Centre for Development Studies.

Mashelkar, R.A. (2007) 'New Geographies of Innovation', The Atlas of Ideas: Mapping the New Geography of Science Conference, London, 17 January.

Nauriyal, D.K. (2006) 'TRIPs-Compliant New Patents Act and Indian Pharmaceutical Sector: Directions in Strategy and R&D', *Indian Journal of Economics & Business*, Special Issue on China and India.

Nelson, R. and Winter, S. (1982) *An Evolutionary Theory of Economic Change*, Cambridge, MA: Harvard University Press.

OPPI (2001) 'OPPI Pharmaceutical Compendium', Mumbai, India: OPPI and Monitor Company Group L.P.

OPPI (2003) 'Outsourcing Opportunities in Indian Pharmaceutical Industry', Mumbai, India: OPPI and Monitor Company Group L.P.

Penrose, E.T. (1959) *The Theory of the Growth of the Firm*. Oxford: Basil Blackwell.

Pharmabiz.com, 'Schedule Y amendment to help India tap US$ 1 billion by 2010: Experts', http://www.pharmabiz.com/article/detnews.asp?articleid=29467&sectionid=50, retrieved December 2006.

Piramal, S. (2007) 'Next People, Next Places, Next Science', The Atlas of Ideas: Mapping the New Geography of Science Conference, London, 17 January.

Porter, M.E. (1991) 'Toward a Dynamic Theory of Strategy', *Strategic Management Journal*, 12: 95–117.

Rosenberg, N. (1969) 'The Direction of Technical Change: Inducement Mechanisms and focusing Devices', *Economic Development and Cultural Change*, 18(6): 1–24.

Rumelt, R.P. (1984) 'Towards a Strategic Theory of the Firm', in R.B. Lamb (ed.), *Competitive Strategic Management*, Englewood Cliffs, NJ: Prentice-Hall.

Srinivas, S. (2004) 'Technological Learning and the Evolution of the Indian Pharmaceutical and Biopharmaceutical Sectors', PhD Thesis, Cambridge, MA.

Teece, D. (1998) 'Capturing Value from Knowledge Assets: the New Economy, Markets for Know-how, and Intangible Assets', *California Management Review*, 40(3): 55–79.

Teece, D.J., G. Pisano and A. Sheun (1997) 'Dynamic Capabilities and Strategic Management', *Strategic Management Journal*, 18(7): 509–33.

Tufts Centre for the Study of Drug Development (CSDD) Report (2003) 'The Price of Innovation: New Estimates of Drug Development Cost', *The Journal of Health Economics* 22: 151–85.

7
Knowledge Exchange with Offshore R&D Units: Novo Nordisk, GN Resound and BenQ Siemens Mobile in China

Julie Marie Kjersem and Peter Gammeltoft

7.1 Introduction

It is widely acknowledged that China has evolved into a global manufacturing powerhouse. GDP growth rates close to 10 per cent a year have been sustained through a virtuous combination of high domestic savings and investment rates, macroeconomic stability, and domestic structural and regulatory reforms on the one hand, and mounting foreign direct investment (FDI) inflows and export volumes on the other. Somewhat less advertised but no less significant is the fact that even though China remains a low-income country in the aggregate, a range of high-tech activities and industries are developing quickly and prospering. These activities are not only targeted and supported by government policies and strategies but they also increasingly attract foreign investment, technology, and R&D activities.

In this chapter we will analyse the emerging phenomenon of foreign investments in R&D in China, approaching it both from the supply side of the constitution of the Chinese system of science, technology and innovation (ST&I), and also from the demand side of foreign firms investing in high-tech activities in China. Seen from the perspective of firms, the offshoring of knowledge-intensive activities to China is not without challenges and, more often than not, they will encounter unforeseen difficulties in the exchange and protection of knowledge between geographically- and organizationally-dispersed R&D units.

Thus, we approach high-tech investments in China from two angles. First, in connection with the Chinese ST&I system, we will account for its development and constitution. Secondly, we will account for foreign investments in R&D into China and the firms' motives for undertaking them. A particular challenge and concern of companies locating knowledge-intensive activities abroad is to attain effective and efficient exchange of knowledge with other units within the corporate network without risking the leakage of proprietary knowledge assets. These issues are discussed through a case analysis of the Chinese R&D labs of Novo Nordisk, GN Resound and BenQ Siemens Mobile.

7.2 The Chinese system of innovation

Contemporary Chinese policies and strategies reflect a strong determination to become a world-leading nation in science, technology, and industrial innovation. Furthermore, current government policies, strategies, and plans express a sentiment that economic development over the last couple of decades may have come to rely too much on foreign technology and that a shift towards stronger 'indigenous innovation' capability is needed. In the following, a brief review of some conventional innovation input and output indicators confirm that significant progress has indeed been made towards this goal.

7.2.1 Chinese innovation system reforms

Innovation is generally recognized as taking place within broader 'innovation systems' – that is, within complex networks which span firms, universities, and government organizations. These complex networks form the national innovation system and are influenced by both market and non-market institutions that impact the direction and speed of innovation and technology diffusion (Lundvall, 1992). The quantitative strength and qualitative features of a country's innovation system has a defining influence not only on domestic growth and innovation processes, but also on its ability to attract foreign investments in high-tech and knowledge-intensive activities, and the Chinese system of innovation increasingly acts as a pull factor for high-tech investments into China.

As we will see, cross-organizational linkages and networks is an area where particular progress has recently been made in China, especially in terms of forward and backward linkages and feedback mechanisms between users and producers of knowledge. The 1990s in particular saw the increasing interlinkage of the Chinese national innovation system. Chinese ST&I policy making shifted from merely elaborating R&D policies to focusing on a more modern approach to ST&I development by integrating more tightly government ST&I efforts with R&D efforts of foreign and national enterprises. Extensive science and technology system reforms and policies were made with a focus on economic development and technological advancement. Interactions between firms, universities, and research institutes became more frequent. Firms began to play a more active role in technological innovation and became the nucleus of the innovative system (Haiyan and Yuan, 2006).

In relation to foreign R&D investments, a number of policies have been put in place in order to encourage the inflow into China. These are focused on support structures such as science parks and incubators. By 2002, over 400 business incubators and 53 high development zones were established at the national level through governmental support (Huang et al., 2004). Thus, one important means by which to attract high-technology FDI has been the establishment of high-tech science parks combined with incentives such as free rent, low tenancy cost, favourable lease terms, and tax relief (Gassmann and Han, 2004). The Zhongguancun Science Park in Beijing, which is the base for 40 universities and 130 research institutes, is one of the more well-established scientific zones (UNCTAD, 2005: 142). Given the attractiveness of the Chinese market, the Chinese government is also pursuing a 'technology for market' policy, whereby they encourage foreign investors to transfer technology to China (Gassmann and Han, 2004). In other words, while the Chinese government is supporting the national ST&I development through various domestic initiatives, reforms, and policies it is also supporting it by encouraging inflow of foreign high-tech investments by incentives, inducements, and the establishment of attractive high-tech science parks.

In the following sections we will take a closer look at a set of conventional innovation indicators, viz. the educational system and human resources, R&D expenditure, high-tech exports, and patenting, and discuss the intellectual property regime.

7.2.2   Innovation input indicators

One of the most important inputs to innovation is human resources. The educational system and human resources are significant components of the Chinese innovation system. The current 11th Five Year Plan reflects this and it is an area which receives major and increasing attention from the Chinese government. Great strides have already been made in primary education and profound emphasis is placed on developing the higher educational system. One crude indication is the increase in number of universities and institutions of higher education: there were 1,552 universities and institutions of higher education in 2003, up from 1,396 in 2002 (EIU, 2005: 25).

Where the total number of researchers in the country is concerned, relative to the total workforce China lags behind other countries, but China is so populous that it is still home to one of the largest pool of researchers in the world. In 2004, China had the second highest absolute number of researchers in the world with 918,000, ranking behind the United States but ahead of Japan (UNESCO Institute for Statistics).

The number of students in China is also increasing rapidly and, again, due to the sheer size of the population, these numbers represent large talent pools. Some 15 million students were enrolled in tertiary education in China in 2002/2003, which is comparable to the USA and to the EU, and China produced 885,000 university graduates in 2002. Moreover, China has the highest number of students in the world enrolled in science and technology education, numbering approximately 2.6 million (UNCTAD, 2005: 296). Even though the number and proportion of students with research training expectedly remains significantly lower than in the USA and the EU, almost 15,000 (1.7 per cent) of the university graduates in 2002 were awarded a PhD degree, a number that rose to 19,000 in 2003 (OECD, 2005: 25). While the Chinese educational system suffers from a number of recognised weaknesses improvements are continuously being made.

China has a large number of students studying overseas: 152,000 Chinese students were enrolled abroad in OECD countries in 2002, accounting for 10 per cent of the total number of foreigners enrolled in university education in OECD countries (OECD, 2005: 25). The Chinese government is actively trying to attract these students back and educated Chinese abroad are increasingly returning to the mainland as an important supplement to domestic human resources.

Another commonly-deployed innovation input indicator is expenditure on R&D. China's gross domestic expenditure on R&D grew by more than a factor of five over the period from 1995 to 2004 to reach 1.23 per cent of GDP (OECD, 2006). R&D expenditure can be split into source of funds and sector of performance. In low-income countries, business expenditure on R&D is usually relatively low and government carries the bulk of both R&D funding and implementation. The need to economize on scarce resources to achieve scale and scope economies and the need for coordination across different sectors and activities also tends to imply a larger involvement of government. Nevertheless the majority of R&D activities in China today are both funded and performed by business rather than government. This is to a large part explained by reforms of the public S&T system around the turn of the millennium during which previously government-run institutes re-registered their business type. More than 300 institutes were spun into an enterprise, more than 600 became profitable firms by themselves, and a few integrated with universities (Gu and Lundvall, 2006). As a consequence, the percentage of R&D performed by business leaped from 49.6 per cent in 1999 to 60.3 per cent in 2000. In 2004 it reached 67 per cent, compared to 70 per cent in USA (OECD, 2006).

Considerable media attention was generated when OECD estimates indicated that China became the world's second largest aggregate R&D spender in 2006, measured in purchasing power parities, thus overtaking Japan. For 2006, R&D intensity was estimated at 1.3 per cent of GDP and spending is projected to increase further to 2.5 per cent of GDP by 2020 in the national 2006–2020 Medium- and Long-Term Program for Science and Technology Development. However, if we look beneath the aggregate numbers, R&D activities are still focused primarily on applied research and development rather than basic research. Compared to the USA and Japan, China still lags behind when it comes to investments in more advanced research. In 2002 a modest 5.7 per cent of R&D expenditures were on basic research, 19.2 per cent on applied research, and 75.1 per cent on experimental development. In comparison, the numbers in the USA were 18.1 per cent on basic research, 20.8 per cent on applied research, and 61.1 per cent on experimental development (OECD, 2004).

7.2.3 Innovation output indicators

Turning now from innovation input to innovation output indicators, the boom in China's manufactured exports has been driven predominantly by the low cost structure. Nonetheless, the increase in foreign firms located in China has also ensured a continual upgrading of the goods exported by China. Today, most of the world's advanced consumer electronics are manufactured in China. In addition, China is increasingly becoming competitive in a number of important high-tech industries (Walsh, 2003), 'high-tech industry' being defined as an industry with high R&D intensity, such as aerospace, computers, pharmaceuticals, scientific instruments, and electrical machinery. One of the indicators is the growing trade in high-tech goods, which made up almost 30 per cent of the manufactured exports from China in 2004 – as compared to just 6 per cent in 1992. The increase in high-tech exports can, to some extent, be viewed as an indicator for the advancement of the industries in China through science, technology, and innovation improvements. This data should be interpreted with caution, though, since FDI in high-tech industries in developing countries is mostly confined to lower-tech and labour-intensive activities within the value chain and very dependent on imports of intermediate goods, with modest local value-added. Furthermore, in China most of these high-tech exports (as much as 85 per cent by some estimates) originate from foreign-invested enterprises highly concentrated in specific geographical locations.

The extent of patenting is often used as an output indicator of innovative activities and over the past decade patenting activity in China has increased significantly, testifying to a strengthening ST&I system. In 2004, 354,000 national patent applications were submitted, up from 78,000 in 1994 (Ministry of Science and Technology of the Peoples Republic of China, 2005). Different factors mitigate the record, however; strong formal protection of intellectual property rights is a relatively recent phenomenon in China and thus increases in patenting activity does not exclusively reflect increases in innovative activities but also the transition and tightening of the property rights regime itself. Furthermore, only a minority of applications submitted by domestic companies are for the more advanced 'invention' patents while the large majority are for the simpler 'utility model' or 'design' patents. Applications submitted by foreign companies in China on the other hand were almost exclusively for invention patents in 2004. More generally, a large number of patent applications and grants in China are by foreign rather than domestic companies (Walsh, 2003: 67) and between 1999 and 2001 almost half of all domestic inventions were foreign owned.

In terms of international patenting, China's share in patenting at the US Patent and Trademark Office and the European Patent Office is still very small

(Schaaper, 2004), but there appears to be a greater propensity to register international patents through the World Intellectual Property Organization where China became the tenth largest applicant in 2005 with 1.8 per cent of the applications.

China is widely criticized for not sufficiently complying with and enforcing intellectual property laws. Formal legislation is adequate by most accounts while implementation and enforcement of the laws is lacking. Administrative or litigative pursuit of infringements remains a highly complicated and uncertain process. Whether a tight WTO-like IP regime is conducive to economic catch-up remains contested, yet awareness is increasing in China that a well-functioning and predictable intellectual property regime is closely connected to accommodation of inward foreign direct investment and smooth integration into the world economy.

7.3 The offshoring of R&D to China

As we saw in the previous section, the Chinese innovation system is advancing rapidly, in both quantitative and qualitative terms, and today China attracts not only resource-seeking and labour-intensive investments, but increasingly also investments in high-tech activities. In the remainder of this chapter we will focus on the 'demand side' of foreign firms investing in high-tech in China. First, we will give a brief account of the general motives of companies to offshore R&D to China, then turn to the analysis of three foreign companies with R&D labs in China.

The motives behind high-tech investments and offshore R&D units can be viewed as more complex in comparison to other forms of FDI (that is, the relocation of production facilities to low-cost countries). Even though motives are many and varied they can be generalized into the generic motives outlined in Table 7.1. They reflect a combination of push and pull factors, resulting from both internal changes and motivations in the firm but also external drivers in the increasingly global arena for conducting business (Gammeltoft, 2006).

This leads us to the analysis of three companies who have already begun to offshore some of their R&D activities to China.

7.3.1 The cases of Novo Nordisk, GN ReSound and BenQ Siemens Mobile

The following case analysis of three companies with R&D activities in China, Novo Nordisk, GN ReSound and BenQ Siemens Mobile, is based on information collected from the companies' R&D units in both China and Denmark. Accordingly, it provides a fuller and more balanced perspective than most prior analyses of companies' international R&D. Explorative open-ended face-to-face interviews with 38 respondents were conducted in the R&D units in both Denmark and China. All of the interviews were conducted in 2005

Table 7.1 Motives for internationalizing R&D

Market-driven	Exploit existing company-specific assets more widely; motivated by market size and proximity; support local sales, closeness to lead customer, improve responsiveness in terms of both speed and relevance
Production-driven	Supporting local manufacturing operations
Technology driven (pull)	Tapping into foreign S&T resources, technology monitoring (especially competitor analysis), acquire/monitor local expertise, knowledge and technologies
Innovation-driven (push)	Generating new company-specific assets; attaining a faster and more varied flow of new ideas, products and processes; capitalise on location-specific advantages through an international division of labour between R&D labs
Cost-driven	Exploiting factor cost differentials
Policy-driven	National regulatory requirements or incentives, tax differentials, monitoring and exploitation of regulations and technical standards

Source: Gammeltoft, 2006, p. 186.

with interviewees employed that year, thus titles and positions may have changed subsequently (Kjersem, 2006).

For each of the three companies we describe and account for the R&D units and their historical evolution. Moreover, we discuss their motives for offshoring R&D to China and the strategic purposes with the R&D units.

7.3.1.1 Novo Nordisk

Novo Nordisk is a focused healthcare company. It is headquartered in Denmark, employs approximately 22,000 full-time employees in 79 countries, and 99 per cent of its sales are outside the Danish market. Novo Nordisk focuses on four core areas – diabetes care, haemostasis management, growth hormone therapy, and hormone replacement therapy – and is the global market leader within diabetes care with an insulin volume market share of 50 per cent (Novo Nordisk website).

Globally, more than 3,000 employees are working on R&D activities in Novo Nordisk, but the R&D activities are mainly located in Denmark. However, the company is commencing on an internationalization strategy with regards to R&D (Share, 2005). They established an R&D unit in China focused on protein expression and purification in bacteria and they have commenced set up of an R&D unit in the USA, focusing on haemostasis management (the stopping of bleeding). The two R&D centres are focused on very different areas, and are not intended to collaborate directly.

Novo Nordisk established its Chinese R&D centre in Beijing in January 2002 and it was the first R&D centre established in China by an international bio-pharmaceutical company with the focus on biotech. The R&D centre moved to greater facilities in the Beijing Zhongguancun Life Science Park in July 2004, which allows the centre to grow from 25 employees up until about 60 employees in 2008. The R&D centre is an integrated part of the Microbiology Department at Novo Nordisk's Globe Discovery organization and they take part in various drug discovery projects with teams in Denmark. The Chinese R&D unit takes part in the initial stages of a drug discovery process, before it is developed and tested.

The main motive for Novo Nordisk to offshore R&D to China can be considered to be policy-driven. The Chinese market is highly attractive for Novo Nordisk, as it is located on Novo Nordisk's top 10 list of sales turnover. In this setting, the policy-driven motive has become particularly important. Mr Ron Christie, managing director of Novo Nordisk in China, expresses it thus: 'As for the government – it doesn't want foreign companies to use China as a low cost source of labour. If you want to make money here you are expected to give something back' (People, 2004: 6).

However, in effect, Novo Nordisk has chosen a strategy that is based on the technology-driven and innovation-driven motives regarding the R&D unit in China and created an advanced objective for the R&D unit: 'Basically the Chinese government wished for us to set up and R&D unit in China [. . .]. So my assignment became to set up an R&D unit in China, but we have to get the best out of them and make sure that they seriously contribute to our research' (interview with the Vice President of Microbiology). The company counts on the R&D unit being able to reversibly contribute innovative ideas back to the home-base R&D site in the near future.

The R&D centre has evolved into a centre of excellence, that is, a uniquely specialized centre within Novo Nordisk, in research in protein expression and purification in the bacteria E. coli. The projects they are involved with centre on a variety of different therapeutic areas such as growth hormone, haemophilia (bleeding disorder), and cancer. It is called R&D, but they do not develop drugs in China. They only conduct research, predominantly applied research, while the development phase is located in Denmark. Completing a project in the discovery phase requires a team, with input and expertise from many different fields and the Chinese R&D unit works closely with the Novo Nordisk discovery research projects. Thus far, the projects they work on have been initiated in Denmark, but the assignments that they receive must be solved independently. The R&D unit is still in its infant stage; however the long-term strategy for the unit is to increasingly contribute to the firm's knowledge base.

7.3.1.2 GN ReSound

GN ReSound is a hearing healthcare company, which develops and produces hearing instruments. With headquarters in Denmark, GN ReSound employs approximately 3,800 employees and has production facilities in five countries, as well as subsidiaries in 20 countries and distributors in 80 countries. At present, the GN ReSound hearing aid division encompasses approximately 200 development engineers.

GN ReSound defines itself as the leading manufacturer of advanced-technology hearing instruments. The company offers a full range of hearing instruments, including software-based digital instruments and digitally programmable and traditional analogue products in all sizes and models. Hearing instrument manufacturers are competing in four different price segments: (1) top; (2) plus; (3) basic; and (4) budget.

Historically, GN ReSound has come about through a series of mergers and acquisitions and consequently they have obtained a patchwork of R&D units around the world. Today they have units in Copenhagen and Præstø in Denmark, Chicago in the USA, Xiamen in China and a small R&D satellite in Eindhoven in the Netherlands. The R&D functions in GN ReSound are divided into six main areas: digital sound processing technology, audiology, algorithm software, fitting software, product development, and product development services. In China, they focus only on development. The satellite centre in Eindhoven is part of the algorithm group. In Chicago they are specialized in development, audiology, and fitting software, whereas they are specialised in all six areas in Denmark.

GN ReSound's R&D unit is located in Xiamen, in the south of China. When the company acquired a manufacturing plant in 1986 in Xiamen, a small R&D unit followed. GN ReSound chose to keep and expand the plant. Initially, the R&D unit functioned as a support and problem-solving unit for production. However, within the past few years it has increasingly gained strategic importance. Now the main tasks are to support the manufacturing and to develop low- and middle-end products. About 30 per cent of their work revolves around assisting manufacturing and the rest centres on product development (interview with the R&D manager in China). In 2002, the Chinese R&D unit became part of the global R&D organization and at present the unit includes 16 employees, with plans to hire up until 30 new employees in 2006.

The main motives for GN ReSound to offshore R&D to China have been cost-driven and production-driven. The Senior Vice President of Research and Core Technology states that: 'First of all we chose China, generally because of cost orientations. It is cheap to develop and it is cheap to produce in China. A great amount of our product production is in China, which also makes it convenient to have a development team nearby the production site.' Furthermore, the market-driven motive is also becoming relevant; the Asian markets are increasingly important for GN ReSound. In addition, the technology-driven motive is depicted in the necessity for GN ReSound to tap into a supplementary pool of human resources (interview with the Vice President of Development).

The Chinese R&D concentrates mainly on developing products for the plus, basic, and budget segments in the markets, whereas products in the top segment are developed in Denmark and the USA. Thus, GN ReSound seeks to advance the use of their existing technological competencies in the lower-cost location of China. Basically, the Chinese unit is re-branding what might have been a top product four years ago, and then they make a derived product for a lower price segment. In this way the unit is to a great extent re-utilizing the existing knowledge from the home-base R&D site. The Chinese R&D unit also has its own models to work on so they do not work on cross-border projects with the Danish unit.

7.3.1.3 *BenQ Siemens Mobile*

In 1985 Siemens became one of the first companies in the world to develop mobile phones. However, the mobile phones division did not prove particularly profitable in 2004. The division sold about 51.1 million handsets during the year, but the mobile phones division posted a loss of 152 million euros on sales of 4,979 billion euros (Siemens Annual Report, 2004: 60). Consequently, Siemens chose to pay the Taiwanese company BenQ, a company spun off from Acer Group in 2001, to take over their mobile phone division. In order to seal the deal, Siemens management agreed to pay BenQ $300 million and to buy $60 millions worth of stock in BenQ (Time Magazine, 2005). BenQ has officially taken over the division in October 2005 and it has taken over all of Siemens Mobile Phones' production sites and R&D facilities. In 2005 BenQ Siemens Mobile was estimated to have a sixth place in the global market with a market share of 5.2 per cent (InfoWorld, 2005).

BenQ Siemens Mobile has its global headquarters in Munich, Germany, and it serves more than 70 markets around the globe. With the takeover of the Siemens division, BenQ Siemens Mobile employs over 7,000 employees worldwide, among which about 2,500 work in R&D units. Worldwide, BenQ Siemens Mobile encompasses eight main development centres for mobile phones, which were acquired from Siemens. These include R&D units in Aalborg (Denmark), Beijing, Shanghai (China), Kamplintfort, Berlin, Munich, Ulm (Germany), and Manaus (Brazil). For all intents and purposes, then it is the former Siemens R&D network which we are analysing. Moreover, we are particularly focussing on the two R&D units in Denmark and China.

For this analysis it has been chosen to mainly focus on the Java technology R&D units. We chose this specific department because they focus on the most advanced kind of mobile phone development. Therefore, the research is mainly restricted to interviews with key employees in the Java unit in China and in Denmark. The Danish R&D unit is responsible for the advanced Java technology development for the mobile phones, where they collaborate with the Chinese R&D unit. The Chinese R&D unit provides human resources and support for the Java projects that are carried out in Denmark.

The mobile phones R&D unit in Beijing was established in 2000 and it had about 1,000 employees in 2006. The location of the R&D unit is in the north of Beijing. The site was chosen because of its proximity to the university and other mobile communication companies, which was thought to make it easier to attract qualified human resources. In addition obviously, it is located in the capital of China, close to a large and advanced consumer base. This unit develops products for the Chinese market and it works on present generation mobile phones (whereas the R&D unit in Denmark works on next generation mobile phones). In addition, they provide engineers for cross-border projects. The main motives for establishing an R&D unit in China is market-driven, as the market entails a huge potential for mobile phones companies; the Chinese market is the single largest market for mobile phone subscribers in the world and it is still far from being saturated. Henceforth, it also becomes relevant to customise the products: 'China is a huge market for us and it is not going to be very successful to have German engineers sitting in Germany, trying to figure out what Chinese people want [...] I don't think it would be possible to get the particular Chinese flavour in Germany', explains the Line Manager for the Java Team in China. Moreover, the cost- and technology-driven motives are also present, as the lower cost and the high recruitment potential also have motivated the establishment of an R&D unit in China.

The strategy of the Chinese R&D unit is to take over the GSM and GPRS platforms and software. This means that the existing platform is being transferred to Beijing. Consequently, the focus in Europe is on the development of a new platform where they will focus on the next 3G and 4G generation mobile phones, which incorporate multimedia features. In Europe, they also develop the high-end products, whereas they in China focus on the lower-end segment.

7.3.2   Cross-company analysis: managing offshore R&D in China

In this section we focus on the management of offshore R&D in China through the analysis of the Danish and Chinese R&D units of Novo Nordisk, GN ReSound, and BenQ Mobile. In alignment with several other scholars, we found the managerial aspects of R&D offshoring to China to be a neglected area of research (von Zedtwitz & Gassmann, 2002; von Zedtwitz, 2004; Gassmann & Han, 2004). To be able to reap the benefits of offshoring R&D to China, knowledge has to be successfully exchanged across borders. Consequently, we will focus on the exchange of knowledge between the geographically dispersed R&D units in Denmark and China.

In order to analyze the exchange of knowledge between the geographically distant R&D units, we have chosen to focus on five dimensions of the knowledge exchange process (see Figure 7.1): the characteristics of the knowledge being exchanged, the Danish unit's motivational disposition towards knowledge sharing, the motivational disposition of the Chinese unit towards

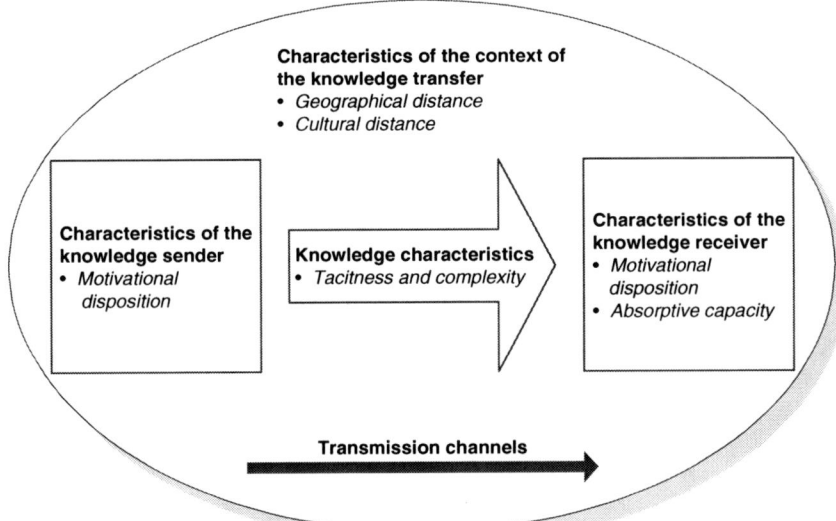

Figure 7.1 Framework for determinants of knowledge exchange
Source: Created by authors based on Szulanski (1996), Simonin (1999) and Gupta & Govindarajan (2000).

knowledge reception, the transmissions channels for exchanging knowledge, and the protection mechanisms associated with knowledge exchange. Finally, we will discuss how the three companies perceived a small set of general features of the Chinese innovation system, which emerged in the interviews as important influences on the knowledge exchange process.

These five dimensions were identified as appropriate to explore in connection with the knowledge exchange process. Since all of the case companies' R&D units in China still can be characterized as young and developing, a large number of the initial knowledge flows have been directed from Denmark to China, as the Chinese R&D units needed to be developed and upgraded. Obviously, this chapter only covers a certain area and does not embrace the many other determinants in the knowledge transfer process, which may very well exist. Moreover, the focus is on the interactive exchange of knowledge rather than on one single delimited transfer.

7.3.2.1 Knowledge characteristics and type of R&D

The knowledge characteristics are important determinants for the transfer of knowledge, as they influence the ease with which it can be exchanged across borders. The difference between tacit and explicit knowledge hinges on whether or not the knowledge can be codified and transmitted in a documented format. When the knowledge to be transferred is codified its transfer is also simpler and less costly. Conversely, the more tacit the knowledge, the more difficult it is to transfer it (Zander & Kogut, 1995; Simonin, 1999; Cummings & Teng, 2003).

The tacit and codified knowledge characteristics are present to different extents in the case companies. In the R&D units of both GN ReSound and BenQ Siemens Mobile, they work with well-known technology platforms. Moreover, they have an ISO 9000 certification for their development processes. That means that all their work is documented and they follow clearly described and standardized work processes. Within GN ReSound, this involves specific information about the procedures, the designs, the measurements, the drawings, and so on. Within BenQ Siemens Mobile, the documentation entails writing down all the software code. They have a test department that tests all the documents and software code. Consequently, the work tasks they carry out in both of the R&D units are often very well defined, formalised and to a great extent, documented.

As outlined in the previous section, the Novo Nordisk R&D unit engages in applied research where they are advancing the discovery of new and improved research processes and therapeutic drugs. The work tasks in the R&D unit can best be described as novel and complex. Moreover, there is a great deal of uncertainty connected to the research process. Out of approximately 45 projects in the discovery organization, only three or four projects will be developed into finished products. Thus, the R&D unit does not engage in extensive documentation: 'We do not have an ISO certificate in the discovery organisation, since we have to operate freely and with so many projects that an ISO certification would kill many forms of creativity and innovation, because the level of documentation for what you are working on will increase rapidly' (interview with Danish scientist). Henceforth, a large part of the knowledge, which is created in the R&D unit in Novo Nordisk is not codified or documented.

7.3.2.2   Motivation of the Danish R&D units to share knowledge

It is noted by many scholars that the behaviour of the knowledge sender with regards to knowledge sharing is key in the knowledge transfer process. The main focus is on the sender's motivation for transferring knowledge (Szulanski, 1996; Simonin, 1999; Gupta and Govindarajan, 2000). Thus, in this section we focus on the challenges related to the knowledge-sharing propensity of the Danish R&D unit. During the case study a range of obstacles were found. These have been divided into what we identify as: (i) the China threat; (ii) the lack of priority of the Danish colleagues; and (iii) the diverse levels of skills. On a positive note, however, the Chinese R&D units also bring additional human resources to aid the Danish R&D activities.

The China threat constitutes a challenge for the knowledge-sharing propensities of the Danish R&D unit. The choice to utilize human resources at a low cost in China instead of in Denmark can be perceived as a threat to their career prospects by the Danish employees and the conception of the 'China threat' has been an issue in all of the companies under examination. The Danish employees in all three of the Danish R&D units agree that in the beginning they did, to some extent, perceive the Chinese unit as a threat, but that this was no longer the case. One of the Danish project leaders from GN ReSound contends that: 'When we started it was difficult to help them, if you were cutting the branch from which you are yourself sitting. But people don't feel like that today.'

Lack of priority of the Danish colleagues is another challenge. From the perspective of the Chinese R&D units, it is emphasized that the willingness to share knowledge relies on the individual person in Denmark. Some are better and more willing than others to share their knowledge. In both of the R&D units it is implied that there may be reluctance in Denmark to spend time on knowledge sharing, which is probably because the Chinese unit is not given a high priority: 'They have so many projects, so we never know when they will finish ours, so we have to wait [...] They don't give our projects such a high priority' one of the Chinese engineers at GN ReSound concludes.

We also found that the *diverse skill levels* create problems for knowledge sharing. It was emphasized that since the colleagues in Denmark often have more experience and knowledge in the area, they find it burdensome to work with colleagues in China that are from a younger R&D unit, and thus they have less experience. One of the Danish scientists from Novo Nordisk states that: 'In the beginning it was annoying that they were not more creative and that they could not come up with better solutions.'

However, the Chinese R&D units also bring *additional resources* with which to conduct R&D activities. In GN ReSound the Chinese R&D unit works on development aspects that are not of great interest for Danish engineers: 'We focus on the top-end products and the Chinese focus on the more standardised products. It is very important for the company that we develop these products, but there are many engineers in Denmark that find that they are not as exciting as the new products [...] The Chinese engineers are extremely good and motivated to make these type of assignments' (interview with the Vice President of Development in GN ReSound). Thus, there is a greater incentive to share knowledge when there is a clear need for the Chinese R&D resources.

7.3.2.3 Motivation of the Chinese R&D units to receive knowledge

The characteristics of the knowledge receiver of the target unit for the knowledge transfer have been found to be of significant relevance. Primarily, the motivation and the ability to absorb new knowledge (absorptive capacity) of the receiver have been the most emphasized impediments to knowledge transfer (Szulanski, 1996; Gupta and Govindarajan, 2000; Cohen and Levinthal, 1990).

The *motivation to receive knowledge* is important in order to get the R&D units up to speed, increase the knowledge base, and accumulate experience. As one of the positive points for conducting R&D in China, all the Danish R&D units described their Chinese colleagues as extremely motivated and eager to learn: 'They are very keen to learn and develop so they can move up the career ladder. Therefore I also think that they are very willing to accept knowledge from other sources or from abroad' (interview with line manager from BenQ Siemens Mobile). Moreover, it is also emphasized that the Chinese employees work very hard and long hours: 'There is not so much "coffee and cake" in China as in Denmark; they really work hard' as one of the Danish scientists at Novo Nordisk stressed.

While the motivation to receive knowledge is high, the *absorptive capacity* of the Chinese R&D unit can constitute a challenge for the exchange of knowledge, since the Chinese employees are lacking the same extent of experience and education as their well-established Danish counterparts. In GN ReSound and BenQ Siemens Mobile it is noted that the Chinese R&D units sometimes take on assignments that they do not yet have the adequate capabilities and experience to carry out. One of the Danish project leaders from BenQ Siemens Mobile states: 'The only thing I actually experience is that they sometimes are a little bit too optimistic. They readily take some assignments that they cannot solve, and then we find out that they could not solve them [. . .] and then we have to start all over again.' Particularly in the case of GN ReSound, this inequality was perceived to be a challenge: 'If you look at it through the company glass then it is much cheaper to have R&D in China. If you look at the competence situation, then we all would say it is too much hassle. That it will be a hassle until we reach the same level. It requires a huge effort to train the unit' (interview with Danish project leader in GN Resound).

7.3.2.4 *Transmission channels for exchanging knowledge*

With the management of innovation around the globe, a firm must be able to coordinate activities and link them in an efficient manner, in order to fully leverage the potential of the offshore R&D unit. Persuad et al. (2002) argue that key challenges within global R&D is about finding the best way to coordinate corporate R&D activities in order to accelerate the pace of innovation. Thus, with significant geographical distance between the R&D units, extensive and efficient coordination mechanisms – particularly within headquarters – become key to making R&D activities flow and function optimally. The existence and richness of the transmission channels for exchanging knowledge are analysed in terms of formal and informal mechanisms in the case companies.

The *formal transmission channels* centre on formalized structures and institutions. GN ReSound and BenQ Siemens Mobile use advanced computer tools and databases to exchange knowledge between Denmark and China. They utilize resource-planning tools, which help plan the projects across borders from the beginning to the end: the activities, the schedule, and the time. Furthermore, the documentation for the work processes are uploaded on shared databases between the R&D sites and the shared intranet is also utilized for information sharing. At Novo Nordisk, they cannot plan their research in such a stringent manner. It is difficult to evaluate and measure the work. However, the management team makes an overall plan each year with regards to what goals they should reach within the different activities and projects. In all of the case companies, employees hold web meetings or phone meetings with their Danish colleagues once a week to update and keep track on the status of the projects. In addition to emails, they also utilize instant messaging. Thus, the day-to-day exchange of knowledge also takes place through the use of collaborative communication tools: i.e. the internet, databases, web/telephone conferences.

The *informal transmission channels* are not bound by formal contractual agreements and institutions, but encompass a person-oriented mechanism and socialization. Since the Novo Nordisk R&D unit does not utilize a formal mechanism to exchange knowledge, they rely heavily on informal transmission channels. In Novo Nordisk they make extensive use of both short-term and long-term assignments abroad. The importance of face-to-face meetings was profoundly emphasized both to start-up projects, but also to transfer the work and solve problems. Since they started the R&D unit in China, they have had expatriates stationed there for a year or more. They also make use of wide-ranging rotation programmes and training for the Chinese employees. To keep up with the increasingly advanced projects, the Chinese employees receive extensive training either by Danish employees who come to China or the Chinese employees will travel to Denmark, typically for three months. BenQ Siemens Mobile also makes use of short-term and long-term assignments, but to a lesser extent. They utilize expatriates widely both in Denmark and in China. In addition, the short-term assignments, usually two weeks in duration, are utilized as a way to enhance the collaborative work and exchange of knowledge. GN ReSound did not place as great an emphasis on the informal channels as the other case companies. They have not engaged in any job rotations or expatriation programmes. They assert it is difficult to get Danish people to go to Xiamen. It is moreover stressed that they do not perceive wide-ranging travelling as a viable solution: 'The idea is that the employees out there teach the new ones. It is about waterfall learning. We are not an educational institution, so if they cannot work independently in China, then much of the point of having them is lost' emphasizes one of the Danish engineers.

7.3.2.5 *Knowledge protection strategies*

This section discusses strategies for the protection of knowledge flows between the R&D units. In relation to knowledge exchange between offshore

R&D units, concerns about IP infringement may constitute an additional barrier. When a firm's IP is pirated, a major element of its competitive posture and advantage may be jeopardized and the management of IP becomes crucial (Teece, 1998).

The IPR system in China has been subject to much scrutiny. Chinese enforcement of IPRs is not yet at the same level as in the West. All of the companies in the case study are aware of the difficulty of protecting their IP assets in China. BenQ Siemens Mobile and Novo Nordisk have not experienced any problems, but GN ReSound has encountered considerable difficulties. GN ReSound has found that employees have left the R&D unit in China to start a low-cost company competing with a product which ReSound perceived as too similar to its own. Thus, IPR protection is something upon which the company increasingly focuses its attention. Nevertheless, none of the case companies found their intellectual property to be easy to infringe upon, due to the complexity of their knowledge and products and the experience needed to copy them. In spite of the problems with IP rights, GN Resound still did not perceive IP infringement as a major threat: 'If some take our hearing aids, then it is easy to copy all the mechanical parts, but it is still not easy to make it function, because there are many factors of stability involved in it. So they will still have a good bit of work to make it function', explains one of the Danish engineers.

There are a variety of strategies which the firm can utilize in order to protect the flow of knowledge between the dispersed R&D units. In the following we will focus on IP strategies based upon patents, secrecy, lead-time advantage, and complementary assets as outlined by Levin et al. (1987) and Cohen et al. (2000).

All of the three case companies focus on *patents* in general to protect their innovations. However, IPRs may not work in practice as they do in theory, since many patents can be invented around at modest costs. Moreover, they often provide little protection in practice because litigation involves cost and resources (Teece, 1998). In Novo Nordisk and BenQ Siemens Mobile they have well-established IPR departments. In GN ReSound the IPR department has just recently been founded (early 2005). Moreover, the Chinese R&D units in Novo Nordisk and BenQ Siemens Mobile work on filing patents (only abroad), whereas the R&D unit of GN ReSound is not yet at that stage. Nonetheless, none of the companies count on utilizing the patents in China. They are aimed at the worldwide market as a means to block or exchange innovations. BenQ Siemens Mobile and GN ReSound – both companies from industries in which the exchange of complementary innovations is important – utilize the patents as thickets to exchange innovations. 'Patents are of enormous importance to BenQ Siemens Mobile. It is not so much to protect our products, as it is to get access to other things that we would like to have in our product portfolio. So we can trade around' (interview with the line manager from BenQ Siemens Mobile).

However, in Novo Nordisk the patents are more used as 'fences' to block the competitors.

Trade secrecy is another way to prevent unwanted appropriation and unintentional technology transfer to rival firms. All of the case companies use secrecy to protect the knowledge and write confidentiality agreements with their employees and suppliers. One of the German project leaders elaborates: 'There are always some holes where information is dropping out, that is why you need to take care that not everybody gets access to everything. We have a single entry point for that.' However, limited access to the knowledge does not relate solely to Chinese employees. It is also an issue for all of the other employees in the case companies. With regards to the more core technologies of the company, the risk of infringement is a challenge in the knowledge exchange. Consequently, the choice in all of the case companies is to keep strategically close to headquarters.

Lead-time advantage does, to a considerable extent, depend upon the industry characteristics and it refers to a firm having a first mover advantage and engaging in frequent technical improvements in order to stay ahead, whereby the competitors should be left behind in the innovation process. Particularly for the mobile phone industry, and for a company like BenQ Siemens Mobile that develops 35 new phones a year, lead-time advantage is seen as pivotal in sustaining the competitive advantage: 'On the other hand, then it goes really fast with the development of mobile phones, so you can say that if they copy something that has already been developed, then they would still lag behind, because then we are working on something new' (interview with the line manager in BenQ Siemens Mobile). In contrast, the development cycles in GN ReSound and to a large extent Novo Nordisk (where it takes approximately 15 years to develop a product) are slower, meaning that lead-time advantage is not as significant.

One effective way to deter the piracy of key assets is to hold *complementary assets* at different R&D sites, which are difficult to imitate. In countries with weak IPR regimes, effective strategies to overcome hold-up, imitation, or piracy of key assets is to own or control key complementary assets, so even if imitation should occur, the total pirated value is limited (Anand & Galetovic, 2004). We found that internal complementary assets within the international R&D organization were utilized to a high extent in the case companies. Nevertheless, it did not seem to be a conscious IP strategy on the part of the companies. Since the Chinese R&D unit of GN ReSound works on their own models, there is a lower extent of cross-border complementary assets involved.

7.3.2.6 *Features of the Chinese innovation system influencing knowledge exchange*

Next, we return to the Chinese innovation system and examine those features that influence the international exchange of knowledge. We will discuss perceptions of the geographical distance; differences of time zone, language, and culture; the issues of 'saving face' and hierarchy; and perceptions of creativity and entrepreneurship of Chinese labour.

The *geographical distance* is perceived as holding the greatest challenges for the Novo Nordisk R&D unit. One of the most often-cited challenges from a Chinese perspective was the lack of synergy effects with the home-based R&D unit and the difficulty in getting an overview of the whole project on which they collaborate: 'Sometimes we make mistakes in our part of the project because we don't know about the new developments or change of goals' (interview with a Chinese scientist in Novo Nordisk). Since the Novo Nordisk R&D unit is focused on applied research and the scientists work closely on cross-border projects generating new knowledge, the physical distance can more easily become a challenge for the successful exchange of knowledge. In GN ReSound and BenQ Siemens Mobile this challenge did not appear prevalent as they work on their own models and specified components.

The *time zone difference* is not seen as a particular challenge since the Chinese employees will wait until the afternoon to call Denmark, where it is morning. For BenQ Siemens Mobile and, to a lesser extent, GN ReSound the time difference is seen as an advantage due to the fact that they can delegate problems to China. The following morning, Chinese colleagues will return the solution to Denmark. In this way the R&D units can engage in around the clock and year-round development.

The *language differences* are perceived to be a relatively substantial challenge, particularly in the cases of BenQ Siemens Mobile and GN ReSound. The lack of adequate English skills for some of the Chinese colleagues makes it difficult to exchange knowledge. In all of the case companies interviewees state that it is difficult to find employees with a high-quality education and a good command of the English language. A number of the Chinese employees in all of the R&D units emphasize that they prefer to write English than to speak on the phone.

In general, the *cultural distance* between the Danish and Chinese cultures did not seem to create significant challenges to the exchange of knowledge, not least because the Chinese employees adapted easily to working with Danish colleagues. Differences in personalities seem to be a more profound challenge. 'I think that it is a very easy cultural interaction. I actually think that there are bigger problems within Europe' (interview with the Vice President of Microbiology in Novo Nordisk). Another cultural aspect that was observed was the internal cultural difference between the Chinese employees that have been studying and working in the USA and those who have not. It was noticed that the Chinese employees with extensive experience from USA also found it difficult to adapt back to the Chinese culture.

Challenges to an open and unfettered exchange of knowledge are the cultural issues of *saving face and a hierarchical mentality* in China. These challenges were encountered in all three case companies. Face saving in the

Chinese culture is often perceived as a problem for innovation, because it implies complying with the preferences of a superior. This perception found support in the interviews in the case companies: 'It can actually decelerate innovation a bit that they are so authoritarian. There is a great loyalty and belief in what the grey haired is saying is probably true [...] In the Danish management culture everything is constantly challenged' (interview with the Vice President of Microbiology in Novo Nordisk).

A *lack of creative knowledge* was reported as a challenge. The Chinese educational system is based on a very rigid structure, which leaves less room for innovative thinking or creativity. For Novo Nordisk, it has been a problem to find creative scientists to be part of the discovery department: 'It is difficult to find creative people. I see it as culture related. I don't think it is possible in one area to have a top-governed system where everything is controlled and then a freely thinking in another area.' As a consequence, the Chinese R&D units tend to receive well-defined assignments. On the other hand, the case companies emphasize the high technical skills and stringent manner of working as positive aspects that contribute to the workflow across borders. For example, in Novo Nordisk it is perceived as a good complement to the Danish less structured way of working. Accordingly, the cultural differences can constitute both a challenge and an advantage for the management of offshore R&D.

7.4 Conclusion

In this chapter we have analysed the phenomenon of high-tech investments to China from two different angles. In the first part of the chapter we focused on the supply side of the Chinese ST&I system. China is in the process of reforming a national innovation system that still reflects weaknesses from the planned economy period. During this process China has made impressive gains in some key ST&I areas, such as increasing the percentage of high-tech exports, increasing R&D investments made by businesses, attracting foreign high-tech investments, improving the legal IPR environment, and revitalizing the educational system.

Seen from the demand side of foreign firms investing in high tech in China there has been an increase in more advanced FDI inflows that are not just focused on low-cost human resources but also are attracted to the Chinese market, skilled human resources, and scientific clusters. Through an analysis of the Chinese R&D units of Novo Nordisk, GN ReSound and BenQ Siemens Mobile, it was shown that the motives for offshoring R&D to China were varied, ranging from mainly cost-driven and market-driven motives in GN ReSound and BenQ Siemens Mobile, to policy-driven, technology-driven, and innovation-driven motives in Novo Nordisk. Furthermore, the strategies and organization of the R&D units depend to a large extent on the type of R&D conducted.

We also investigated the management of offshore R&D to China. The results reveal that there are differences within the case companies depending upon which type of R&D is conducted by the unit. Since GN ReSound and BenQ Siemens Mobile focus on development work that builds on existing knowledge platforms, and Novo Nordisk focuses on research where the work tasks are highly complex, the companies are faced with different challenges. Furthermore, BenQ Siemens Mobile has a global R&D network, in which we found that they do not experience the same extent of challenges, due to an increased incentive to exchange knowledge in order to make the global workflow function.

With regards to the knowledge-sharing propensity of the Danish R&D unit, we found that the managerial challenges were seen to be the 'China threat' and fear of losing work, lack of priority of the Danish colleagues, and diverse skills levels. One aspect that positively influenced the knowledge-sharing propensity was the additional resources that the Chinese R&D unit bring to R&D activities. The Chinese employees were highly motivated to receive transfers of technology and knowledge but absorptive capacity and lower level of experience were seen as a challenge to knowledge exchange.

In spite of the lax IPR regime in China the case companies have still off-shored knowledge to China. Admittedly, all of the case companies place an importance on patents, however they serve as 'thickets' to exchange complementary knowledge or as proprietary assets for the global market. In addition to a patent strategy, all of the companies employed secrecy as a mean to deter infringement risks. However, particularly for GN ReSound and Novo Nordisk, they were mindful about transferring core knowledge to China. Moreover, the lead-time advantage strategy was only noticed in BenQ Siemens Mobile, due to the extremely short development cycles. Finally, both BenQ Siemens Mobile and Novo Nordisk are, more or less consciously, employing a strategy based on complementary assets.

With respect to the features of the Chinese innovation system influencing the knowledge exchange process, the geographical and cultural distance did present some challenges. For GN ReSound and BenQ Siemens Mobile, where the R&D units are focused entities, geographical distance did not prove to be a barrier. Furthermore, they are able to take advantage and exploit time zone difference. For Novo Nordisk the geographical distance does constitute a challenge. In particular, the Chinese employees found it difficult to get an overview of the projects and to obtain the synergy effects of being in a larger R&D organization. In all of the R&D units the language and lower English skills in China was perceived as an obstacle. Concerning the cultural distance, it was emphasized that differences in personalities rather than culture is a greater challenge. Nevertheless, there are a number of Chinese cultural traits, which were believed to be a challenge to the knowledge exchange between the R&D units. These reflect issues of saving face, hierarchical mentality, and the reported less creative mindset in the Chinese culture.

In conclusion, as the People's Republic of China is liberating its market and FDI is flowing to the country in ever-increasing amounts, the ST&I system is also evolving. One of the parameters for developing a successful ST&I system has been to attract foreign high-tech investments to China. For firms, the offshoring of R&D is a non-trivial and risky endeavour. Nevertheless, the Chinese R&D units included in this case study analysis have proven to be very successful and growing ventures.

References

Anand, Bharat and Alexander Galetovic (2004) 'Strategies That Work When Property Rights Don't', in Gary Libecap (eds), *Intellectual Property and Entrepreneurship*, Greenwich, CT: JAI Press.

Cohen, Wesley M., Richard R. Nelson and John P. Walsh (2000) 'Protecting Their Intellectual Assets: Appropriability Conditions and Why US Manufacturing Firms Patent (or Not)', *NBER Working Paper 7552*, Cambridge, MA: National Bureau of Economic Research.

Cohen, Wesley M. and Daniel A. Levinthal (1990) 'Absorptive Capacity: a New Perspective on Learning and Innovation', *Administrative Science Quarterly*, 35: 128–52.

Cummings, Jeffrey L. and Bing-Sheng Teng (2003) 'Transferring R&D Knowledge: the Key Factors Affecting Knowledge Transfer Success', *Journal of Engineering Technology Management*, 20: 39–68.

Economist Intelligence Unit (EIU) (2005) *China, Country Profile 2005*, London.

Gammeltoft, Peter (2006) 'Internationalisation of R&D: Trends, Drivers and Managerial Challenges', *International Journal of Technology and Globalization*, 2(1/2): 177–99.

Gassmann, Oliver and Zeng Han (2004) 'Motivations and Barriers to Foreign R&D Activities in China', *R&D Management*, 34(4): 423–37.

Gu, Shulin and Bengt-Åke Lundvall (2006) 'China's Innovation System and the Move Toward Harmonious Growth and Endogenous Innovation', *Innovation: Management, Policy & Practice*, 8(1–2).

Gupta, Anil K. and Vijay Govindarajan (2000) 'Knowledge Flows Within Multinational Corporations', *Strategic Management Journal*, 21: 473–96.

Haiyan, Wang and Zhou Yuan (2006) 'The Evolving Role of Universities in the Chinese National System of Innovation', *National Research Center for S&T for Development, Ministry of S&T Peoples Republic of China*.

Huang, Can, Celeste Amorim, Mark Spinoglio, Borges Gouveia, and Augusto Medina (2004) 'Organization, Programme and Structure: an Analysis of the Chinese Innovation Policy Framework', *R&D Management*, 34(4): 367–87.

InfoWorld (2005) 'New BenQ, Siemens Mobile Phone Company Opens', http://www.infoworld.com/article/05/10/03/HNbenqsiemens_1.html, accessed 20 February 2007.

Kjersem, Julie Marie (2006) *The Internationalisation of R&D – Offshoring Knowledge to China Viewed Through Case Studies of Novo Nordisk, GN ReSound and BenQ Mobile*, Master's Thesis, Copenhagen Business School.

Levin, Richard C., Alvin K. Klevorick, Richard R. Nelson and Sidney G. Winter (1987) 'Appropriating the Returns from Industrial Research and Development', *Brookings Papers on Economic Activity*, 3: 783–831.

Lundvall, Bengt-Åke (ed.) (1992) *National Systems of Innovation; Towards a Theory of Innovation and Interactive Learning*, London: Pinter.

Ministry of Science and Technology of the Peoples Republic of China, 2005: http://www.most.gov.cn/eng/statistics/2005.

OECD (2004) *OECD Science, Technology and Industry Outlook 2004*, Paris: OECD.

OECD (2005) *OECD Science, Technology and Industry Scoreboard 2005 – Towards a Knowledge-based Economy*, Paris: OECD.

OECD (2006) *OECD Science, Technology and Industry Outlook 2006*, Paris: OECD.

People (2004) Another First in China, 19 of June, Novo Nordisk.

Persuad, Ajax, Vinod Kumar and Uma Kumar (2002) *Managing Synergistic Innovations Through Corporate Global R&D*, Westport, CT: Greenwood Press.

Schaaper, Martin (2004) 'An Emerging Knowledge-Based Economy in China? Indicators from OECD Databases', STI Working Paper 2004/4.

Share, quarterly investor update from Novo Nordisk (2005) 'Scientific Satellites', May.

Siemens Annual Report (2004) http://www.siemens.com.

Simonin, Bernard (1999) 'Transfer of Marketing Know-How in International Strategic Alliances: An Empirical Investigation of the Role and Antecedents of Knowledge Ambiguity', *Journal of International Business Studies*, 30(3): 463–90.

Szulanski, Gabriel (1996) 'Exploring Internal Stickiness: Impediments to the Transfer of Best Practices Within the Firm', *Strategic Management Journal*, 17: 27–43.

Teece, David J. (1998) 'Capturing Value from Knowledge Assets: the New Economy, Markets for Know-how and Intangible Assets', *California Management Review*, 40(3): 55–79.

Time Magazine (2005), 'Taiwan Steps Up', 25 June.

UNCTAD (2005) *World Investment Report 2005: Transnational Corporations and the Internationalization of R&D*, New York and Geneva: UN.

UNESCO Institute for Statistics http://www.uis.unesco.org.

von Zedtwitz, Maximilian (2004) 'Managing Foreign R&D Laboratories in China', *R&D Management*, 34(4): 439–52.

von Zedtwitz, Maximilian and Oliver Gassmann (2002) 'Market versus Technology Drive in R&D Internationalisation: Four Different Patterns of Managing Research and Development', *Research Policy*, 31: 569–88.

Walsh, Kathleen (2003) *Foreign High-Tech R&D in China*, Washington, DC: The Henry L. Stimson Center.

Zander, Udo and Bruce Kogut (1995) 'Knowledge and the Speed of Transfer and Imitation of Organizational Capabilities: an Empirical Test', *Organization Science*, 6(1): 76–92.

8
Learning from the Bangalore Experience: The Role of Universities in an Emerging Regional Innovation System in Asia

Jan Vang, Cristina Chaminade and Lars Coenen

8.1 Introduction

This chapter is about the role of universities and public research organizations in initiating, maintaining and sustaining the development of regional innovation systems in high-tech sectors in Asian countries, exemplified by Bangalore, India. Over the past two decades researchers and policy makers have increasingly acknowledged the importance of universities and other publicly financed research institutions as engines of knowledge-based growth and enhanced innovative performance in developed economies. Especially in the context of regional economic development and fuelled by the Silicon Valley success story, expectations on the presence and contribution of universities to regional high-technology agglomerations have been high (OECD, 1999).

Universities are conceptualized as creators and providers of knowledge spillovers for industrial innovation and thus as key actors in the national and regional innovation systems. In particular, the triple helix narrative has been widely heralded as the new policy paradigm that puts universities at the heart of knowledge-based regional economic development (Etzkowitz and Leydesdorff, 2000; Jacob, 2006). It explicitly seeks to reform academia into entrepreneurial universities and to strengthen industry–university–state interaction. Under this paradigm, it is believed that in order to harness scientific knowledge for innovation, industry–university linkages have to be stimulated through various mechanisms such as the promotion of academic entrepreneurship, the establishment of science parks and incubator centres, and the development of technology transfer support infrastructure.

The discussion on the third task that was grounded in experiences in California, or the USA more generally, is swiftly disseminating to other regions such as Asia. Policy makers all over the world are now discussing the importance of the entrepreneurial university as an engine for growth, particularly in high-tech regions. Similar debates are increasingly taking place in policy circles in India. We argue, however, that this envisaged function of universities runs the risk of misplaced policy learning by ascribing universal truths to western practices (Amin, 1989; Said, 1993; Yeung, 2003) ignoring the specific context – including the historical trajectories – in which this interaction between the university and the industry has taken place. This chapter attempts to contribute to the current discussion in India about the role of universities in the development of high-tech clusters, particularly IT. As the flagship of the IT industry in India, Bangalore is often referred to as the Silicon Valley of Asia. This chapter therefore scrutinizes the role played by universities and research institutes for the emergence of the Bangalore IT cluster. As such, we critically discuss the adequacy of readily-imported, 'one-size-fits-all' models on the role of universities in stimulating high-tech regional clusters. Instead, we endorse a non-deterministic evolutionary perspective that emphasizes firms', regions' and nations' degrees of freedom in strategizing, i.e. contextually shaped strategy in action (Nygaard, 2001). Bangalore has been chosen as it represents one of the few cases in Asian countries, which have come close to having constructing a full-scale regional innovation system (Chaminade and Vang, forthcoming, 2008; Vang and Chaminade, 2006).

The structure of the remainder of the chapter is as follows. First, we introduce the dominant perspectives, positions and findings on the role of universities in (regional) innovation systems. Then we turn to the case, Bangalore, where we analyze the different roles of universities and publicly funded research in the different phases of Bangalore's development into a regional innovation system (Chaminade and Vang, forthcoming, 2008) by analyzing the strategic needs of indigenous firms. The chapter is rounded off with more general conclusions from the study and the challenges they pose to established insights on the role of universities in innovation systems research.

8.2   Universities in innovation systems: the current debate

While having its origin in OECD economies, innovation systems research has become increasingly interested in Asian, Latin American and African economies. This has entailed a stronger focus from innovation, defined narrowly as R&D-related activities to a broader perspective that also encompasses competence building and upgrading to higher value-added activities in global value chains (Chaminade and Vang, 2006; Vang and Asheim, 2006). Central to the innovation systems approach is the claim that upgrading is possible when there is *an environment* that supports interactive learning and innovation. Firms' isolated efforts to make this transition tend to fail in the longer term. The literature claims that the interaction often takes place with other firms and organizations co-located in the same regional area (Lundvall and

Borrás, 1999). The importance of the local interactions for firms, particularly SMEs, holds for developed (Asheim et al., 2003; Cooke and Morgan, 1998; Cooke and Will, 1999; Schmitz, 1992) as well as developing countries (Albu, 1997; Giuliani, 2004; Giuliani and Bell, 2005; Pietrobelli and Rabellotti, 2006; UNIDO, 2001 and 2004). Firms located in a region might benefit from static and dynamic externalities supporting their ability to compete in local and global markets.

Our theoretical vantage point for this chapter is the regional innovation systems (RIS) approach. RIS is defined as a 'constellation of industrial clusters surrounded by innovation supporting organizations' (Asheim and Coenen, 2005). The approach puts the emphasis on the systemic dimension of the innovation process, with the focus on the dynamic interaction between the different elements of the system. Four related system-elements can be identified (Doloreux, 2002):

- Firms within a cluster (constituting the knowledge exploitation subsystem).
- Knowledge infrastructure (constituting the knowledge exploration subsystem) in which universities are included.
- Institutions (the 'rules' regulating the behaviour of the actors in the RIS and their interaction).
- Policy (intended to improve the overall innovative performance of the RIS).

Thereby, the regional innovation system is boiled down to two main knowledge-related sub-systems, the interactions between them as well as a governance system underpinning it (see Figure 8.1). The first type of sub-system, involved with knowledge exploitation, concerns the companies in the region's main industrial clusters, including their customers and suppliers. Industrial clusters are defined as the geographic concentration of firms in the same or related industries (Porter, 1998; Pietrobelli and Rabellotti, 2004; for a critique, see Martin and Sunley, 2003). In this sense, industrial clusters represent the production component of the regional innovation system. The second sub-system, involved with knowledge exploration and support, includes research and higher education institutes (universities, technical colleges, and R&D institutes), technology transfer agencies, vocational training organizations, business associations, finance institutions, etc. (Asheim and Coenen, 2005). It provides the infrastructure backing up the innovative performance of the first type of actors. The knowledge-creating and -diffusing organizations bestow the resources and services (knowledge, capital, and so on) to support innovation among the local firms.

As a third element, institutions are an important factor that shapes the territorial context of the RIS and, thus, the ways that actors in the region create,

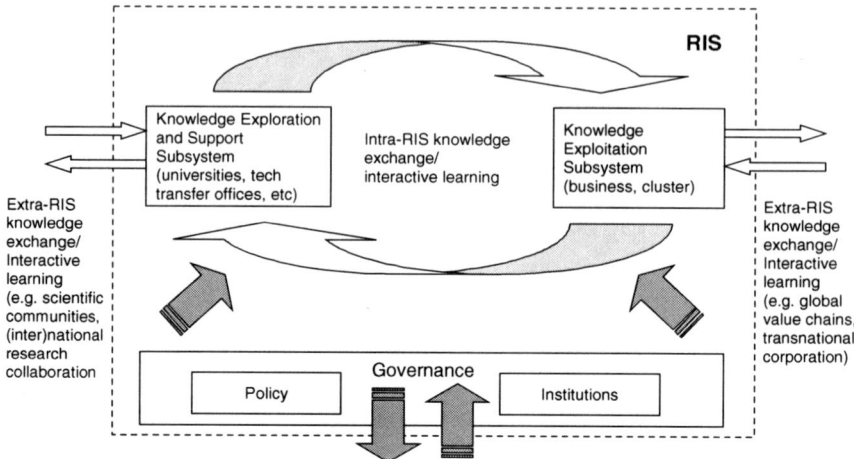

Figure 8.1   Model of a RIS
Source: Authors.

exchange, exploit and forget knowledge. As formal regulations, legislation, and informal societal norms, they produce (and are reproduced by) the structures and meanings that regulate (but not wholly determine) the actions and interactions of firms and other organizations (Gertler, 2004; Hollingsworth, 2000; Nooteboom, 2000; North, 1990). Fourthly, policy plays an increasingly important role, not the least due to the rising popularity and diffusion of the RIS concept into policy-making circles. However, the functioning of the RIS is also influenced by policy frameworks and decision taken outside the boundaries of the region (Isaksen, 2003), for example, through national science and technology policy and central decisions about the extent and level of regional administrative devolution. On a general level, RIS policy seeks to improve the overall system by increasing learning capabilities and knowledge diffusion (Doloreux, 2002), but the way this policy is specifically shaped can take different forms dependent on the region's characteristics (Asheim and Isaksen, 2003).

Although universities have always been considered a crucial element in the system of innovation, there has recently been a rising interest in the specific role that they should play supporting the development of different innovation systems (Lundvall, 2002; Asheim et al, 2006) where special attention has been placed on the so-called 'third task' or mission (Goddard and Chatterton, 2003). The third task (after teaching/education and research) refers to direct interaction between universities and society. This can be interpreted in a variety of ways. The third tasks range from creating new high-technology firms, consulting for local industry, delivering advice to politicians and policy makers and informing the general public and shaping the national spatial distribution of social opportunities and services. Although, historically, universities have been engaged with society in a variety of manners (Benneworth and Arbo, 2006), most innovation system researchers tend to privilege direct economic engagement over other potential roles (Molas-Gallart et al., 2002). Thus, the third task often refers to direct collaboration between university and the industry. Such direct interaction between actors in the knowledge exploration and the knowledge exploitation subsystem fits very well in a RIS framework. It should, therefore, not be seen as a surprise that university's third task often is advocated from a regional innovation policy perspective.

However, concern has also been raised that this emphasis on direct collaboration with industry might divert attention and resources away from university's core activities, i.e. (public) research and teaching (Lundvall, 2002; Martin and Etzkowitz, 2006). This debate has resulted in polarized views on the role(s) of universities in regional systems of innovation, which tend to ignore the territorial and historical context in which the university is embedded. Such contextualization is especially important when discussing the role of university in Asian regional innovation systems and the case of Bangalore in particular. This debate shall be outlined and an effort is made to propose a more nuanced framework to understand the role of universities in RIS in Asia, taking Bangalore as a flagship example. It appears that the discussion has been split in two camps that either emphasize university's generative or developmental role (Gunasekara, 2006). While the generative role refers mainly to knowledge creation, diffusion and exploitation processes, the developmental role is more pre-occupied with the governance dimensions that regulate the interaction of university in the regional innovation system.

8.2.1 The generative role of universities

The generative role of universities underlines the contribution of academia to knowledge-based regional development through the production of advanced basic research and trained personnel. The knowledge outputs that are produced can take different forms, for example, as scientific and technological information, equipment and instrumentation, skills or human capital, networks of scientific and technological capabilities and prototypes for new products and processes (Mowery and Sampat, 2005). Such discrete outputs have varying potential across industry to become commodified knowledge. It is not the objective of the generative role to supply the industry with knowledge solutions (in the sense of applied knowledge) but to produce science (basic knowledge) and to train human resources.

Obviously, science-based and high-technology industries benefit more from industry–university linkages compared to, for example, service providers. The generative role of universities is often couched in terms of knowledge spillovers. Academic knowledge spillovers, measured by the location of inventors citing university patents, have a tendency to be localized in the university's region (Adams, 2002; Trajtenberg et al., 1997). Such mea-surements do not say very much about the mechanisms by which knowledge spillovers are realized. Zucker and Darby (1996) have highlighted the role of so-called 'star scientists' that drives the commercialization of break-through discoveries. Others find that the average level of human capital conditions the ability to develop and implement new technology (Glaeser et al., 1992). This, in part, relates to Cohen and Levinthal's (1990) argument that the appropriability of new knowledge by a firm is dependent on its absorptive capacity, i.e. the firm's level of prior related knowledge and related intensity of efforts in acquiring new knowledge. The consensus in the literature seems to be that 'knowledge spillovers are geographically bounded within lim-ited space over which interaction and communication is facilitated, search intensity is increased, and task coordination is enhanced' (Feldman, 2003). While the localization of knowledge spillover is often explained through institutional similarities, this literature remains rather silent about which institutions matter and how, as well as how firms or regions located outside these areas can access and benefit from the localized knowledge spillovers.

Many (national) innovation system researchers, and particularly its early proponents, subscribe to the generative role of university in innovation sys-tems. They tend to treat universities as a more or less autonomous systems adhering to norms of academic research and teaching. Their research sup-ports the claim that university research plays a small role in industrial R&D projects (Cohen et al, 2002; Fagerberg, 2004). It implies that the importance of direct, purposeful industry–university interaction, in other words the third task, should not be exaggerated. Lundvall (2002) even goes so far to state that university's 'most significant contribution to society and the economy will remain well-educated graduates with critical minds and good learning skills'. He acknowledges that universities can be involved in 'third task' activities but sees it both as dangerous and mainly relevant for specific industries such as science-based industries, biotech and software production. While sceptical of the constraints implicated in direct co-operation with industry, he stresses the importance of openness to the environment or surrounding society 'to ensure that the long-term, creative and critical aspects of academic research can survive'. In sum, the generative role of universities in innovation sys-tems ascribes to traditional activities in the universities – that is, scientific research and teaching. In contrast, the developmental perspective takes a broader outlook on university's contribution to innovation.

8.2.2   The developmental role of universities

While acknowledging the knowledge-generating and -disseminating activi-ties underlined in the generative role, the developmental role puts a stronger emphasis on university's impact on the governance of the regional innova-tion system and on the close interaction between university and industry in the development of what has been called 'economically useful' knowledge.

Table 8.1 University roles

	Generative role	Developmental role
Role of university	Supply of qualified human capital and basic research	Driver of regional growth. Entrepreneurial university
Research type	Basic research and non-industry specific applied research	Prominently applied research, although some basic research also takes place
Functions	Education and research	Third task (but also training and research)
Networks with local actors	Weak. Independent institution.	Strong. Blurring boundaries in the role of the different organizations as knowledge providers
Equivalent mode	Mode 1	Mode 2
Theoretical influences	Research policy, (national) innovation systems, standard economics	Triple helix, regional innovation systems
Advantages	Research excellence. Autonomy in research and training and focus on basic research. Clear boundaries between roles of the different organizations. Long-term orientation of research	Coordination between the different actors in the generation of knowledge and its transformation into commercial outputs
Inconveniences/ Critics	Research might be disentangled from industry and thus never turn into new products and services that impact growth	The university might lose its core "competence" resulting in an identity crisis. There is a risk of concentrating too much efforts in applied research and abandon basic research (for which firms have less incentives). Potential problems of recruiting researchers if they become reduced to cheap consultants.
Policy implications	Emphasis on basic research and education	Emphasis on applied research that suits industry needs

Source: Authors.

As such, it conceptualizes the university not only as an autonomous player strictly involved in knowledge generation, but also as an actor that participates, both formally and informally, in shaping regional institutional and social capacities through close interaction with the industry, as summarized in Table 8.1. While having been a central concern in regional studies for quite some time, this perspective has lately been increasingly dominated by the triple helix model.

The triple helix finds its foundation in a spiral model of interaction between university–industry–government where knowledge production and innovation transcends organizational boundaries (Etzkowitz and Leydesdorff, 2000). According to Etzkowitz (2002), it refers to a move towards a new global model for the management of knowledge and technology. In this model, universities should be rethought in ways that allow for new interfaces with industry and, even, a shift towards roles that were traditionally allocated to private industry such as the exploitation of intellectual property rights, technology transfer, the establishment of science parks and spin-off firms. In fact, Etzkowitz criticizes the innovation system approach for adopting a static and essentialist perspective:

> The National Systems of Innovation (NSI, [read: generative role]) approach is especially well suited to analysis of bounded phenomena, within nations or individual firms. Although other sources are taken into account, incremental innovation is viewed as primarily occurring within the firm, through various forms of learning (Lundvall, 1988). A different model of the sources of innovation is required to account for discontinuous as opposed to incremental innovation (Etzkowitz, 2002 p. 1).

While the generative role treats universities as more or less independent units, the triple helix model emphasizes the emergence of hybrid, recursive and cross-institutional relations between university, industry and government. In other words, the institutional boundaries between university and industry are blurring. According to Etzkowitz and Klofsten (2005), this has resulted in three fundamental changes in the role of universities in stimulating innovation. Firstly, universities have become one of the drivers of innovation in a knowledge-based economy through its research activities. Secondly, there is a shift towards purposeful, collaborative relationships between university and industry as well as government. Thirdly, each sphere has started to take the role of the other. This has lead Etzkowitz and Klofsten (2005) to argue that the entrepreneurial university is *the* core institution in an innovating region. The triple helix model seeks to spur innovation of the more radical kind by conceptualizing the university as an incubator or seedbed that provides support structures for overlapping networks of academic research groups and start-up firms.

In contrast to the generative role, the developmental role is, mainly, concerned with how university-industry interaction in a regional innovation system is organized. As a result, it has a stronger focus on novel associative structures and modes of governance (Cooke and Morgan, 1998) compared to research that deals with the generative role of universities.

The triple helix approach has been rapidly disseminated worldwide and it is increasingly being used in policy circles, including several Asian countries, as the new paradigm for the design of innovation policies. However, in the dissemination process the ideas have been de-contextualized. The model is based on studies conducted in the Boston region with MIT (Massachusetts Institute of Technology) as the exemplar of an entrepreneurial university. While case studies have been conducted in other regions and universities, the triple helix model has been criticized for its totalizing framework. Very little has been reported on the disadvantages and conflicts of interests related to academic entrepreneurialism (Gunasekara, 2006), or on the adequacy of the model to the specific territorial and historical context of the RIS.

This issue is particularly relevant for developing countries. As recent research has pointed out (Juma et al., 2001 cf Sardana and Krishna, 2006; Turpin and Martinez-Fernandez, 2003), the use of the triple helix concept in a developing country context is problematic. In most of the cases, the interaction between the government, the university and the industry does not materialize due to the lack of resources and the weaknesses of the different actors involved in the system. Despite the relevance of their critique, what this literature tends to implicitly assume is that a more proactive role of the university and a closer interaction with the industry is desirable (Krishna, 2001; Basant and Chandra, 2006) and that policy makers should be more actively supporting the transition from the generative role to the developmental role of the university (Sardana and Krishna, 2006). This is a critical issue when discussing the future of high-tech clusters in India such as the Bangalore cluster. Hence, attention is now turned to analyze the role played by the Indian universities – especially those in Bangalore – in facilitating the emergence of the IT cluster in Bangalore.

8.2.3 Contextualizing the debate

The first step in this process is contextualizing the debate on the role of universities in (regional) innovation systems. In this sense, the following three dimensions need to be taken into account.

Time matters: There is abundant documentation supporting that most Asian countries (apart from Japan) – or rather their firms – should be conceptualized as imitators, as they tend to rely on second or third mover advantages. That is, they compete on producing cheaper products than their competitors in the developed world by copying, re-engineering and imitating the first movers' products without having to bear the R&D costs or they simply serve as sites for outsourcing or offshoring production because of Ricardian or absolute cost advantages. This challenges the notion of firms in both dominant approaches to the role of universities in systems of innovation. Both assume that innovation per se is pivotal. However, the literature on developing countries emphasizes the importance of imitation. It is by no means clear that universities are important when innovation is not so crucial among the clustered firms nor has the literature provided systematic answers for which of the three different university tasks (training, research and third task) become relevant in the evolution of the industries in developing countries. It does not provide clear answers to whether there are generic evolutionary propensities attached to, respectively, generative and developmental roles (for example, whether generative activities are more important earlier that later). In this context the importance and trade-off between different types of interaction with universities for Asian countries has not been investigated (see Mowery and Sampat, 2005 for an attempt in the USA).

Position in the value chain matters: The sectoral and regional innovation systems literature emphasizes the industry-specific dimension of innovation. However, most of the studies often ignore that firms in the developed and developing worlds tend to operate in different segments of the value chain of the same industry. Most of the studies tend to focus on the knowledge-intensive (read R&D-intensive) activities in the value chain and thus one cannot mechanically generalize their findings to other parts of the value chain (that is, the role occupied by Asian firms). The identified sector specificities are thus unlikely to reflect industry differences but specific positions in the value chain. Hence the different roles of universities discussed in the developed world might not be relevant in Asian countries.

Strategy matters: In the same vein it should be emphasized that the literature ignores the importance of the specificities of firm strategies. It assumes that all firms pursue global leadership in a certain segment of an industry. Hence, it applies a hierarchical concept of knowledge that is detached from firms' strategic aims and visions. Alternatively, one could think of 'knowledge-levels' as context and strategy specific. The demand for particular types of public knowledge provision (that is, the role of public universities) is contingent on firms' strategies. Public knowledge provision can both support existing strategies and create the conditions underpinning new alternative strategies. In other words, whether universities should aim at becoming developmental or generative reflects a balance between developmental goals, institutional resources and firm's strategies and capabilities. Hence, it is not *a priori* assumed that supporting global leadership based on a developmental university should be privileged compared to a strategy - based on generative universities - exploring the advantages of, for example, targeting southern markets (D'Costa, 2006).

While other issues could have been emphasized, the pivotal importance of these dimensions illustrate the need for an inductive study that discusses the role of the university, taking into account the specificities of the regional innovation system.

8.3 The role of universities and public research in Bangalore's software 'RIS'

Bangalore has emerged as one of the largest and fastest-growing software clusters outside the USA (Nadvi, 1995; Parthasarathy, 2004a). Bangalore is not only a hub for software-related industries, but also houses several high-tech clusters (such as defence and aeronautics), and is considered to be the scientific and engineering centre of India in terms of research and training and, partly, manufacturing. Despite the weight of the TNCs in the Bangalore IT sector, the large majority of firms are SMEs (NASSCOM, 2005).

Bangalore has attracted the attention of scholars around the world for its impressive software growth export rates, which are superior to those of competing IT hubs in Israel, Brazil or China (Arora and Gambardella, 2004; Athreye, 2005). The value of export, for example, typically grows more than 30 per cent annually, while revenues grow at 30 to 40 per cent (www.bangaloreit.in). The growth of the software industry in India is based on exports to global markets, principally, to the US. This export-led development trajectory or model has important implications for the industrial structure of the RIS and the possibilities for upgrading of the indigenous firms (D'Costa, 2006). India has an estimated share of 65 per cent of the global IT services offshoring segment and around 46 per cent of the global BPO market (NASSCOM-McKinsey Study, 2005a).

Bangalore is considered to be a success story and, as such, is receiving increasing attention from Indian policy makers that are discussing how to support the growth of the IT sector through a variety of policies in the education, science and technology realms (Indian Ministry of Human Resource Development, 2006). As in many other countries, the open debate is whether to support a more generative role of the university or a more developmental role. To understand which of the two options is more adequate for Bangalore, it is necessary to analyze how the RIS evolved over time and also to identify the different needs of the firms located in Bangalore.

8.3.1 Emergence of the RIS and fragmentation of the knowledge exploitation subsystem

There is a heated debate on the role of public research in the initial phase of Bangalore's development. As many authors have acknowledged (Arora and Gambardella, 2004; Athreye, 2005), the early development of Bangalore as a specialized hub in the software industry was due to the location in the region of some of the best educational institutions such as the world-renowned Indian Institute of Science, the Indian Institute of Information Technology, the National Institute of Mental Health and Neuro-Sciences, the Central Food Technological Research Institute, the Indian Space Research Organisation, the National Aeronautical Laboratory and so on. The high concentration of knowledge providers in the region resulted in a critical mass of highly qualified, yet cheap labour force which could explain the initial interest of numerous US firms in locating their outsourcing activities to the region. Direct research spillovers from university research seem somewhat unimportant.

Instead, the interaction between the indigenous firms and the TNCs seems to be major drivers behind the development of certain firms such as TCS, Wipro and Infosys in the early phase.[1] Through interacting with the TNCs, these firms became more familiar with the work organization and requirements of the US firms (delivery times, quality, reliability) while the US firms started to outsource tasks to be performed entirely in Bangalore. Co-operation was facilitated by the role of the Indian transnational community in the USA (Saxenian, 2001), particularly those that held important positions in US firms (Vang and Oberby, 2006).

As the Bangalore software RIS matured, both Bangalore and US firms improved their competences in managing outsourcing and offshoring, and built up inter-cultural competencies and created their own local networks. Employee attrition and wage increases forced the firms to introduce advanced management techniques (Arora et al, 1999; Athreye, 2005) alluding to the importance of managerial education as an added value to the existing engineering training capacity. The broader knowledge base combined with the existence and building up of reputation as reliable suppliers in the US market plus an aggressive certifying strategy among most Indian firms have permitted a handful of firms to move up the global value chain (to the provision of R&D services for TNCs) and, even in some cases, develop their own innovation strategy and enter new niche markets with their own final product (Parthasarathy and Aoyama, forthcoming).

Only a small group of firms has benefited in terms of knowledge spillovers from the interaction with the TNCs and has effectively moved to higher added-value activities. As acknowledged by D'Costa (2006), most of the SMEs located in Bangalore provide standardized services; therefore, the incentives for the TNCs to create long-term arrangements with the indigenous SMEs are low.[2] Their absorptive capacity also remains low. Only a small group of firms that has been able to build an absorptive capacity and create distinctive capabilities are benefiting from the interaction with TNCs. The growth model that the indigenous firms have adopted (i.e. export- and TNC-driven) has created a fragmented industry with very weak local linkages (D'Costa, 2006). Table 8.2 plots the distribution of the industry by segments. Basically the top 15 Tier I and Tier II companies account for 70 per cent of the IT services software revenues. The vast majority of firms (emergent players) are still responsible for only 15 per cent of the IT services. Furthermore, while the market share of the Tier I firms has increased from 32 per cent in 2001–02 to 45 per cent in 2004–05, Tier II lost ground from 35 per cent to 16 per cent.

Table 8.2 India's IT structure and performance

Category	No. of players	Share of India's total IT/BPO export revenues	Performance
Tier I Players	3–4	• 45% of IT Services • 4-5% of BPO	Revenues greater than USD 1 billion
Tier II IT Players	7–10	• 25% of IT Services • 4-5% of BPO	Revenues USD 100 million-USD 1 billion
Offshore operations of Global IT majors	20–30	• 10-15% of IT Services • 10-15% of BPO	Revenues USD 10 million-USD 500 million
Pure play BPO providers	40–50	• 20% of BPO	Revenues USD 10 million-USD 200 million (Excluding top provider with USD 500 million)
Captive BPO units	150	• 50% of BPO	Revenues USD 25 million-USD 150 million (top 10 units)
Emerging players	>3000	• 10–15% of IT Services • 5% of BPO	Revenues less than USD 100 million (IT) Revenues less than USD 10 million (BPO)

Source: NASSCOM-McKinsey (2005a).

The question that these numbers pose is if it is possible to talk about the role of the university for the whole Bangalore system of innovation with such a fragmented and polarized industry.

8.3.2 The knowledge exploration subsystem

The situation of the universities in India is extremely heterogeneous in terms of performance and tasks undertaken. Four Indian universities are listed in the top 50 universities in the world (THES-QS, 2006): the Indian Institute of Technology, the Indian Institute of Management, the Indian Institute of Science and the Jawaharlal Nehru University. These universities are considered to be world-class in both training and research as well as for collaboration with the industry (Basant and Chandra, 2006).

In a recent report on the IT industry, the Indian Ministry of Human Resource Development classifies the IT-related higher education institutions into three categories (2006):

• Category I embraces the six IITs (Indian Institute of Technology), the Indian Institute of Science, two IIIT (Indian Institute of Information

Technologies) and six IIMs (Indian Institute of Management). Of the 15 institutions, three are located in Bangalore (IIIT, IISC, IIM). Category I institutions supply postgraduate education and advanced research.

- Category II includes the 17 Regional Engineering Colleges (RECs) and 33 other established universities and technical institutions. The RECs, now renamed as National Institutes of Technology (NITs) provide high-level undergraduate and postgraduate education in the IT field, among other technical fields. The University of Vishveshvarayya, located in Bangalore, offers courses in the IT field.
- Category III: Other government and self-financing institutions (include approximately 200 government-supported IT training institutions and 550 self-financing institutions). This is an extremely heterogeneous category embracing both low-end academies and the corporate training centres such as the Infosys Training Centre which is the largest IT training institution in Bangalore. The two top IT service firms located in Bangalore (Infosys and Wipro have established their own training centres).

If one eliminates the handful of world-class technical institutions, the picture is one of shortages of high-quality staff (Arora and Gambardella, 2006; NASSCOM-McKinsey 2005a, 2005b), and under-investment in research facilities. According to a recent NASSCOM-McKinsey report (2005a), only one-quarter of the technical graduates and 10-15 per cent of the general college graduates are suitable for employment in the IT and BPO industries. This fragmentation of the university system, together with a shortage in the supply of human capital, explains why some TNCs and also local firms have started to build their own training centres in Bangalore as the recent examples of Infosys, Wipro or TCS show.

The shortage of human capital is also reaching the university. Almost 25 per cent of the IT and computer science faculty positions remain vacant, due to the heavy demand from industry and the high salaries offered by the industry (Indian Ministry of Human Resource Development, 2006). With few exceptions, universities are almost exclusively devoted to the provision of (qualified) manpower to the local firms (Basant and Chandra, 2006). Research is often more basic and, as a consequence, universities are not playing a significant role in supporting innovation and generating research suitable for local firms. Faculty with postgraduate education is hard to find (Indian Ministry of Human Resource Development, 2006). As Basant and Chandra (2006) indicate, only a handful of institutions provide both high-quality undergraduate and postgraduate teaching and research. These are the universities considered as Category I, which are actively engaged in research in collaboration with the industry.[3] Table 8.3 provides some examples of research collaboration between the top five IT service firms and a selection of universities located in Bangalore and the surrounding region.

Table 8.3 Sample of collaboration in research

Top 5 IT services companies located in Bangalore	University partner in Bangalore (area of collaboration)	University partner elsewhere (area of collaboration)
TCS	IISc (Advance product design and prototyping)	IIT Delhi (Lab for intelligent internet research) IIT Chennai (Computational engineering) University of California (Internet quality of service) University of Wisconsin (Business components) Carnegie Mellon University (Center for the Study of Software Industry)
Infosys	IIIT Bangalore (Banking, C-BIT center)	
Wipro	Institute of Bioinformatics and applied biotechnology (Bioinformatics)	
HCL Tech	–	IIT Delhi (Supercomputing facility for bioinformatics and computing biology)
Satyam		Center for cellular and molecular biology (Hyderabad) (Bioinformatics) John Hopkins University (Health care)

Source: information compiled from the websites of the firms and universities.

As an example, the International Institute of Information Technology (IIIT) located in Bangalore hosts research laboratories from different companies such as Honeywell, Intel, HP or Siemens as well as several research centres such as C-Bit (research on banking systems), C-Ait (research on automotive software) or C-Hit on healthcare and IT. Similarly, the Indian Institute of Science has contributed significantly to the growth of the biotech cluster in Bangalore, generating some important spillovers in terms of bioinformatics research for the IT industry (Basant and Chandra, 2006). But outside those top higher education institutions, interactive learning with universities is thus weak (D'Costa, 2006).

In sum, the Bangalore system of innovation is fragmented with large disparities in both the knowledge generation (universities and research institutions) and exploitation subsystems, with organizations that are competing glob-ally coexisting with low-quality education institutions and firms operating in the low end of the value chain. Under these circumstances, the discussion on whether to invest the scarce resources in improving the general IT education of a larger majority of population (generative role), to invest in more elite institutions that can better support the research requirements of the most advanced firms, or to maintain the current distribution is a crucial, albeit highly complicated one. The analysis of the possible future scenarios in Bangalore might shed some light to the issue.

8.3.3   A dual university–industry system?

The combined analysis of the potential future strategies for the knowledge exploitation subsystem as well as the type of university and education insti-tutions available to the Bangalore firms might illustrate the open debate between the generative and developmental role of the university. There are five possible strategies for the IT firms located in Bangalore (NASSCOM-McKinsey, 2005a; D'Costa 2006) (Table 8.4):

- *Global Champions*: Become global IT service and software players, compet-ing in the same market segment as IBM, Accenture and other such players.
- *IT specialist in a segment:* Become an IT specialist in three or four major industry verticals or cross-industry service lines targeting OECD markets.
- *IT specialist in secondary markets:* Become an IT specialist targeting niche markets (such as Japan, Latin America or other Asian countries).
- *ADM specialist:* Turn into an ADM factory (specialized in advanced design manufacturing), providing low-cost application development and ser-vice maintenance, linking with smaller business consulting and system integrators.
- Or become a *BPO specialist*, focusing on process re-engineering or other specialized BPO services.

The selection of a particular strategy is a function of the existing capabil-ities of the firm. It follows that the most advanced IT firms will be the ones prepared to adopt the first strategy, while the others will be more or less compelled to adopt a different one.

One factor that emerges clearly from this table and the historical evolu-tion of Bangalore is that, initially, the leading institutes had only a positive indirect spillover to the private industry and even today the provision of high-quality specialized research benefits only a small proportion of the firms. These are the firms that are performing better in terms of exports and revenues. The long-term derived spillovers on the regional innovation system remains to be seen. Different scenarios are possible. The university–industry collaboration can turn out to be successful and support the general upgrading of the cluster by generating spillovers into the SMEs and spin-offs

Table 8.4 Strategy and university roles

Strategy	Actors	Number	Role of university	Possible potential university partners
Global Champions	Tier I Indian firms (TCS, Wipro, Infosys, Satyam and HCL Technologies)	5–7	Developmental university: Training, advanced research and third task	Category I (ex. IISc, IIT IIIT. Total 15 institutions in the country)
IT specialist in a segment	Tier II suppliers (ex. Patni computers, NIIT, Mastek, i-flex, Polaris, CMC, Birlasoft, Mindtree)	10–15	Generative role (training and specialized research)	Category I (ex. IISc, IIT IIIT. Total 15 institutions in the country)
IT specialist in secondary markets (ex. Other Asian countries, Latin America)	Emerging players	>3000	Basic generative role (training of HR), Language training required!	Category II -Regional Engineering Colleges (17 institutions) and other University or Technical colleges (33)
ADM (advanced design manufacture)	Emerging players	>3000	Basic generative role (training of HR). Language training required	Regional Engineering Colleges
BPO specialists	Pure play BPO providers	2000	Basic generative role	REC and other government and self-financed institutions approx 50 in total)

from the global champions. These spillovers might even support other strategies (that is, *IT specialist in secondary markets*, for example) and reduce the entry costs for firms to upgrade and change strategy in this direction. This might be needed due to the increased global competition from other low-cost countries and/or labour shortage and increased salaries in Bangalore. Positive spillovers to firms aiming at becoming *IT specialist in a segment* are likely to be limited. Due to their technological gap and lack of absorptive capacity, spillovers into *ADM-* and *BPO-oriented* firms are not likely to happen either. In a less optimistic scenario the investments in a developmental role of the universities will either lead to further technological diversity among the indigenous firms in the cluster or benefit mainly the TNCs.

Extrapolating the experience of Bangalore suggests that at best the increased technological inequality among the indigenous firms is the most likely outcome. Lack of social capital and absorptive capacity (Chaminade and Vang, 2006; Vang and Chaminade, 2006) does not support the idea that we might expect great spillovers to SMEs (albeit successful collaboration in embedded software suggests that it might be possible). Currently, the TNCs located in Bangalore (and firms in the OECD countries) attract almost all the IT candidates from the best universities. Hence, the construction of localized knowledge spillover from universities to indigenous firms – especially SMEs – is an uncertain process and requires policy measures targeting these challenges.

Investing in increasing the quality of a greater number of training institutions (thus supporting the generative role of the university) could impact a larger number of firms in the regional innovation systems, thus contributing to a general upgrading of the RIS. Important, for example, for the third strategy, there is a need to invest in languages. Apparently, one of the reasons that India is losing market to China and other countries is their lack of knowledge of Japanese, French and Spanish. Yet the uncertainty with respect to this strategy is the future competitiveness of the Bangalore firms as their competitiveness is likely to be gradually eroded by increased wages and/or competition from other countries. While Tier I firms continue to grow, Tier II and the rest are losing markets. This suggests that the traditional generative role of the universities requires to be complemented by other policy measures supporting and creating incentives for the upgrading of the non-first-tier firms. Upgrading for these firms – at least from third tier and down – is not likely to emerge without public support. One particular task generative universities should focus on in education and research is innovation management, as this might provide the competencies needed for upgrading the indigenous SMEs.

Seen from a policy perspective one might also argue that the discussion on generative versus developmental universities is interesting from a 'distribution perspective'. Meaning focusing on 'how many of each' as opposed to 'either–or' is the main question. While being explorative in nature, our arguments suggest that attention should be paid to upgrading the promising mid-level universities with the highest direct contact with indigenous SMEs. This could imply a modified developmental role (that is, applied research of relevancy for SMEs upgrading). This, however, requires fundamental changes in budgets, internal incentive structure, salaries and formalized competency building.

8.4 Conclusions

Innovation systems research focusing on the role of universities in generating economic development is divided into two camps with radically opposing claims: proponents of the generative versus the developmental university. The current debate is based on experiences of the developed world and not well-suited to the reality of developing countries and regions such as Asia. In this vein the particularities of the firms' strategic choices (for upgrading and global expansion) and derived requirements for university-based support are often neglected. The literature draws on a hierarchical knowledge concept, which is less relevant to the strategies applied by catching-up Asian countries. Finally, the positions are polemic and constructed as 'either–or' positions that seldom take into account the specificities of different innovation systems.

Translating the discussion to the Bangalore regional system of innovation provides some interesting insights. The case suggests that in the initial phases, when the regional system of innovation is still in its emerging phase, it is sufficient for the universities to focus almost exclusively on the supply of qualified human resources with general and industry-specific skills (i.e. generative role). In the later phases, universities' roles become more complex and tied to the specific strategic choices made by the firms (their strategies are again conditioned by the innovation system and changes in the global competitive landscape). The IT firms located in Bangalore should follow one of these five different strategies if they want to stay in the market: aiming at becoming global champions, IT specialist in a segment, IT specialist in secondary markets or specialized in ADM or BPO.

Different strategies require different responses from the universities and call for different types of corrective policy measures. Firms aiming at becoming global leaders are most likely to benefit from universities taking a developmental role. Yet it is not unlikely that the TNCs will harvest the main benefits unless targeted policy measures are implemented. Second-tier firms and SMEs, in general, are not likely to acquire benefits. Generative universities can mainly support the upgrading of the firms in Bangalore by focusing on specific types of education of relevancy for their upgrading (i.e. innovation management). Reliance on generative universities is increasingly likely to require alternative supportive policy measures for third-tier firms. Second-tier firms are likely not to encounter major problems as competitiveness even among industry leaders within their field (i.e. IT services) is based on 'innovation without research'. Global leaders can maintain their current position and expand in IT service, but it is not likely that they will be able to compete in the science-based parts of the industry if the leading universities maintain a generative role. Concerning the distribution of different types of universities, we suggest that focus should be paid to the 'generative' role of mid-level performing universities. This is likely to lead to spillovers for the indigenous firms – and especially the SMEs. Yet, much research remains to be done in this area.

Notes

1. Yet, it should be acknowledged that changes in the domestic banking sector also played a central role in fuelling the development of the IT industry.
2. TNCs seem more concerned with protecting intellectual property rights. This applies not only with respect to external alliances, but also by applying highly modular R&D research strategies with minimum knowledge dissemination to their Indian subsidiaries.
3. Not all universities are present in Bangalore. The Indian firms need to develop subsidiary strategies that facilitate physical proximity to the relevant universities when the nature of the university-industry collaboration requires this. The strategies should also encompass subsidiary R&D mandate.

References

Adams, J.D. (2002) 'Comparative Localization of Academic and Industrial Spillovers', *Journal of Economic Geography* 2: 253–78.

Albu, M. (1997) 'Technological Learning and Innovation in Industrial Clusters in the South', SPRU electronic working papers, SPRU, University of Sussex.

Amin, S. (1989) *Eurocentrism*, London: Zed Press.

Arora, A., V.S. Arunachalam, J. Asundi and R. Fernández (1999) 'The Indian Software Industry. Carnegie Mellon Heinz School Working Papers. www.heinz.cmu.edu/wpapers/author.jsp?id=ashish.

Arora, A. and A. Gambardella (eds) (2005) *From Underdogs to Tigers: the Rise and Growth of the Software Industry in Brazil, China, India, Ireland and Israel*, New York: Oxford University Press, pp. 7–40.

Arora, A. and A. Gambardella (2004) 'The Globalization of the Software Industry: Perspectives and Opportunities for Developed and Developing Countries', NBER Working paper series 10538.

Asheim, B.T. and L. Coenen (2005) 'Knowledge Bases and Regional Innovation Systems: Comparing Nordic Clusters',*Research Policy*, 34(8): 1173–90.

Asheim, B.T., L. Coenen and J. Vang (forthcoming) 'Face-to-face, Buzz and Knowledge-bases: Socio-spatial Implications for Learning and Innovation Policy', *Environment and Planning C*.

Asheim, B.T. and A. Isaksen (2002) 'Regional Innovation Systems: the Integration of Local "Sticky" and Global "Ubiquitous" Knowledge', *Journal of Technology Transfer* 27: 77–86.

Asheim, B.T., A. Isaksen, C. Nauwelaers and F. Tödtling (eds) (2003) *Regional Innovation Policy for Small-Medium Enterprises*, Cheltenham: Edward Elgar, pp. 21–48.

Asheim, B.T. et al. (2006) *Constructing Regional Advantage: Principles, Perspectives, Policies*, DG Research Report, Brussels: European Commission.

Athreye, S. (2005) 'The Indian Software Industry', in A. Arora and A. Gambardella (eds), *From Underdogs to Tigers*. New York: Oxford University Press, pp. 7-40.

Basant, R. and J. Chandra (2006) 'Role of Educational and R&D Institutions in City Clusters: Bangalore and Pune Regions in India', mimeo.

Benneworth, P.S. and P. Arbo (2006) 'Understanding the Regional Contribution of Higher Education Institutions: a Literature Review', *IMHE 'The Regional Contribution of Higher Education' Project Report*, Paris: OECD.

Chaminade, C. and J. Vang (2006) 'Innovation Policy for Asian SMEs: an Innovation Systems Perspective', with J. Vang in H. Yeung, *Handbook of Research on Asian Business*, Cheltenham: Edward Elgar.

Chaminade, C. and J. Vang (2008) 'Globalisation of Knowledge Production and Regional Innovation Policy: Supporting Specialized Hubs in Developing Countries', *Research Policy*, 37(10), in press.

Cohen, W.M., R.R. Nelson et al. (2002) 'Links and Impacts: the Influence of Public Research on Industrial R & D', *Management Science* 48(1): 1–23.

Cohen, W. and D. Levinthal (1990) 'Absorptive Capacity: a New Perspective on Learning and Innovation', *Administrative Science Quarterly* 35: 128–52.

Cooke, P. and K. Morgan (1998) *The Associational Economy: Firms, Regions and Innovation*. Oxford: Oxford University Press.

Cooke, P. and D. Will (1999) 'Small Firms, Social Capital and the Enhancement of Business Performance through Innovation Programmes', *Small Business Economics*, 13: 219–34.

D'Costa, A. (2006) 'Exports, University–Industry linkages and Innovation Challenges in Bangalore, India', World Bank Policy Research Working Paper 3887, April 2006. http://econ.worldbank.org.

Doloreux, D. (2002) 'What We Should Know About Regional Systems of Innovation', *Technology in Society*, 24: 243–63.

Etzkowitz, H. and M. Klofsten (2005) 'The Innovating Region: Toward a Theory of Knowledge-based Regional Development', *R&D Management*, 35(3): 243–55.

Etzkowitz, H. (2002) *MIT and the Rise of Entrepreneurial Science*, London and New York: Routledge.

Etzkowitz, H. and L. Leydesdorff (2000) 'The Dynamics of Innovation: From National Systems and "Mode 2" to a Triple Helix of University–Industry–Government Relations', *Research Policy* 29(2): 109–23.

Fagerberg, J. (2004) 'What Do We Know About Innovation? Lessons from the TEARI Project'. Paper presented the TEARI final conference 'Research, Innovation and Economic Performance - What Do We Know and Where Are We Heading?', Brussels.

Feldman, M. (2003) 'The Locational Dynamics of the US Biotech Industry: Knowledge Externalities and the Anchor Hypothesis', *Industry and Innovation*, 10(3): 311-29.

Gertler, M. S. (2004) *Manufacturing Culture: The Institutional Geography of Industrial Practice*, New York: Oxford University Press.

Glaeser, E.L., H.D. Kallal et al. (1992) 'Growth in Cities', *Journal of Political Economy* 100(6): 1126–52.

Goddard, J. B. and P. Chatterton (2003) 'The Response of Universities to Regional Needs', in F. Boekema, E. Kuypers and R. Rutten (eds), *Economic Geography of Higher Education: Knowledge, Infrastructure and Learning Regions*, London: Routledge.

Giuliani, E. (2004) 'Laggard Clusters as Slow Learners, Emerging Clusters as Locus of Knowledge Cohesion (and Exclusion): a Comparative Study in the Wine Industry. LEM Working papers, Pisa, Laboratory of Economics and Management – Santa Anna School of Advanced Studies: 38.

Giuliani, E., and M. Bell (2005) 'When Micro Shapes the Meso: Learning Networks in a Chilean Wine Cluster', *Research Policy* 34: 47–68.

Gunasekara, C. (2006) 'Reframing the Role of Universities in the Development of Regional Innovation Systems', *Journal of Technology Transfer*, 31(1): 101–11.

Hollingsworth, J.R. (2000) 'Doing Institutional Analysis: Implications for the Study of Innovations', *Review of International Political Economy*, 7(4): 595–644.

Indian Ministry of Human Resource Development (2006) 'IT Manpower Challenge and Response', Interim Report of the Taskforce of IT in HRD. http://www.education.nic.in/it_hrd_report/.

Isaksen, A. (2003) '"Lock-in" of Regional Clusters: the Case of Offshore Engineering in the Oslo Region', in *Cooperation, Networks, and Institutions in Regional Innovation Systems*, Cheltenham: Edward Elgar, pp. 247–73.

Jacob, M. (2006) 'Utilization of Social Science Knowledge in Science Policy: Systems of Innovation, Triple Helix and VINNOVA', *Studies of Science*, 45(3): 431–61.

Krishna, V.V. (2001) 'Changing Policy Cultures, Phases and Trends in Science and Technology in India', *Science and Public Policy*, 28(3): 179–94.

Lundvall, B.-A. and S. Borrás (1999) 'The Globalizing Learning Economy: Implications for Innovation Policy', Luxembourg: Office for Official Publications of the European Communities.

Lundvall, B.-A. (2002) 'The University in the Learning Economy', DRUID Working Papers, No. 6, ISBN: 87-7873-122-4.

Martin, B. and H. Etzkowitz (2006) 'The Origin and Evolution of the University Species', *Journal for Science and Technology Studies*, 13(3–4): 9–34.

Martin, R. and P. Sunley (2003) 'Deconstructing Clusters: Chaotic Concept or Policy Panacea?', *Journal of Economic Geography* 3(1): 5–35.

Molas-Gallart, J., A. Salter, P. Patel, A. Scott and X. Duran (2002) 'Measuring Third Stream Activities', Final Report to the Russell Group of Universities, Brighton: SPRU, University of Sussex.

Mowery, D.C. and B. Sampat (2005) 'Universities in National Innovation Systems', in J. Fagerberg, D. Mowery and R. Nelson (eds), *The Oxford Handbook of Innovation.* Oxford: Oxford University Press, pp. 209–39.

Nadvi, K. (1995) 'Industrial Clusters and Networks: Case Studies of SME Growth and Innovation', Vienna: UNIDO.

NASSCOM- McKinsey (2005a) 'Extending India's Leadership of the Global IT and BPO Industries'. http://www.mckinsey.com.

NASSCOM-McKinsey (2005b) 'The Emerging Global Labor Market'. http://www.mckinsey.com/mgi/rp/offshoring/.

Nooteboom, B. (2000) 'Learning by Interaction: Absorptive Capacity, Cognitive Distance and Governance',*Journal of Management and Governance*, 1–2: 69–92.

North, D.C. (1990) *Institutions, Institutional Change and Economic Performance*, Cambridge: Cambridge University Press.

Nygaard, C. (ed.) (2001) *A Reader on Strategizing*, 1st edition, Copenhagen: Samfundslitteratur.

OECD (1999) *Boosting Innovation: The Cluster Approach*, Paris: OECD.

Parthasarathy, B. (2004a) 'India's Silicon Valley or Silicon Valley's India? Socially Embedding the Computer Software Industry in Bangalore', *International Journal of Urban and Regional Research*, 3: 664–85.

Parthasarathy, B. (2004b) 'Globalizing Information Technology: the Domestic Policy Context for India Software Production and Exports', Interactions: an interdisciplinary journal of software industry. http://www.cbi.umn.edu/iteractions/parthasarathy.

Parthasarathy, B. and Y. Aoyama (forthcoming) 'From Software Services to R&D Services: Local Entrepreneurship in the Software Industry in Bangalore, India', *Environment and Planning A*.

Pietrobelli, C. and R. Rabellotti (2004) *Upgrading in Clusters and Value Chains in Latin America: The Role of Policies. Sustainable department Best Practices Series*. New York: Inter-American Development Bank.

Pietrobelli, C. and R. Rabellotti (2006) *Upgrading and Governance in Clusters and Value Chains in Latin America*, Cambridge, MA: Harvard University Press.

Porter, M.E. (1998) 'Clusters and the New Economics of Competition', *Harvard Business Review*, 76(6): 77–90.

Said, E. (1993) *Culture and Imperialism*, London: Chatto & Windus.

Sardana, D. and V.V. Krishna (2006) 'Government, University and Industry Relations: The Case of Biotechnology in the Delhi Region', *Science, Technology and Society*, 11(2): 351.

Saxenian, A. (2001) 'Bangalore: the Silicon Valley of Asia?', Centre for Research on Economic Development and Policy Reform. Working paper 91. http://www.sims.berkeley.edu/~anno/papers/ bangalore_svasia.html.

Schmitz, H. (1992) 'On the Clustering of Small Firms', *IDS Bulletin – Institute of Development Studies*, 23: 64–9.

Trajtenberg, M., R. Henderson and A. Jaffe (1997) 'University versus Corporate Patents: a Window on the Basicness of Invention', *Economics of Innovation and New Technology*, 5: 19–50.

Turpin, T. and C. Martinez-Fernandez (2003) 'Riding the Waves of Policy', *Science, Technology and Society*, 8(2): 215–34.

UNIDO (United Nations Industrial Development Organization) and UUND Program (2001) 'Development of Clusters and Networks of SMEs'. http://www.unido.org/userfiles/PuffK/SMEbrochure.pdf.

UNIDO (2004) *Partnerships for Small Enterprise Development*, New York: United Nations.

Vang, J. and B. Asheim (2006) 'Regions, Absorptive Capacity and Strategic Coupling with High-Tech TNCs: Lessons from India and China', *Society, Science and Technology*, 11(1): 39–66.

Vang, J. and M. Overby (2006) 'Transnational Communities, TNCs and Development: the Case of the Indian IT-services Industry', in B.-Å. Lundvall, I. Patarapong and J. Vang (eds), Asia's Innovation Systems in Transition, Cheltenham: Edward Elgar.

Vang, J. Chaminade, C. (2006) 'Building RIS in Developing Countries: Lessons from Bangalore', CIRCLE Electronic Working Paper 2006/02. http://www.circle.lu.se.

Yeung, H. and G.C.S. Lin (2003) 'Theorizing Economic Geographies of Asia', *Economic Geography*, 79(2): 107–28.

Zucker, L. G. and M. R. Darby (1996) 'Star Scientists and Institutional Transformation: Patterns of Invention and Innovation in the Formation of the Biotechnology Industry', PNAS 12 November, 93(23): 12709–12716.

9
Foreign Talent and Innovation: China and India in the Japanese Software Industry[1]

Anthony P. D'Costa and Tomoko Kobayashi

9.1 Introduction

The purpose of this chapter is to relate innovation dynamics to the international mobility of technical talent in Asia. As the significance of human capital has increased due to the demands of knowledge-based activities, those countries that are confronted with supply shortages of technical talent must compete for such talent in the global marketplace. We posit that Japan, despite its strength in information and communications technologies (ICT), faces impending labour shortages and remains a high-cost producer. Japan has entered the phase of 'demographic crisis' where a declining fertility rate resulting in labour shortages is an increasingly significant issue (see National Institute of Population and Social Security Research, 2003; Nakamura, 2004; Asia Program Special Report, 2003).[2] While labour shortages and cost reductions can be compensated for partly by increasing productivity, and Japan has a good record on this, we believe the domestic availability of technical professionals for its knowledge-intensive sectors will be limited. Furthermore, the historical insulation of Japan's information and communications sector from the global one has prevented Japan from using cheaper, often better talent found elsewhere. Hence, we believe cost reduction through offshore outsourcing is likely to enhance Japan's competitiveness and will become prevalent in Japan in the future.

There is already some evidence of labour shortages. In a 2001 vendor survey almost one-third of the responses listed the supply and demand of IT engineers as a constraint; and one-fifth expressed the need for more high-tech engineers for future growth (Japan Information Technology Services Industry Association (JISA), 2006: 10, 14). Variously, contributors in this volume have indicated the importance of China and India as sources of large skill bases, although India has the advantage of a much younger population in comparison to China's ageing society. References have been also made to talent supply constraints in some of the smaller OECD countries in Europe and Asia and also in some of the larger countries such as Germany and Japan.

216

In this chapter we examine the role of foreign talent, mainly from China and India, in the Japanese software industry to bring out some of the innovation challenges and opportunities faced by the three countries.[3]

We know that the availability of technical talent is critical to innovations in general – and in the software industry in particular. The international mobility of technical talent in an era of global integration complements the domestic supply of talent. Talent mobility enhances knowledge networks and epistemic communities. Both benefit and sustain regional clusters where such talent resides. Using Japan as an entry point we hypothesize that labour shortages under high-cost production in Japan will induce greater movement of Asian technical talent to Japan. We also hypothesize that China and India, for various structural and institutional reasons, are likely to exhibit different degrees of engagement with Japan, with China enmeshed with the Japanese software industry far more than India. We provide partial evidence for both by examining the inflows of technical professionals to Japan. We anticipate few Indian professionals in Japan, given India's cultural distance from Japanese business practices and its stronger links to the US and other Anglophone countries. However, we believe that cost pressures and the shortages of software engineering talent in Japan are likely to create a greater engagement with other Asian countries, including India.

We relate the significance of talent flows to Japan's demographic predicament. The growth rate of population has fallen to a replacement rate. Gender role expectations, despite significant economic and social shifts, remain unfavourable to Japanese women in male-dominated technology sectors. And, similar to many other OECD economies, science and engineering education is becoming less popular in Japan. Consequently, the supply of technical talent as a whole has been constrained. The demographic dilemma is compounded by the fact that Japan, for both cultural and economic reasons, has been wary of large influxes of foreign workers and professionals, while foreigners themselves find living in Japan less hospitable than in many other OECD locations (Papademetriou and Hamilton, 2000; Douglass and Roberts, 2000). The catch is that Japan is an affluent country with the world's second-largest high-technology market, after the USA. Its IT services industry has been steadily increasing since the mid-1990s, accounting for 2.88 per cent of GDP, while its GDP has remained stagnant during this period (JISA, 2006: 1–2). Japan's international competitiveness in manufacturing is legendary but fast-second followers such as South Korea and Taiwan are rapidly narrowing the technological gap in many high-tech sectors. As manufacturing becomes subject to knowledge-intensive, research-driven activity, Japan, by necessity, must find creative solutions to generating technical talent domestically and securing it internationally.

To date, there is little research on foreign talent in Japan. One notable exception is an OECD-sponsored study, although its findings may no longer reflect the current state of the market (see Kobayashi, 2001). Much of the focus on foreigners in Japan has centred on low-wage workers in construction, small manufacturing industries, and the 'entertainment' sectors (see Ahmed, 2000). There are some references to western professionals in the Japanese business services sector, but not to technical professionals (Sassen, 1998; Fuess, 2003), while virtually all of the research on Indian and Chinese professionals abroad is focused on immigration to the US (Saxenian, 1999; Leng, 2002).

For the purposes of this study we assume that Japan has a strong national innovation system, in which inter-firm collaboration (*local* subcontracting and outsourcing) is very high. The *keiretsu* system of business organization and the long-established practice of buyer–supplier relationships in Japan have been strengths for socializing risks and undertaking innovative activities.[4] Over the course of several decades the state has contributed substantially to Japan's innovation dynamics. Japanese businesses, in turn, have fostered internal labour markets through long-term investments in 'on the job training' for workers.

There are, of course, weaknesses in the Japanese system of innovation. Global competition demands substantial organizational flexibility to respond to market developments, especially in the software industry, which could be limited by Japanese long-term inter-firm relations and internal labour markets designed to maintain business stability (see Anchordoguy, 2004). Given high production costs the Japanese economy faces the spectre of increased competition from other Asian countries. On the one hand, Japan must induce innovation-based productivity growth and, on the other, it must find adequate human capital to sustain this effort. Hence, Japan needs to re-evaluate its innovation system with the understanding that competitiveness through innovation-based productivity growth will be inadequate to maintain its leadership position in critical technologies without adequate supplies of technical professionals.

In the next section we discuss the importance of the mobility of technical talent to innovation-based competitiveness. The following section examines Japan's demographic crisis in relation to its weakness of talent supply. The fourth section broadly examines the inflows of technical talent into Japan from elsewhere in Asia. This is followed by an analysis of Chinese and Indian software firms in Japan, bringing out business and innovation strategies of firms in Japan from all three countries. Section six outlines some broader policy challenges and opportunities for all three countries in the software industry and suggests ways to enhance greater integration for innovation among them.

9.2 The significance of international talent mobility to innovation

Following the literature on international migration and social networks, we assume that the movement of technical talent in the global economy is based on the mismatch between the supply of and demand for technical workers, albeit influenced by social, political, and cultural factors. Technical professionals are also tied into social and knowledge networks, which facilitate their global mobility and the sharing of technical and commercial knowledge (D'Costa, 2006; Lam, 2005). For example, social networks comprise relatives, friends, and alumni, while knowledge networks go beyond these to include professional colleagues and peers. The global movement of talent accompanies the movement of goods and services, capital, and technology (Dicken, 1998; Held et al., 1999).

The intellectual and entrepreneurial contribution by foreign and expatriate technical talent in both host and home countries has been recognized in recent studies (Saxenian, 2003, 2006; OECD, 2001; Khadria, 2004a, 2004b; D'Costa, 2006). However, there is as yet little information on either the software sector or the extent of foreign technical involvement in Japan (OECD, 2003a, 2003b). We also do not know much about how they fit in with Japanese innovation strategies in the IT industry. In an otherwise excellent report by the US-based Association for Computing Machinery (2006) on the offshoring of software production, the analysis stuck to a static view of a relatively fixed division of labour between Japan and China and the USA and India. Based on our information from the field, as well as developments at the bilateral level, we can anticipate more dynamic interactions between Japan and India in the medium- to long-term period. Consequently, there are significant gaps in our understanding of how talent-scarce Japan fits into the innovation dynamics of Asia and how its competitiveness in high-tech industry might be impacted in the future.

Technical talent is critical to knowledge-based activities for the simple reason that knowledge workers require an advanced tertiary education – especially in the fields of science and technology. They possess specific knowledge or expertise that places them at an economic advantage when in demand. What is profoundly different for knowledge workers today is that the demand for their skills need not be driven by local or national economies. With increased global economic integration, professional workers can serve far-flung international markets either from the home base or by being in the market (D'Costa, 2003).

Following the logic of cluster dynamics, when a critical mass of technical talent accumulates in a region, we anticipate network externalities (Chaminade et al., Chapter 4; D'Costa, Chapter 3). Economic externalities are those effects that are either not intended or anticipated or which do not result directly from the activity itself. There are both positive and negative externalities associated with economic activities, many of which only show up later. For example, the spending on education may have a favourable impact on the children of the educated; or, conversely, increased spending on automobiles could worsen air quality. Network externalities refer to economic impacts based on the number of users, consumers, or producers and their integration. Thus, as more people use personal computers, the easier it becomes to design a variety of software programs for PCs. There is not only a scale effect but there is also a 'network' effect as connections established between users allow information to be shared and collective problems to be solved (Meyer, 2001). Network effects give rise to 'epistemic communities' (Cowan, 2004: 8). The links between users, suppliers, manufacturers, research and development (R&D) centres, and universities could theoretically support a gamut of interrelated industries that tap into the knowledge pool. For example, interlocking corporate directorships, representing a high density of networks, contribute to knowledge transfer (O'Hagan and Green 2004: 131–2). Networks thus have a 'social' dimension to them as they tend to generate knowledge that spills over from the immediate economic activity and multiplies both within this and across other economic activities (Maskell and Malmberg, 1999).[5] This is facilitated by the inter-firm mobility of professionals, which generates flows of knowledge and fosters social networks. The result of such mobility, whether within a local cluster or internationally to other clusters, acts as a reinforcing mechanism for cluster sustainability (Caspar, 2007).

The more dense the networks in a region and the larger the number of nodes in the global network, the greater is the overall production of knowledge within the network and also the greater the possibility of its diffusion through the numerous channels linking the different nodes. Access to both high volume and high quality of knowledge is critical to network externalities. An integrating economy is likely to have more links to the various knowledge nodes representing technical, organizational, commercial, market, and infrastructural information at the global level (Davenport, 2004). However, it is possible that the links to the nodes are weak, as is often the case with economies that suffer from structural dependence. In the case of technical and professional workers who have worked abroad, the access to knowledge nodes need not be of higher quality if such workers provide peripheral services to the more knowledge-intensive projects carried out at the principal nodes. For example, the Indian IT industry, despite its global prominence, still suffers from weak innovation links (D'Costa, Chapter 3), while Japan, strategically insulated until recently, has fallen behind in innovativeness in the software sector (Anchordoguy, 2004).

Students have an important place in the global movement of talent (Tremblay, 2001: 61). First, there is increased social interaction within the science and engineering peer groups that contributes to technical knowledge at both the individual and collective levels. Secondly, as a high percentage of students tend to remain in the receiving countries, their contribution to the host economies is substantial (Burrelli, 2004). This permanent migration adds to the stock of knowledge in the host country, which in static terms is seen as brain drain for the sending country. However, this stock can be also seen as a 'brain bank' to be tapped to benefit sending countries at a later date (Meyer and Brown, 1999; Saxenian, 2006).[6]

As economies expand and there is an increase in the global demand for technical services, sending countries such as China and India could become viable sites for the offshore development of technical services for global clients. In this scenario, the migration of talent is expected to expand the pool of technical professionals in sending countries and benefit receiving countries (D'Costa, 2006). Conversely, there is also the possibility of expatriate talent returning home to establish businesses for familiar foreign clients.[7] This long-term circulation of talent contributes to knowledge and commercial networks in both sending and receiving countries. The entry of foreign talent creates the basis for a diverse knowledge base, cost reduction opportunities via offshore links created by expatriate talent and ultimately new development centers for the global market. The clustering of talent in general tends to energize innovative activities due to information externalities and reduced transactions costs. Familiarity of receiving countries with sending countries becomes a critical condition for strengthening global partnerships for securing talent in a competitive talent market and establishing entrepreneurial links with the receiving countries abroad. Hence, given varying familiarity of China and India with Japan, their relationships with Japan are also likely to be different.

9.3 Japan's demographic dilemma

There are at least two reasons for the shortages of highly skilled technical professionals in Japan. The first is a shrinking pool of students pursuing technical studies in the OECD economies. Secondly, a declining fertility rate and a concomitant ageing population are contributing to an absolutely smaller talent pool (see Cohen 2004: 38–9). As a result, the dependency ratio (non-working population divided by the working population) is increasing in the OECD economies. The bulk of the population in developing countries is young, which also contributes to high dependency ratios, but countries such as India and China have an absolutely larger pool of talent (D'Costa 2004). However, in contrast to India, China has a slow-growing population and its dependency ratio is on the rise due to an ageing population. While global population has increased by nearly 40 per cent since 1980, the share of population of the high-income countries fell from 19 per cent to 16 per cent (World Bank 2003). Conversely, the share of the low-income countries increased from 36 per cent to 41 per cent. Since technical workers come typically from the upper- and middle-income households from poor and middle-income countries, such aggregate data do not necessarily convey the nature of skilled labour migration. Nevertheless, a rudimentary understanding of expected talent flows to Japan can be extrapolated from its demographic crisis.

By 2015 Japan is expected to witness an absolute decline in its population, with its average annual rate of population growth falling to −0.2 per cent (World Bank, 2003; see also National Institute of Population and Social Security Research, 2003: 18). Currently it has the highest proportion of people in the 65 and over age group (Asia Program Special Report, 2003). Its child dependency ratio has fallen consistently since 1940 and stood at 21.2 per cent (National Institute of Population and Social Security Research 2003: 14). Conversely, its old-age dependency increased from 10.2 per cent in 1970 to 26.5 per cent in 2001. The demographic data for some of the low- and middle-income countries such as India, Thailand, China, and the Philippines show that although population growth rates are expected to decline considerably by 2015, their absolute population growth will keep their dependency ratios high. One-quarter to one-third of their population is currently under 14 years of age, which makes the dependency ratio high by OECD standards. At the same time the population group between 15 and 64 years of age, classified as the labour force, will continue to rise. China and India are expected to increase their workforce by 7 per cent and 18 per cent respectively over the period 2001–10, with a combined labour force of 1.4 billion. For Japan the labour force has been predicted to shrink absolutely from 68.2 million in 2001 to 66 million by 2010 – a decline of over 3 per cent (World Bank 2003). This implies future flows of people to Japan. It also suggests that as more people become dependent on those who have jobs and as the number of jobseekers itself increases in the developing countries, the pressure to find employment will mount drastically. Under such a situation international migration will be one outlet for venting 'excess' labour, assuming that these economies will not be able to absorb all of the growth in the working age population.

Japan has exhausted the alternative sources of labour. For example, it is institutionally impractical in the medium-term to tap rural residents. Many rural areas of Japan are already depopulated. There is great reluctance to induct women into the more male-dominated economic sectors as gender role expectations have not changed substantially (Nakamura, 2004: 18–22) and levels of female participation in Japanese labour markets is already high. There are discussions on how to incorporate the elderly population more, however, for the knowledge-intensive industry this is not a realistic solution at this time. Automation is also not a real solution for coping with labour shortages since Japanese industry has been substantially automated. Under this predicament there are two medium-term complementary options. One is allowing more foreign workers into Japan and the other is to increase the outsourcing of production of goods and services from abroad. This will allow firms to redeploy their workforce and complement it with foreign professionals in areas where larger inputs of human capital are required. Selectively, Japan has already exercised both options. Throughout the 1980s Japanese firms farmed out production to South East Asian economies. Rising business costs, especially wages, the appreciation of the yen, massive foreign exchange revenues, and greater environmental consciousness pushed Japanese businesses to invest overseas. However, a good part of this outsourcing relies on inexpensive industrial labour in mature industries using standardized technologies. In knowledge-intensive activities, where there is greater uncertainty, using foreign talent locally or abroad via outsourcing arrangements has not been used widely in Japan. However, local Japanese businesses are relying on foreign workers, as evidenced by increasing inflows of migrants into Japan. Acutely aware that the small and medium firms in Japan were becoming uncompetitive, the government relaxed somewhat its immigration laws to import labour, often under the guise of industrial training.[8]

9.4 The pattern of foreign talent flows

Historically speaking, Japan has had an unfavourable view of foreigners (Kuwahara, 1998; Iguchi, 1998; Cohen and Zaidi, 2002: 28–9). Cultural homogeneity, linguistic barriers, and the nationalist sentiments of Japanese firms have limited interactions with the international economy and community and precluded large inflows of immigrants. Particular forms of corporate governance and inter-firm arrangements such as the *keiretsu* have effectively excluded foreign firms from the Japanese economy. Long-term employment, local subcontracting relations, and reliance on internal labour markets have essentially excluded the employment of foreign workers. Consequently, the level of migration to Japan has been relatively low. However, due to demographic pressures and shifts in social preferences and attitudes toward work, international migration to Japan is on the rise. Japan attracts mainly Chinese, Koreans, Filipinos, and Brazilians (mostly of Japanese ancestry). Most foreigners in Japan enter labour-intensive industries in small and medium-sized enterprises, for manufacturing, construction, and entertainment. The key difference between Japan and most other OECD countries is that permanent residency and Japanese citizenship are highly restricted (Hanami, 1998; Brody, 2002). Based on data from the Ministry of Justice, the share of permanent residents relative to total registered foreigners shows a consistent decline from a high of 97.1 per cent in 1952 to 38.9 per cent in 2001 (National Institute of Population and Social Security Research, 2003: 115). This trend, however, is not inconsistent with increasing mobility of people since registered foreigners in Japan increased more than threefold during this period. This may suggest tightening of permanent residency after 1950 but flexibility toward short-term inflows of people.

One way of establishing the increasing inflows of foreign talent is to examine the type of professionals entering the country.[9] If we consider foreign entrants in four categories by resident status – professor, researcher, engineer, and intra-company transferee – part of foreign technical talent, we find that between 1992 and 1997 the number of such entrants increased from 9,321 to 11,196 – an increase of 20 per cent over a five-year period (Kobayashi, 2001: 121).[10] In later years (1998–2004), we find the total number of such high-skilled foreign technical talent registered in Japan increasing from 30,000 to 45,000 – an increase of 50 per cent over a five-year period (Table 9.1). Such technical talent represented 50 per cent of the total number of professionals (defined as the total of professionals, engineers, and intra-company transferees). What is incontrovertible is the importance of Asian talent in the Japanese economy, representing nearly 75 per cent of total technical talent. Asians also represent a slightly higher share of technical talent relative to total Asian professionals, 57 per cent compared to 51 per cent for foreigners as a whole.

When we disaggregate the Asian data we find that there is a big divergence between China and India's share of talent resident in Japan (Table 9.1). In 2004, China represented 55 per cent of Asia's technical talent, 58 per cent of Asia's professionals, and 38 per cent of total professionals registered as aliens in Japan. The corresponding figures for India stood at 11 per cent, 8 per cent and 5 per cent respectively. Thus India's share of Asian technical talent is only one-fifth of China's.[11] This is consistent with our hypothesis of the anticipated pattern of flows. Indian technical professionals prefer English-speaking receiving countries, which also draw them as students. The USA is clearly a preferred destination for Indian technical talent and students, displacing the UK over the years (see D'Costa and Parayil, Chapter 1, Figure 1.3). However, notwithstanding the big gap between the Chinese and Indian flows, the former has declined and the latter has increased somewhat. For example, China's share of Asian technical talent fell from 68 per cent in 1998 to 55 per cent in 2004, while India's share increased from 6 per cent to over 11 per cent during the same period. We believe that as demand for technical talent increases globally, and as severe shortages emerge in the large OECD economies, especially Japan, India's presence will become visibly larger. This is likely to be a compound result not only of more Indian firms seeking a share in the large, lucrative Japanese market over the long haul but also because Japanese IT companies, wishing to reduce costs, will engage Indian IT companies more extensively. Since Japanese prefer far greater 'face to face' interaction, the presence of Indian software and technical professionals is also expected to increase in Japan.

Foreign students in technical professions often contribute to the talent pool in the receiving country. This is certainly the case in the USA, where a large number of students elect to stay (D'Costa, 2006). For example, nearly 70 per cent of foreign PhD students in the USA had plans to remain there

Table 9.1 Chinese and Indian Alien registrations in Japan by professional/technical background

	Total			Asia		
	1998	2001	2004	1998	2001	2004
#Professionals**	45076	57646	65313	21318	26617	31681
1.Professor*	5374	7196	8153	2433	3697	4386
2.Research*	2762	3141	2548	1859	2308	1856
3.Engineers**	15242	19439	23210	13029	16727	20397
4.Intra-company**	6599	9913	10993	3331	5221	6810
Total of 4 = (1+2+3+4)	29977	39689	44904	20652	27953	33449
Total talent (TT) (**)	60318	77085	88523	37678	48565	58888
Share of 4/Total talent (%)	50.9	51.5	50.7	54.8	57.6	56.8
Share of Asia's TT to Total TT (%)				62.5	63.0	66.5
Share of Asia's 4/ Total 4 (%)				68.9	70.4	74.5

	China			India		
	1998	2001	2004	1998	2001	2004
#Professionals**	14774	16830	19170	800	1121	1258
1.Professor*	1659	2228	2471	102	246	292
2.Research*	1145	1387	1043	224	234	171
3.Engineers**	9904	11382	11981	709	1286	2298
4.Intra-company**	1367	193	2753	260	674	993
Total of 4 = (1+2+3+4)	14075	15190	18248	1295	2440	3754
Total talent (**)	26045	28405	33904	1769	3081	4549
Share of T of 4/ Total country talent (%)	54.0	53.5	53.8	73.2	79.2	82.5
Share of country 4/ Total Asia 4 (%)	68.1	54.3	54.6	6.3	8.7	11.2
Share of country total talent to total talent (%)	43.2	36.8	38.3	2.9	4.0	5.1
Share of country total talent to Asia total talent (%)	69.1	58.5	57.6	4.7	6.3	7.7

Notes: #Professionals = Professor, Journalist, Investor/Business Manager, Lawyer, Accountant, Medical Staff, Research, Humanities/International Business.
Total talent = sum of professionals, engineers, and intra-company transfers.
Total technical talent (based on resident status) = Total of 4 = sum of Professor, Researcher, Engineer, and Intra-company transfers.
Source: Japan Immigration Association, various years.

after completing their studies (Guellec and Cervantes, 2001: 92). However, in Japan, given the difficulties of assimilation and job prospects for foreigners, the permanent residency of foreigners is not inevitable. Even Chinese students prefer going to the USA and other OECD countries. Furthermore, in contrast to the USA, Japan does not have an educational system that draws heavily on foreign students. In 2005, the total number of students stood at 121,812, the highest to date. In 2006, the number fell to 117,927. In general, foreign students in Japan tend to pursue social sciences and humanities rather than engineering or sciences.[12] However, we anticipate that student populations will increase in the future, particularly when the shortages of technical talent become acute. Curiously, the Japanese government does not seem to regard foreign students as a future pool of technical talent – notwithstanding their recognition of the importance of technical education.

However, as total enrollments in various engineering fields have increased only marginally in Japanese technical colleges since the early 1970s (Figure 9.1), the possibility of increased foreign students must be given due consideration. From 1973 to 2006, the numbers increased from about 48,000 to about 56,000 – an increase of 17 per cent over the two decades (Ministry of Education, Culture, Sports, Science and Technology, various years). Mechanical, electrical, and chemical engineering has consistently declined since the late 1980s and early 1990s, while electronic and information-related subjects grew sharply since the mid-1980s, but stabilized by the early 1990s. There was a sharp increase in engineering under 'others', which covered several fields such as architecture, design engineering, and systems-related engineering and also various types of information and communications technology engineering such as electrical electronical, electrical electronical system engineering, visual information, and management information.

In the face of increased demand for talent this reallocation of engineers towards the IT industry may not be adequate. While total foreign college students doubled from 62,000 in 1994 to 130,000 in 2005 (Japan Immigration Association, various years), the share of foreign engineering students has fallen consistently since 1998. From 18 per cent in 1998, the share declined to 13 per cent in 2004 (Japan Student Services Organization, various years).[13] In recent years, the growth rate of foreign students has slowed down or turned negative, hence the share of foreign engineering students has increased somewhat – to 15 per cent of the total. These developments suggest that the availability of foreign talent in Japan is not assured. Already the competition for global talent suggests Japan must actively seek out such talent (D'Costa, 2006). A recent presentation by an official of METI (Ogawa 2007) indicated that the Japanese labour market in the information services industry is tightening up, with many employers voicing concern over inadequate availability. Some projections made by the World Information Technology and Services Alliance (in Ogawa, 2007) point to Japan's relative decline in

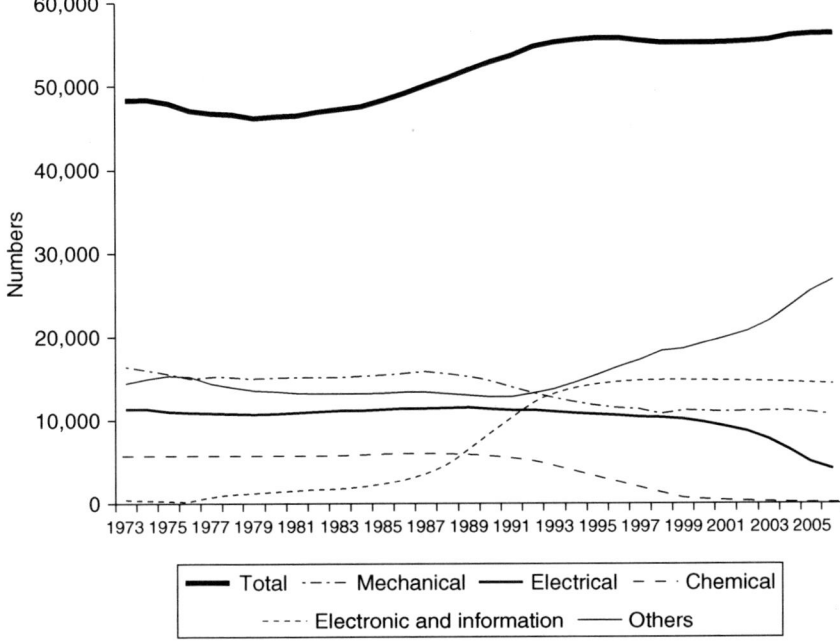

Figure 9.1 Trends in technical student enrollments in Japan in technical colleges

Note: Mechanical = Mechanical Engineering and Mechanical Electric Engineering; Chemical = Chemical Engineering and Industrial Chemistry; Electronic and Information = Electronic, Electronic Control, Control Information, Electronic Information, Information Electronic, Information, and Information Communication; Others = Substance Engineering, Architecture, Design Engineering, Interdisciplinary, System-related, etc.

Source: Ministry of Education, Culture, Sports, Science and Technology, *Gakkou Kihon Chousa Houkoku* (Statistics on School and Education in Japan), various years.

world IT market shares by 2015 compared to China's and not that much larger than India's.[14]

It is evident that at present Japan is not an attractive destination for Indian students. This is also evident from data on foreign students who change their visa or immigration status for employment purposes. Of the nearly 6,000 students who converted their visa status for employment in the fiscal year 2005–06, 96 per cent were from Asia, of which three-quarters were from China (Japan Immigration Bureau, various years). India's share was less than 1 per cent of the Asian visa conversions. Cultural distance, especially language, has been the principal reason why Japanese clients have not been very enthusiastic about offshoring work to India (JISA, 2006: 38–9; see also Kjersem and Gammeltoft, chapter 7) and few Indians, compared to the Chinese, study Japanese language or business culture.[15]

Japan has a long way to go before becoming a large receiving country of foreign talent. Currently, more high-skilled Japanese citizens go abroad than foreigners come to Japan. Intra-company transferees, assumed to be highly skilled professionals, ranged from 5,000 to 6,500 foreigners in the 1990s and between 10,000 and 11,000 in 2001 and 2004 (Japan Immigration Association, various issues). In contrast, nearly 53,000 Japanese nationals went abroad to take up various posts (Kobayashi 2001: 121–2). Similarly, nearly 200,000 Japanese citizens went abroad for studies and technical training and over 100,000 for research in the late 1990s. Given its demographic dilemma, Japan will have to move quickly to attract talent or else face the fierce competition ensuing from shortages of talent in all major IT-producing countries, including China and India.

9.5 Software strategies and the triangular relationship between China, India, and Japan

While it is difficult to obtain estimates of the Japanese software market, given the size of the Japanese hardware industry, one can easily infer the market to be very large. In 2005, the IT services market was estimated at ¥14.56 trillion or roughly $123 billion (Ministry of Economy, Trade, and Industry (METI) 2006).[16] Based on a sample of 58 firms, the Japan Information Technology and Services Industry Association (JISA) reported that the total amount of offshore outsourcing was ¥49 billion or less than $450 million (METI, 2005). Both the share of international outsourcing and also the participation of both China and India has been low in the Japanese market.

This anomaly is reflected in Japan's share of India's software exports, around 2 per cent, compared to the almost two-thirds that went to the US (NASSCOM, 2004). Poorer regions of the world imported more software services from India than Japan. The relative dependency ratio was just 0.2.[17] Among all of the regions Japan had the lowest dependency ratio, which suggests India's penetration of the Japanese market to be extremely low at this time and conversely presents future opportunities for growth provided certain technical and business conditions are met.

Of the total Japanese market, 61 per cent represented customized services, a market in which both China and India are generally competent. The total estimated offshore market was $447 million, of which more than 50 per cent went to China, followed by 10 per cent to the US, and 8 per cent to India (Table 9.2). According to a JISA survey, the Chinese share of offshoring is higher than reported in part because many Chinese IT firms work with Japanese partners (JISA 2006: 38).[18] This logic could also apply to India, raising India's share, since we know that Indian firms worked mainly for US firms and US firms received around 10 per cent of Japanese offshoring projects (see Table 9.2). However, the intermediate links between

Table 9.2 The scale of overseas software outsourcing from Japan

	(Unit: million Yen)					
	2002 (n = 58)	2002 Share (%)	2003 (n = 58)	2003 Share (%)	2004 (n = 77)	2004 Share (%)
China	9,833	48.5	26,280	53.7	32,241	62.4
USA	3,260	16.1	4,988	10.2	5,147	10.0
India	1,908	9.4	6,312	12.9	4,255	8.2
Australia	–		2,626	5.4	3,133	6.0
UK	20	0.1	1,827	3.7	2,126	4.1
Philippines	1,864	9.2	2,494	5.1	2,117	4.1
Korea	1,952	9.6	1,871	3.8	1,415	2.7
France	–		834	1.7	548	1.1
Canada	496	2.4	616	1.3	262	0.5
Vietnam	30	0.1	30	0.1	216	0.4
Others	888	4.4	1,082	2.2	237	0.5
Total	20,251	100.0	48,960	100.0	51,697	100.0

Source: Japan Information Technology Services Industry Association 2005.

India and Japan via the USA are unlikely to be strong for the simple reason that the USA remains India's most lucrative market. This does not negate future possibilities since there are instances of Indian firms hiring Americans (often of Japanese ancestry) to work on projects for the Japanese market, while Singapore has been a gateway to the Japanese for some Indian firms (D'Costa interviews, Bangalore, Delhi, February 2005). Furthermore, several large Indian firms have established development centres in China, mainly in Dalian in the north-east, to enter the Japanese market for cost, talent, and market proximity reasons. Dalian was under Japanese rule and today has a pro-Japanese atmosphere with a large Japanese-speaking pool of technical talent. Chinese firms have also set up software centres in Bangalore, India to tap into Indian technical talent (see D'Costa and Parayil, Chapter 1).

The smaller size of Chinese firms, and services based on the low-end segments of software development requiring less project management skills, make Chinese firms more suitable for Japanese projects, given Japan's penchant for domestic sourcing and in-house development. The propensity to deal with Chinese firms rather than Indian ones goes beyond the cultural affinity, often claimed by Japanese businesses. The particular trajectory of the Japanese software industry has imposed limits on international outsourcing. For example, Japanese hardware producers have traditionally bundled their software, which has limited the development of an independent software industry in Japan (Anchordoguy, 2000). The Chinese are more engaged with the Japanese due to greater Japanese investments in Chinese manufacturing as the result of rising Japanese costs and expanding Asian markets. Hence, Chinese IT service providers have an advantage in supplying Japanese production units in China. In this context, not only has foreign investment in India been low but Japanese FDI has been very low.

Specific to the IT industry, wage costs are an important determinant in the pattern of software outsourcing by Japan (see JISA, 2006: 35). In general, offshoring by Japanese firms is also driven by cost considerations and the shortage of specialized skills (JISA, 2006: 38). As reported by Nikkei Computer (2005: 59), China's man-month cost varies from ¥150,000 to ¥350,000 for a junior software engineer/programmer, whereas the corresponding figures for India and Japan are ¥200,000–450,000 and ¥450,000–900,000 respectively. For a senior-level engineer/programmer the high figure for China was ¥1,000,000 compared to India's ¥1,500,000 and Japan's ¥2,300,000. As Indian software firms entered earlier into the global market, they have developed better project management skills for outsourced projects than Chinese firms (Personal Interviews, Tokyo, Japan 2005 and 2006). This shortcoming is also echoed by a Nomura Research Institute report, which states that the number of 'experienced professionals in specialized work' in China is limited (Konomoto, 2002: 3).

The link between the international mobility of technical talent and knowledge-intensive activities such as in the software industry can be seen in terms of the types of services provided to clients (Figure 9.2). Roughly there are three sources of services – onsite, offsite, and offshore. In the case of a client, services could be provided by a vendor onsite, which could be domestic or international (in Country A). The work could be done offsite by either a domestic or foreign vendor in Country A. The services could also be obtained from offshore sources (from Country B). If services are obtained onsite but provided by a foreign company, which in turn contracts the work to an offshore vendor then we have a movement of technical talent. Talent from Country B will go to Country A temporarily and develop the service either offsite in their own company or onsite at the client's location. In the case of offshore supply, talent in Country B will develop the services and export to Country A. There is, of course, a link between a foreign company working onsite/offsite in Country A and offshore development in Country B. The trade-off between onsite, offsite, and offshore can be seen in terms of cost reductions being small or large respectively. Most service projects in the IT industry have some combination of onsite, offsite, and offshore talent involved, although in the case of Japan offshore is still limited. The general trajectory for both Chinese and Indian software firms in Japan is to provide services onsite/offsite, and, if successful, offshore development thereafter.

The cost of providing services onsite in Japan is much higher due to the higher cost of living and salaries compared to services provided through offshore arrangements from China and India. However, the general institutional preference of Japanese firms to interact closely with their suppliers

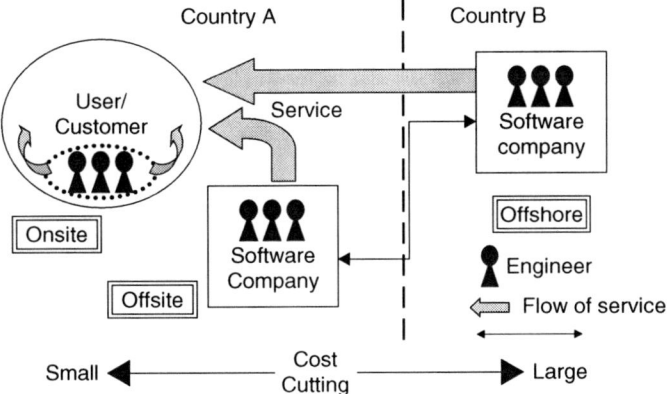

Figure 9.2 Types of software service

suggests that Japanese IT companies would prefer more foreign personnel onsite/offsite rather than fewer. The foreign origin of personnel is less important in the case of onsite work. It is the unfamiliarity of Japanese firms in long-distance outsourcing of services that calls for closer interactions on an everyday basis. This lack of trust naturally leads to higher costs but it is a trade-off, which Japanese companies are willing to make. Anecdotal evidence suggests a ratio of 20:80 for onsite and offshore personnel respectively for Japanese projects, while for American firms it is 10:90. Thus, assuming the cost of doing business in Japan and the US to be similar, Japanese costs of outsourcing of software services to offshore sources in India is likely to be higher.

In any outsourcing model there is a considerable partitioning of the project into discrete components, some of which are carried out by the client (or client's local partners) and others by the overseas vendor. What is the distribution of value creation in this division? In other words, what is the scope of innovation and learning under an international outsourcing arrangement? We present some estimates of Japanese IT companies on the breakdown of project components by onsite and offshore development (Figure 9.3). It is evident that the project requirements are determined by the Japanese client at the client's site no matter who the supplier is. As we move down the value chain it becomes clear that a larger share of onsite work is outsourced. For example, about 70 per cent of the architecture is done onsite by the client (or under exceptional cases by foreign talent onsite) compared to 30 per cent in module design, coding, and unit testing. While Figure 9.3 does not reveal the division of work between Japanese and foreign firms, we can infer that the bulk of Chinese and Indian professionals will fit into customized coding work. In the areas of testing and integration the client takes on most of the responsibility – 80 per cent and 100 per cent respectively. Japanese firms are

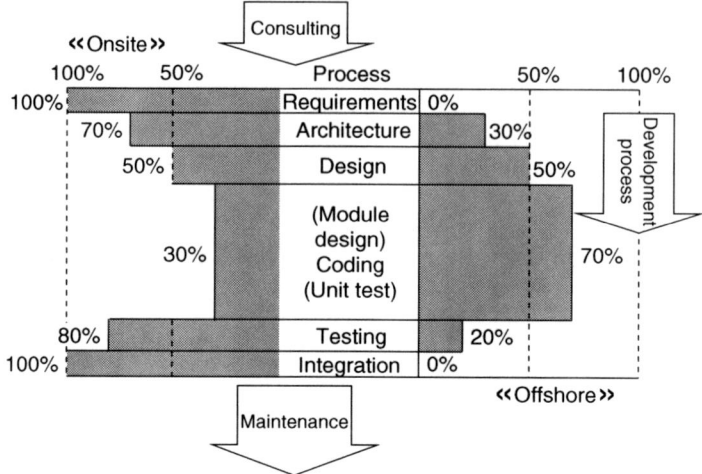

Figure 9.3   Offshoring of software development
Source: Kobayashi 2006.

reputed for their stringent testing requirements; hence it is expected that the onsite share of this would be higher than offshore. However, integration of the different components of the software development process is a highly skilled activity and since users of the software need to ensure its functioning, the integration process has to be done onsite. What is also not revealed here is the particular Japanese institutional setting in which specialized firms or divisions of large Japanese firms, also known as systems integrators, have considerable control over this segment of software development. The big five systems integrators are NTT Data, Hitachi, NEC, Fujitsu, and IBM Japan. Their control stems from their large size, their long-term relationship with their clients, and their familiarity with business practices in the domestic market. Foreign firms, who may have systems integration skills, find it difficult to enter this segment.

There are other factors that also contribute to Japan's low level of international outsourcing in the IT industry. Japanese institutional and corporate structures limit outsourcing arrangements to familiar, local firms. There are also social and cultural differences between China, India, and Japan, especially at the corporate level. Most importantly, however, it is the particular trajectory of the Japanese IT industry that has imposed limits on outsourcing. The lack of an independent software industry in Japan due to its emphasis on hardware manufacturing suggests entry barriers for foreign firms wishing to enter the Japanese IT industry (see Anchordoguy, 2000). Also, Japan's strong presence in hardware-intensive software development raises its own entry barriers. The Japanese are known for their design and embedded software, areas that the Indian industry is only beginning to develop. Projects from Anglophone countries still entail customized software services that are technically different and perhaps less challenging than Japanese manufacturing-based software needs (D'Costa 2002). Given the nature of software development and specific Japanese institutional practices, it is evident that a Japanese offshore model limits significant opportunities for learning for foreign vendors. In the area of design activities both China and India are making considerable inroads and hence could capture part of the design work estimated to be split 50 per cent each onsite and offshore (Figure 9.3). More importantly, the spheres of architecture, testing, and integration remain limited in scope for offshore development and hence the learning opportunities for talent supplying countries also remain truncated.

Notwithstanding structural and institutional barriers to greater participation by Chinese and Indian technical professionals, there is evidence of entrepreneurial initiatives by both Chinese and Indian professionals in the Japanese software market (Table 9.3). In surveys (see Table 9.3) conducted by Tomoko Kobayashi of Chinese and Indian software firms, we find that most Chinese entrepreneurs entered Japan as college students, whereas Indian entrepreneurs came to Japan initially as part of an employer's project assignment. Five of the six Indian entrepreneurs surveyed came under the 'engineering' visa category, with one as an intra-firm transferee. The Chinese entrepreneurs entered Japan in the 1980s, the Indians in the 1990s. Most Chinese entrepreneurs obtained advanced degrees from Japanese technical institutions, with 13 of the 32 getting doctoral degrees. This is in contrast to one Indian entrepreneur who came as a student and another Indian entrepreneur who obtained a degree from a Japanese university. This difference between the Chinese and Indians may be due to alternative educational and entrepreneurial options for Indian students in English-speaking countries such as the USA and the UK. Evidence also suggests that IT professionals from India tend to terminate their studies at the master's level, whereas Chinese students tend to pursue PhD degrees more than Indians (see D'Costa Chapter 3). Unlike Indian entrepreneurs, many Chinese entrepreneurs either became Japanese nationals (4) or permanent residents (8). Only one of the six Indian entrepreneurs surveyed became a permanent resident. Since it takes ten years of continuous residency in Japan to become a permanent resident and Chinese entrepreneurs arrived earlier than Indians, the presence of more Chinese permanent residents does not come as a surprise. More importantly, as engineer and business manager visa categories are now easier to obtain, the presence of more Chinese and Indian talent in Japan can be expected. However, both Chinese and Indian entrepreneurial software firms in Japan typically locate at the low end of the value chain. Their Japanese clients may not be either end users or system integrators but rather second or third tier software subcontracting houses.

Preliminary findings from other surveys of Indian software firms in Japan indicate the presence of large Indian firms emphasizing marketing and sales

Table 9.3 Profile of Chinese and Indian software entrepreneurs in Japan

	No. of Entrepreneurs/ Professionals Surveyed	Time of Entry of Professionals	Status of Initial Entry	Education Prior to Entry	Advanced Degrees in Japan	Initial Job Experience	Completion of Ph.D.
China	32	1980s	Mostly College Students	Technical Universities in China	Mostly Yes	Mostly in Japan, some in China	High Share
India	6	1990s	Mostly as Engineers, on project assignment	Technical Universities in India	Mostly No	Mostly in India	None

Note: Full details of the survey have been suppressed due to confidentiality agreements.
Source: Kobayashi, T. Tokyo, 2003, 2006.

with small shares of sales in Japan. There are also a few small, but dedicated service providers to the Japanese software industry (D'Costa, Firm Interviews, Bangalore, Chennai, Delhi (2005), Tokyo (2005, 2006)).[19] The representatives of large Indian firms in Japan are occupied mostly in finding new business, with some staff devoted to maintenance and post-sales services. Indian entrepreneurs, either in Japan or in India working for Japanese clients, have had some professional contact with the Japanese market in the past and many of them have found the technical challenges associated with the Japanese market attractive. The attractiveness of the US market for Indian companies outweighs some of the intangible benefits that the Japanese market offers over the long haul. However, it is clear that Indian participation will increase in the future even if the Chinese have had a head start in the Japanese software sector.

We anticipate divergent engagements by Chinese and Indian firms with the Japanese industry. Chinese software firms are generally small, many spun-off from universities by academic entrepreneurs. Some of these firms have Japanese-language capabilities and better Japanese connections than Indian software firms. After working for Japanese firms these Chinese companies launch their own small enterprises in Japan. The Indian industry, although fragmented, has several very large firms with revenues exceeding one billion dollars. Project management capabilities and experience with large projects make Indian firms more competitive in general. However, as software process development becomes standardized and as Japanese firms adopt such international standards, some of the unfamiliarity claimed by Japanese firms is likely to disappear. There is already some evidence of successful Indian firms in the Japanese market. For example, the turnaround of the bankrupt Long Term Bank of Japan to Shinsei Bank was made possible in part by the large deployment of Indian software professionals in Japan and in India. The fact that 61 per cent of Japan's IT services market of $120 billion is in customized services, a market in which Indians excel, suggests that both China and India have future opportunities in Japan.

9.6 Policies for new Asian software dynamics

Based on the analysis of talent flows in Japan and the characteristics of the Japanese software industry, we find the following factors contributing to the current division of labour between China and India for the Japanese outsourcing market:

1. Japanese firms are unfamiliar with the structural and institutional logistics of offshore outsourcing;
2. the Chinese vendors have a clear advantage over all countries, especially India, in terms of talent supply and outsourcing volume, while India is heavily engaged with the US market;

3. the demographic crisis in Japan suggests a fundamental rethinking of the role of foreigners in Japan and offshore development by Japanese companies is needed;
4. the difference in production costs in the three countries and continued increases in costs in India and China are likely to induce a realignment in Japan's integration with both China and India.

We identify several policy areas that will benefit all three countries collectively in their innovation efforts. First, we emphasize the importance of the movement of technical talent among the three countries but especially for Japan, which clearly has an impending shortage. Secondly, enhanced mobility of Chinese and Indian talent is likely to impact at least three spheres: formation of epistemic communities, transfer new knowledge, and access to new markets, which will benefit all three countries. Japan can foster the movement of foreign professionals and increase its competitiveness by facilitating the movement of talent. Three interrelated policy areas are: immigration to attract foreign students and professionals for the IT industry as a whole; industry policy to encourage international outsourcing; and education to raise the quality of its software engineering talent through university reforms, including university–industry linkages. For China and India, the policy interventions also include education to create an internationally competitive talent pool, fostering more inclusive domestic development, and establishing bilateral and trilateral exchanges among the three countries.

The promotion of foreign talent should not pose a significant challenge since the presence of foreign workers in Japan is no longer a new phenomenon. However, the gravity of its demographic imbalances must be politically and socially acknowledged so that adjustment to a more internationalized Japanese economy becomes easier. Immigration reforms will be critical especially if circulation of talent is entailed and if Japanese business preference for greater face-to-face interaction is to be met. The other strategy will be to attract foreign students with the intent to create a locally-trained foreign talent pool in Japan, not dissimilar from the US approach.

A long-term focus on revamping engineering training in Japanese universities, especially for software development, will be important for Japan. This strategy could be complemented by drawing on foreign talent from China and India to energize local student learning. There is some evidence of the latter, but they are selective and confined to provincial government initiatives rather than part of a national innovation strategy. For example, the Kitakyushu Science Research Park (KSRP) and Fukuoka city in Japan have sought Chinese technical talent and students. The geographical proximity of several East Asian countries, including the Shanghai region of China, make it particularly attractive for Japanese semiconductor and other high-tech companies in Kyushu island to recruit foreign talent. Several universities have established training and research facilities in the science park with the aim of securing Asian technical students for long-term needs (D'Costa, Field Visit, KSRP and Fukuoka, June 2005). As we have shown earlier most Chinese software entrepreneurs came as students to Japan, obtained advanced degrees, and set up their businesses in Japan.

Japan must also turn its attention to its university system and nurture industry–university partnerships for generating new talent, an issue which the Japanese government has correctly identified (JISA, 2006: 96). Some initiatives have been taken by central government ministries such as the Ministry of Economy, Trade, and Industry (METI) and provincial authorities such as Kitakyushu and Fukuoka. Recently, METI and the Ministry of Education, Culture, Sports, and Science have launched IT-education-related programmes. But for this effort to succeed Japan will have to attract foreign students as well as foreign technical professionals in a highly competitive global labour market, evident from the high demand for Indian talent in the US.[20]

In addition, there must be some reforms of Japanese business practices. It will be necessary to introduce incentives for international outsourcing as well as hiring foreign professionals. Current business practices have tended to limit the degree of internationalization. For example, of the $120 billion IT services market in Japan less than 5 per cent is outsourced from abroad. However, since face-to-face interactions are preferred by Japanese companies and shortages of talent imply greater engagement with foreign talent, businesses will have to alter their strategies by making international talent more acceptable. One intermediate possibility could be working through intermediate 'gateways' such as Singapore, the USA, and Australia. Rather than dealing directly with India, Japanese companies could go through intermediate companies located in these gateways. For example, Philips of the Netherlands in India works through Singapore for Japanese projects, while Infosys of India has been utilizing Japanese Americans for the Japanese market (D'Costa Field Visit, Bangalore, February 2005). Similarly, China could act as a gateway for Indian software firms, using Chinese professionals and Indian project managers in China for the Japanese market. This is already taking place in Dalian, in north-eastern China, while in Bangalore, India major Chinese IT companies are tapping into Indian technical capabilities.

No doubt Japanese firms are experiencing competitive pressures, as revealed by a recent survey. Of the 77 firms surveyed, 51 and 17 companies expressed that cost reduction and availability of qualified foreign partners/engineers is a necessary part of a growth strategy (JISA, 2006). This suggests interactions among the three countries at the regional and global levels will reinforce not only the formation of epistemic communities but free up Japanese talent at home for alternative innovation-based activities. Given the large differential between Japanese and Indian costs and the imminent shortages of technical professionals worldwide and in Japan, we anticipate that in the future there will be collaborative offshoring strategies with both

China and India. The same survey in JISA revealed that the most important element in selecting an outsourcing partner was the quality and quantity of engineering talent. This obviously calls for a complementary strategy on the part of both China and India to create a large, technical talent pool for multiple markets in the global economy.

There are three benefits that would arise from the creation of a talent pool. First, it would diversify markets for all, especially India's, which exports a mere 2–3 per cent of its total software exports to Japan. Secondly, given Japan's strengths in setting high-technology standards, both China and India can create new opportunities for learning by servicing the Japanese market. Thirdly, and related to the second benefit, greater interaction among the three countries would not only reinforce technical networks but also provide a long-term basis for reinforcing the return of expatriate talent. The latter is likely to accelerate new entrepreneurial initiatives in China and India and generate new export niches in the Japanese market.

Both China and India are actively expanding their talent pool. For example, in 1998 China had 486,000 person years of scientists/engineers engaged in R&D, whereas by 2005, the figure stood at 1.119 million person years (Cao et al., Chapter 10). In India, the promotion of technical talent has gained wide popularity (NASSCOM, 2003) with IT employment increasing nearly tenfold from 56,000 in 1990 to 522,250 a decade later (NASSCOM, 2002: 63). The number of Indian professionals with an engineering degree has increased from 43,000 in 1997 to nearly 450,000 students enrolled for engineering degrees in 2006 (see Bound, 2007: 62). Notwithstanding such expansion there are lingering problems associated with the quality of talent in India and China (see D'Costa, Chapter 4; Cao et al., Chapter 10). Not only should the average quality of engineering education improve in all three countries but more importantly as global demand for talent increases both India and China are likely to lose the momentum of their growth, should they fail to transform their raw talent into technologically competent professionals.

One way to accomplish this is to create an R&D ecosystem, which is currently weak in both countries (D'Costa Chapter 4; Cao et al. Chapter 10). Revamping university education generally – and doctoral training in engineering in particular – will go a long way to fill the research dimension of a national innovation ecosystem. The Japanese high-technology market is driven significantly by manufacturing and hardware development. Both are areas in which Indian firms have had relatively low levels of exposure to international competition, hence it would take considerable technical competence and commercial wherewithal to make a dent in the Japanese market. Some large firms in India such as TCS, Wipro, Satyam, Infosys, and HCL, and a few small, but highly specialized ones such as Soft-jin, HTL, and Nihon-Indussoft have a presence in Japan. Most Indian firms are currently ill-prepared to tackle this dynamic market. Those Indian firms that could offer solid engineering-oriented solutions are most likely to succeed in Japan.

In this context, focusing on the embedded software market in Japanese manufacturing will be highly lucrative (see Kojima and Kojima, 2007).

Furthermore, the Japanese market poses substantial cultural barriers, hence it behooves Chinese and Indian engineers to narrow the cultural gap. Few Indian professionals and students study Japanese language and business culture compared to the Chinese, though some of the larger Indian firms, such as Wipro, TCS, Satyam, and HCL, have begun extensive in-house Japanese language training programmes. However, Japanese companies also ought to bridge the gap by studying English and making communication among all parties easier. Although large investments have been made by both the government and individual citizens there are significant shortcomings of English-language training in Japan (Ministry of Ministry of Education, Culture, Sports, Science and Technology, 2003). Remedying this will not only enhance inter-cultural business interactions, but will also bring Japan out of its relative insularity.

9.7 Conclusion

This chapter has demonstrated the significance of human capital in the global software industry. It was argued that the mobility of talent and subsequent circulation led to the formation of social networks and epistemic communities. These are crucial for the transfer of commercial and technical knowledge from one location to another in order to sustain growth. Given Japan's serious demographic predicaments and its position as the world's second largest IT market, a closer examination of Japan's engagement with global talent flows is warranted. Japan is a particularly interesting case because of its relative insulation from global developments in the internationalization of services. As both China and India are emerging as visible players in the high-technology industries, and especially in services, understanding the extent of Japan's participation in this development will be indicative of its future innovative trajectories.

Japan's insertion into global talent flows and outsourcing arrangements will expand its knowledge networks and membership in global epistemic communities. For China, India, and other Asian economies emigration, education, and experience in Japanese industries promises a stream of commercial and technical knowledge flows back to home countries in the future. The professional networks and the new learning opportunities in new geographic, product and service, and technical markets far outweigh the static losses associated with talent outflows. Moreover, offshore development in general suggests a greater portion of the development work in China and India, thereby limiting the loss of talent over the long haul.

Following the social network framework we can anticipate that both China and India will deepen their engagement with Japan. Assuming Japanese firms increase the international outsourcing of software services in general, conditional on reforms of Japanese immigration and business policies, the flows of technical talent to Japan from China and India are likely to increase. Depending upon the length of the projects, the temporary flows of talent could result in a Japan-based brain bank of foreign talent. This, like its US counterpart, could facilitate the transfer of a qualitatively different kind of technical and commercial knowledge to Japan and to the talent-generating countries through offshore development. The net result is the expected overlap of epistemic communities straddling three large economies and IT sectors. Over time, it is likely to lead to greater tripartite interaction and, more importantly, large technological spillovers through network externalities. In order to make the transition successfully all three countries will have to collaborate in the broader areas of immigration, macro-level bilateral agreements, and industry-specific areas of technical cooperation. There are visible signs of improved India–Japan bilateral relations with a free trade agreement in the offing and cooperating at new levels and areas of mutual interests.

Asia's need for technical talent is growing. Singapore, in its quest to establish a high-tech, service and financial hub, has been attracting Asian talent. Even countries such as Taiwan and South Korea, not known for labour shortages, are at a juncture where the need for IT service professionals is acutely felt. Consequently, China cannot be the sole labour reservoir for Asia since it is also facing the spectre of an ageing society. Given these emerging developments, when placed in the larger context of Mode 4 trade in services under the General Agreement on Trade in Services (GATS) negotiations,[21] we can anticipate higher temporary flows of high-technology service workers worldwide, with active involvement of India, which has the most to gain from this development (World Bank, 2004; OECD, 2003c). Under this scenario it will be extremely difficult for Japan to remain insulated from such global flows of talent or resist the advantages that come with such talent for international competitiveness. All three countries could translate the challenges of talent supply to historic opportunities for increased market access, greater technological learning, and economic growth.

Notes

1. A preliminary version of this research was carried out by Anthony D'Costa as a Sabbatical Fellow of UN University's World Institute of Development Economics Research in Helsinki. It is available under 2004 research papers at www.wider.unu.edu. He thanks WIDER for its congenial professional environment. He also collected some of the Japan-based data as an Abe Fellow of the Japan Foundation, to which he is grateful. We are also grateful to Janette Rawlings for her substantive and editorial support. The usual disclaimers apply.
2. A survey by the Japan Information Service Industry Association revealed that Japan in 2001 needed 800,000 IT workers (Hindu, 2001).

3. We, of course, recognize brain drain concerns of sending countries, which arguably lead to a heavy cost burden for sending economies and substantial economic and technological benefits for talent-receiving countries (see Kar and Beladi 2003:38). But for large dynamic talent-generating economies such as China and India the movement of talent does not pose a big risk.

4. *Keiretsu* is a corporate governance structure in which firms such as manufacturers, suppliers, trading companies, and banks are interlinked by cross-equity holdings. Although no majority control is exercised by any one firm the ties that bind these firms are inclusive and hence exclude non-members.

5. There is a growing recognition that technical networks are important first to innovative capability, and second to bringing back expatriate scientists who are perceived to enhance national competitiveness (see Carvajal, 2004).

6. Saxenian in her recent volume (2006) correctly notes that the previous approaches to economic development (state-led structuralist models) are less applicable in a globalizing environment. Her account of global entrepreneurship, mostly US-led, pulled in by Silicon Valley ('stickiness'), suggests that at the global level there are structural impediments to overcome. This naturally begs the question: how can other regions break the monopoly of Silicon Valley? Or, to put it another way, what is the role of the state in a globalizing economy? And, for that matter, what will the innovation world look like in the future, given that China, India, and other Asian countries are displaying greater abilities in science and technology matters?

7. Among the six Indian entrepreneurs surveyed by Tomoko Kobayashi, one of them came to study in Japan, and five of them later went back to India to establish a software business for the Japanese market. See the discussion of Chinese and Indian entrepreneurs in Japan in section 9.5.

8. Since trainees are not workers this category of foreign workers did not come under the purview of Japanese labour laws. Workers from Thailand, China, Philippines, Brazil, Peru, and Bangladesh are employed in low-wage factory work, construction, and entertainment sectors in Japan. The selectiveness of Japanese policy toward foreign workers is evident from the fact that Japan is quite open to the entry of "entertainers" but highly restrictive toward the much-needed nurses and other health care givers (Ito, 2004). In 2002 there were 123,000 entertainers (or 85 per cent of total foreigners allowed to work) compared to 4 'medical service professional' (Ito, 2004). Brazilian and Peruvian spouses and children of Japanese nationals are also hired as trainees and exceed the number of formally recognized trainees as they are free from work restrictions. Such workers complement rural-based manufacturing industries in Japan. Most trainees come from China, with the rest largely from South East Asia.

9. The data are collected by a number of Japanese government ministries. Hence, the data are not always consistent. Moreover, there are a variety of definitions under which the entry of foreigners is recorded.

10. There are of course other categories, which fall under the definition of 'Professionals' such as journalists, investors, lawyers, medical professionals, and humanities and international business personnel (Japan Immigration Association, various years). These professionals largely fall under the OECD definition of human resources in science and technology (HRST) but with our focus on the software and IT industry, we omit these categories.

11. It is also noteworthy that India sends more professionals to Japan (5 per cent) than it exports software (2–3 per cent of total exports).

12. According to the Japan Student Services Organization (JASSO), for the fiscal year ending March 2006, the distribution of foreign students was as follows: social sciences 39.6 per cent, humanities 23.3 per cent, engineering 15.0 per cent, and science 1.3 per cent (JASSO, 2006). Asia contributed nearly 93 per cent of the international students, with China's contribution of 63 per cent of total students, followed by South Korea's 13.5 per cent and Taiwan's 3.4 per cent (Japan Student Services Organization, 2007). Indian students numbered 525, representing a mere 0.4 per cent of the total. For example, in 1997–98 only 2.2 per cent of the Indian students going abroad under government programmes went to Asia as a whole (Khadria, 2004a: 28). This is consistent with Indian students' preferences for Anglophone countries such as the US. In 2001, nearly 47,000 Indian students went to the US, accounting for 78 per cent of all Indian students enrolled in OECD countries (Khadria, 2004a: 29). Nearly 14,000 doctoral degrees were granted to Indian students in the US (National Science Board, 2006: A2-123).

13. The total number of foreign engineering students in Japan has increased, with nearly 18,000 foreign students for 2006.

14. The estimated shares are 2.8 per cent for India, 19.3 per cent for China, 3.9 per cent for Japan, and 41.5 per cent for the US. The corresponding shares for 2001 were less than 1 per cent, less than 1 per cent, 11.7 per cent and 50.5 per cent (Ogawa, 2007: 5). Japan's share of employees in the information services industry was 570,000 in 2005 compared to India's 130,000 and China's 90,000 (Ogawa, 2007: 6).

15. A survey of the Japanese Language Proficiency Test in 2004 showed that there were 90,356 Chinese applicants, compared to India's 3,869 (JISA, 2006: 39). Similarly, China had more than 111 times the teachers for Japanese language than India.

16. According to METI's Current Survey of Selected Service Industries for December 2006, Japan's IT services market was ¥1095 billion or roughly $9.3 billion.

17. This ratio is computed by taking the share of Indian exports to Japan (2 per cent) and divided by the region's share in world IT services spending ($34.9 billion/$349.1 billion).

18. As Indian firms worked mainly for US firms and US firms received 10 per cent of Japanese offshoring projects, India's share of Japanese outsourcing could be actually higher than reported (see Table 9.2). This has been confirmed independently by both authors in their respective surveys of Indian firms in Japan.

19. The analysis of this survey is left for another paper.

20. India garnered 49 per cent of the H1-B visas in 2001 granted to foreign employees in the USA, with 92 per cent of them IT-related (Hira, 2004: 842). The H1B quota allocated for 2006 was exhausted on the very first day.

21. Mode 4 is a pending proposal under the WTO, which is designed to enhance exports of services by allowing temporary migration of workers from the developing world to the rich countries.

References

Ahmed, I. (2000) *The Construction of Diaspora: South Asians Living in Japan*, Dhaka: University Press Ltd.

Anchordoguy, M. (2004) *Reprogramming Japan: The High Tech Crisis under Communitarian Capitalism*, Ithaca, NY: Cornell University Press.

Anchordoguy, M. (2000) 'Japan's Software Industry: A Failure of Institutions?', *Research Policy*, 29(3): 391–408.

Asia Program Special Report (2003) 'The Demographic Dilemma: Japan's Ageing Society', Washington, DC: Woodrow Wilson International Center for Scholars, No. 107.

Association for Computing Machinery (2006) *Globalization and Offshoring of Software: A Report of the ACM Job Migration Task Force*, New York: ACM.

Bound, K. (2007) *India: The Uneven Innovator*, London: Demos.

Brody, B. (2002) *Opening the Door: Immigration, Ethnicity, and Globalization in Japan*, New York: Routledge.

Burrelli, J.S. (2004) 'Emigration of U.S.-Born S&E Doctoral Recipients', National Science Foundation InfoBrief, June, NSF04-327.

Carvajal, D. (2004) 'The Workplace: Calling Scientists Back Home', *International Herald Tribune*, July 7, 2004, www.iht.com. Accessed 7 July 2004.

Caspar, S. (2007) 'How do Technology Clusters Emerge and Become Sustainable? Social Network Formation and Inter-firm Mobility within the San Diego Biotechnology Cluster', *Research Policy*, 36(4) 438–55.

Cohen, B. (2004) 'Urban Growth in Developing Countries: A Review of Current Trends and a Caution Regarding Existing Forecasts', *World Development*, 32(1): 23–51.

Cohen, M.S. and Zaidi, M.A. (2002) *Global Skill Shortages*, Cheltenham: Edward Elgar.

Cowan, R. (2004) 'Network Models of Innovation and Knowledge Diffusion', Maastricht: MERIT-Inonomics Research Memorandum Series, 2004-016, http://www.merit.unimaas.nl.

D'Costa, A.P. (2006) 'The International Mobility of Technical Talent: Trends and Development Implications', *UNU World Institute of Development Economics Research*, Research Paper No. 2006/143 (www.wider.unu.edu).

D'Costa, A.P. (2004) 'Globalization, Development, and the Mobility of Technical Talent: India and Japan in Comparative Perspectives', *UNU World Institute of Development Economics Research*, Research Paper No. 2004/62 (www.wider.unu.edu).

D'Costa, A.P. (2003) 'Catching Up and Falling Behind: Inequality, IT and the Asian Diaspora', in K.C. Ho et al., *Asia.com: Asia Encounters the Internet*, London: Routledge, pp. 44–66.

D'Costa, A.P. (2002) 'Export Growth and Path-Dependence: The Locking-in of Innovations in the Software Industry', *Science, Technology and Society*, 7(1): 51–89.

Davenport, S. (2004) 'Panic and Panacea: Brain Drain and Science and Technology Human Capital Policy', *Research Policy*, 33(4): 617–30.

Dicken, P. (1998) *Global Shift: Transforming the World Economy*, New York: Guilford Press.

Douglass, M. and G.S. Roberts (eds) (2000) *Japan and Global Migration: Foreign Workers and the Advent of a Multicultural Society*, London: Routledge.

Fuess, S.M. (2003) 'Immigration Policy and Highly Skilled Workers: The Case of Japan', *Contemporary Economic Policy*, 21(2): 243–57.

Guellec, D. and M. Cervantes (2001) 'International Mobility of Highly Skilled Workers: From Statistical Analysis to Policy Formulation', in OECD, *International Mobility of the Highly Skilled*, Paris: OECD Proceedings, pp. 71–98.

Hanami, T. (1998) 'Japanese Policies on the Rights and Benefits Granted to Foreign Workers, Residents, Refugees and Illegals', in M. Wiener and T. Hanami (eds) *Temporary Workers or Future Citizens? Japanese and US Migration Policies*, New York: New York University Press, pp. 211–37.

Held, D. et al. (1999) *Global Transformations: Politics, Economics and Culture*, Stanford: Stanford University Press.

Hindu (2001) 'Software: Focus Shifts to Japan', 29 November 2001, http://www.hindu. com 9/8/2004.

Hira, R. (2004) 'U.S. Immigration Regulations and India's Information Technology Industry', *Technological Forecasting and Social Change*, 71(8): 837–54.

Iguchi, Y. (1998) 'What We Can Learn from the German Experiences Concerning Foreign Labor', in M. Wiener and T. Hanami (eds), *Temporary Workers or Future Citizens? Japanese and US Migration Policies*, New York: New York University Press, pp. 203–318.

Ito, T. (2004) 'Vitalize Japan's Economy with Foreign Workers – Nurses or Entertainers?', Glocom Platform, Japanese Institute of Global Communications, 28 September 2004, http://www.glocom.org 9/18/2004.

Japan Immigration Association (various years) *Statistics on The Foreigners Registered in Japan*, Tokyo.

Japan Immigration Bureau *'Ryugakusei to no nihon kigyou to heno shushoku joukyou nit-suite'* (The Statistics of Converting of Status of Residence by Foreign Students in Order to Work in Japan), various years, http://www.immi-moj.go.jp/toukei/index.html.

Japan Information Technology Services Industry Association (JISA) (2006) *IT Services Industry in Japan*, Tokyo: JISA.

Japan Information Technology Services Industry Association (JISA) (2005) *Computer Software bunya in okeru kaigai torihiki oyobi gaikokujin shurou to ni kansuru jittai chousa* (Statistics on International Trading and Foreign Labor in Computer Software Business), http://www.jisa.or.jp/statistics/download/Findings2005.pdf.

Japan Student Services Organization 'International Students in Japan', various years, http://www.jasso.go.jp/statistics/intl_student/data04_e.html.

Kar, S. and Beladi, H. (2003) 'Skill Formation and International Migration: Welfare Perspective of Developing Countries', *Japan and the World Economy*, 16(1): 35–54.

Khadria, B. (2004a) 'Human Resources in Science and Technology in India and the International Mobility of Highly Skilled Indians', Paris: OECD, STI Working Paper 2004/7, DSTI/DOC(2004)7.

Khadria, B. (2004b) 'Migration of Highly Skilled Indians: Case Studies of IT and Health Professionals', Paris: OECD, STI Working Paper 2004/7, DSTI/DOC(2004)6.

Kobayashi, S. (2001) 'International Mobility of Human Resources in Science and Technology in Japan: Available Data, Quality of Sources, Concepts and Proposals for Further Study', in OECD, *International Mobility of the Highly Skilled*, Paris: OECD Proceedings, pp. 109–24.

Kobayashi, T. (2006) 'International Division of Labor in Software Industry and Indian High-tech Immigrants in Japan', Presented at the International Geographical Union Commission on the Geography of Information Society, Sydney, Australia, 26–30 June.

Kobayashi, T. (2003) *The Business Operations and Human Networks of Chinese Immigrant Entrepreneurs in Software Industry*, unpublished Master's Thesis, Human Geography Division, Department of General Systems Studies, University of Tokyo, Tokyo.

Kojima, S. and Kojima, M. (2007) 'Making IT Offshoring Work for the Japanese Industries'. Paper presented at the First International Conference on Software Engineering Approaches for Offshore and Outsourced Development (SEAFOOD), ETH Zurich, Switzerland, 5–6 February.

Konomoto, S. (2002) 'China's Rapidly Growing Infocom Industry and Approaches by Japanese Companies', *Nomura Research Institute*, NRI Papers, No. 53, 1 August.

Kuwahara, Y. (1998) 'Japan's Dilemma: Can International Migration be Controlled?', in M. Wiener and T. Hanami (eds), *Temporary Workers or Future Citizens? Japanese and US Migration Policies*, New York: New York University Press, pp. 203–318.

Lam, A. (2005) 'Work Roles and Careers of R&D Scientists in Network Organizations', *Industrial Relations*, 44(2): 242–75.

Leng, T. (2002) 'Economic Globalization and IT Talent Flows Across the Taiwan Strait: The Taipei–Shanghai–Silicon Valley Triangle', *Asian Survey*, 42(2): 230–50.

Maskell, P. and Malmberg, A. (1999) 'Localized Learning and Industrial Competitiveness', *Cambridge Journal of Economics*, 23(2): 167–85.

Meyer, J.-B. and M. Brown (1999) 'Scientific Diasporas: A New Approach to the Brain Drain', UNESCO Management of Social Transformation (MOST) Discussion Paper No. 41, 1–18.

Meyer, J.-B. (2001) 'Network Approach versus Brain Drain', *International Migration*, 39(5): 91–110.

Ministry of Economy, Trade, and Industry (METI) (2006) 'Current Survey of Selected Service Industries: Information Service Category', http://www.meti.go.jp/statistics/downloadfiles/hv14301j.xls.

Ministry of Economy, Trade, and Industry (METI) (2005) 'METI Annual Survey 2005', http://www.jisa.or.jp/english/statistics/meti_an2005.html.

Ministry of Education, Culture, Sports, Science and Technology (various years) 'Gakkou kihon chousa houkoku' (Statistics on School and Education in Japan), Tokyo.

Ministry of Education, Culture, Sports, Science and Technology (2003) 'Regarding the Establishment of an Action Plan to Cultivate "Japanese with English Abilities"', http://www.mext.go.jp/english/topics/03072801.htm, 2003/03/31. Accessed 10 May 2007.

Nakamura, M. (2004) 'Baby Boomers Face Retirement', Nomura Research Institute, NRI Papers, No. 77, 1 July.

NASSCOM (2004) www.nasscom.com. Accessed 9/8/2004.

NASSCOM (2003) *Strengthening the Human Resource Foundation of the Indian IT Enabled Services/IT Industry*, New Delhi: NASSCOM (Report by KPMG Advisory Services Private Ltd. In association with NASSCOM under the aegis of the Department of IT, Ministry of Information Technology and Communications, Government of India).

NASSCOM (2002) *Strategic Review 2002*, New Delhi: NASSCOM.

National Institute of Population and Social Security Research (2003) 'Population Statistics of Japan, 2003', Tokyo: National Institute of Population and Social Security Research.

National Science Board (2006) *Science and Engineering Indicators 2006*, vol. I, Washington, DC: US National Science Foundation.

Nikkei Computer (2005) *Miangiru China, India IT Power* (Overflowing Chinese and Indian IT power), *Nikkei Computer*, 59.

O'Hagan, S.B. and M.B. Green (2004) 'Corporate Knowledge Transfer via Interlocking Directorates: A Network Analysis Approach', *Geoforum*, 35(1): 127–39.

OECD (2003a) 'Current Regimes for Temporary Movement of Service Providers, Case Study: United States of America', Paris: OECD, Working Party of the Trade Committee, TD/TC/WP(2002)23/FINAL, 6 February 2003.

OECD (2003b) 'Current Regimes for Temporary Movement of Service Providers, Case Study: Australia', Paris: OECD, Working Party of the Trade Committee, TD/TC/WP(2002)22/FINAL, 6 February 2003.

OECD (2003c) 'Service Providers on the Move: The Economic Impact of Mode 4', Paris: OECD, Working Party of the Trade Committee, TD/TC/WP(2002)12/FINAL, 19 March 2003.

OECD (2001) *International Mobility of the Highly Skilled*, Paris: OECD, OECD Proceedings.

Ogawa, K. (2007) 'Japan's IT Professionals Development Policy: Towards the Development of High-end IT Professionals', METI: Information Services Policy Division, March 15, 2007 (Power Point presentation).

Papademetriou, D.G. and K.A. Hamilton (2000) *Reinventing Japan: Immigration's Role in Shaping Japan's Future*, Washington, DC: Carnegie Endowment for International Peace.

Sassen, S. (1998) *Globalization and Its Discontents*, New York: New Press.

Saxenian, A. (2006) *The New Argonauts: Regional Advantage in a Global Economy*, Cambridge, MA: Harvard University Press.

Saxenian, A. (2003) 'The Silicon Valley Connection: Transnational Networks and Regional Development in Taiwan, China, and India', in A.P. D'Costa and E. Sridharan (eds), *India in the Global Software Industry: Innovation, Firm Strategies and Development*, Basingstoke: Palgrave Macmillan.

Saxenian, A. (1999) *Silicon Valley's New Immigrant Entrepreneurs*, San Francisco: Public Policy Institute of California.

Tremblay, K. (2001) 'Student Mobility Between and Towards OECD Countries: A Comapartive Analysis', in OECD, *International Mobility of the Highly Skilled*, Paris: OECD Proceedings, pp. 39–67.

World Bank (2004) *Sustaining India's Services Revolution: Access to Foreign Markets, Domestic Reform and International Negotiations*, Washington, DC: World Bank, pp. 23–5.

World Bank (2003) World Bank Development Indicators Database, http://go.worldbank.org/BDEXK5OE00. Accessed 15 September 2004.

10
Success in State-Directed Innovation? Perspectives on China's Plan for the Development of Science and Technology

Cong Cao, Richard P. Suttmeier and Denis Fred Simon[1]

10.1 Introduction

In January 2006, China announced the initiation of a new 15-year 'Medium to Long-Term Plan for the Development of Science and Technology' (hereafter MLP). The plan calls for China to become an 'innovation-oriented society' by the year 2020, and a world leader in science and technology by 2050. Since the plan was announced, a variety of detailed implementing policies have been introduced and major projects have been initiated in the context of China's 11th Five-Year Plan (2006–10). In this chapter, we review the objectives of the plan and some of the subsequent implementing measures, and explore whether state-directed innovation is likely to be a winning strategy for twenty-first-century technological development.

The plan commits China to the development of capabilities for 'indigenous innovation' (*zizhu chuangxin*) and to the 'leapfrogging' into leading positions in new science-based industries by the end of the plan period. China intends to invest 2.5 per cent of its increasing GDP into R&D by 2020, up from 1.42 per cent in 2006, to raise the contributions to economic growth from technological advance to over 60 per cent, and to decrease its dependence upon imported technology to less than 30 per cent. It also calls for China to become one of the top five countries in the world in terms of the number of invention patents granted, and for Chinese-authored scientific papers to become among the world's most often cited. The plan is likely to have important impacts on the trajectory of Chinese development and thus warrants careful attention from the international community.

10.2 Background and rationale

The plan is a remarkable piece of policy in a variety of ways. It builds on important policy initiatives over the past 25 years, including the 1995 commitment to 'strengthen the nation through science, technology and education' (*kejiao xingguo*) and the more recent notion of 'empowering the nation through talent' (*rencai xiangguo*). Under rubrics such as these, China has made great efforts in recent years to push forward in science and education, including increased expenditures on R&D and enlarged enrollments in higher education (see Table 10.1). It also had begun to take the notion of technological innovation as a complex, systemic problem seriously, with new initiatives pertaining to intellectual property law, venture capital, and standards emerging over the past decade.

In spite of the many signs of progress in China's science and technology development, the MLP comes at a time of serious concern about the nation's overall development experience. At the 17th National Congress of the Chinese Communist Party held in October 2007, China's leaders have reaffirmed to make China an 'overall well-off society' (*quanmian xiaokang shehui*), with per capita GDP by 2020, quadrupling the level in 2000 (revising the target of quadrupling of total GDP – in the past 25 years GDP has grown faster than per capita GDP). This goal will require continued rapid economic growth. However, the leadership is also aware that the high-speed growth of the past two decades – with its overinvestment, excessive dependency on exports, inefficient utilization of resources, and devastating effect on

Table 10.1   China's science, technology, and education: some indicators

	1998	1999	2000	2001	2002	2003	2004	2005	2006
R&D expenditures									
Gross Expenditures on R&D (GERD) (US$1billion)	6.65	8.20	10.80	12.60	15.56	18.61	27.75	29.91	36.79
GERD/GDP (%)	0.69	0.83	1.00	1.07	1.22	1.31	1.23	1.34	1.42
R&D performance									
Enterprises (%)	44.83	49.59	59.96	60.43	61.18	62.37	66.83	68.32	71.08
Basic Research (%)	5.25	4.99	5.22	5.33	5.73	5.69	5.96	5.36	5.19
Human resources									
Scientists/Engineers Engaged in R&D (1,000 person-year)	485.5	531.1	695.1	742.7	810.5	862.1	926.2	1119	1224
Graduate Student Enrollment (1,000)	198.9	233.5	301.2	393.3	501.0	651.3	819.9	978.6	1100
Undergraduate Student Enrollment (1 million)	3.41	4.09	5.56	7.19	9.03	11.09	13.33	15.62	17.39

Sources: National Bureau of Statistics and Ministry of Science and Technology (comps.), *China Statistical Yearbook on Science and Technology 2006* (Beijing: China Statistical Press, 2006).
National Bureau of Statistics, Ministry of Science and Technology, and Ministry of Finance, *2006 Statistical Bulletin of National R&D Expenditure*.
National Bureau of Statistics, *2007 China Statistical Yearbook*.

the environment – cannot be sustained. Rapid growth towards creating the 'overall well-off society' necessarily will have to follow a different path, one characterized by greater efficiency and productivity gains and based on new knowledge and rationalized institutions. Along with continued institutional reforms, China's capacity for research and innovation, therefore, requires a new sense of urgency.

From a strategic perspective, the MLP can be thought of as addressing four critical problems in China's scientific and technological development in particular. First, in spite of China's remarkable economic performance, its record of innovation in commercial technologies has been quite disappointing, recent improvements in its patenting performance notwithstanding. Instead, its dependence on foreign technology had grown consistently over the past 20 years, in part, because of the state's 'market for technology' strategy which was intended to entice MNCs to transfer technology in return for market opportunities. This policy, arguably, was quite successful in helping to make China the manufacturing centre of the world and for stimulating China's impressive rapid growth of high-technology exports. However, Chinese leaders have concluded that this policy may have run its course.

With its accession to the WTO, China gave up some of the policy tools it had used to leverage foreign interest in Chinese investment opportunities for access to technology. Foreign corporations, in short, can no longer be counted on to transfer technologies, especially advanced technologies needed by more sophisticated Chinese manufacturers. In addition, China has become increasingly dissatisfied with the relative gains it has received from its role in the international division of labour. The royalties Chinese firms pay for foreign technology cut into already slim profit margins and are often perceived as excessive. It has become increasingly obvious to China's political and business elites that those who own the intellectual property, and who control technical standards, enjoy privileged positions in international production networks and profit most from them. In addition, as a result of continued conflicts over IPR and standards with the USA and other countries, China has concluded that the international IPR and standards regimes may not serve China's interests, but work instead to serve the international leaders in innovation who control the architecture of the global technological systems. Hence, the Chinese industrial economy of the twenty-first century should, in this view, set its own standards as well as generate and incorporate its own IPR (Suttmeier, Yao, and Tan, 2006). This requires an enhanced capability for technological innovation – thus leaving China no choice but to become an 'innovation-oriented society'.

Chinese technological capabilities were also failing to meet the nation's needs in such areas as energy, water and resource utilization generally, environment protection, and public health. The negative environmental consequences of two and a half decades of rapid economic growth cannot be overestimated, and continued environmental degradation over the next 15 years would make a mockery of the idea of an 'overall well-off society', whether or not per capita GDP targets were reached. China's quest for energy will only increase in the coming years and call for new technologies for conservation and new energy sources, as well as the procurement of more conventional energy supplies, will be needed. In short, these broad areas of social needs, characterized by both technical *and* institutional problems, cannot possibly be managed without increasingly sophisticated technological capabilities.

Finally, the technological challenges of meeting national defence needs are also powerful drivers for the introduction of the MLP. In spite of China's nuclear weapons and space achievements, its overall capabilities for defence-related technological innovation have, until recently, not been formidable. As with civilian production technology, the modernization of Chinese military technology has largely depended upon imports from abroad. China, of course, also has come to realize the importance of dual-use technology, and its relevance for twenty-first-century high-technology warfare, and has begun to exploit the opportunities which dual use offers. But imported dual-use technology, especially more sophisticated know-how, is subject to export restrictions, especially from the United States. Hence, the international environment would be an unreliable source of critical technologies; again, the need for indigenous capacities for innovation seemed self-evident.

The state of Chinese science, as well as technology, has also had its disappointments. In spite of the swelling ranks of research personnel and increasingly generous funding, the performance of the research system has not lived up to expectations. Many of China's best and brightest have sought career opportunities abroad, and despite an array of incentives that have been offered by various national- and local-level entities, it has been difficult to attract them back; by most accounts, the brain drain has slowed the development of high-level scientific leadership. Simply stated, the human resource base, qualitatively, is not commensurate with its quantitative dimensions. The resources committed to scientific research have, in fact, led to a rapid expansion in Chinese-authored papers in *Science Citation Index* catalogued journals, but the actual contribution – measured by citations – of Chinese papers have been disappointing (Kostoff et al., 2006). Accordingly, Chinese science continues to face important challenges as it seeks to move into international research frontiers. The PRC has yet to establish a research tradition that is both conducive to creative activities and tolerant of failure (Qiu, 2007). Research management is still notoriously top-down and inefficient. Scientists have been preoccupied with quick outcomes and immediate returns. Research too often is derivative in nature which has become a form of 'soft corruption' which not only wastes resources, but also discourages creativity. Scientific misconduct seemingly is widespread, and often covered up and protected.

Table 10.2 Topics of strategic research on China's MLP

- Overall strategy for medium- to long-term S&T development
- S&T system reform and national innovation system
- S&T in modern manufacturing development
- Agricultural S&T
- S&T in energy, resources, and ocean
- S&T in transportation
- S&T in modern services industry
- S&T of population and health
- S&T of public security
- S&T in ecology, environment protection, and recycled economy
- S&T in urban development and urbanization
- National defence S&T
- Strategic high-tech and industrialization of high- and new-technology
- Basic science
- Conditions, platforms, and infrastructures for S&T development
- S&T human resources
- S&T input and management model
- Law and policies for S&T development
- Culture for innovation and S&T popularization
- Regional innovation system and S&T innovation system

Source: http://www.people.com.cn/GB/keji/1059/2592822.html (accessed 14 January 2005).

Thus, when China began preparations for the MLP in 2003, it did so from a position of both great progress in science and technology and, simultaneously, progress which was filled with numerous weak spots in areas of critical national need. As the importance of science and innovation for twenty-first-century China gained high-level political attention, the Chinese turned to their own legacy of science planning as the way to move forward towards its 2020 aspirations. In particular, they were inspired by the most celebrated of the past science plans, the 12-year plan (1956–67) which helped lay the foundation for modern science in the People's Republic and which included among its achievements China's successes in nuclear weapons and space technology (or so-called *liangdan yixing*). This planning model was characterized by the identification of important centrally selected priority projects and the mobilization of resources to work on them, a feature which also characterizes the MLP.

Preparation for the MLP began in 2003 when more than 2,000 scientists, engineers, and corporate executives were mobilized into a programme of 'strategic research' in order to identify critical problems and research opportunities in 20 areas considered to be of central importance for China's future. As seen in Table 10.2, these ranged from S&T management reforms, to human resources, to basic research, national defence, etc. In contrast to earlier planning process, at the outset this one was remarkably open and involved social scientists (mainly economists) and foreign scholars. The openness of the 'strategic research' phase eventually gave way to a more secretive process during which time the bureaucracy massaged the reports of the twenty working groups and attempted to reach compromises in what was by most accounts a contentious and unusually drawn-out process of drafting the public version of the MLP. Narrowing the focus of the MLP as well as setting priorities proved to be a very onerous process that at one point required direct intervention by Premier Wen Jiabao.

10.3 Structure and content

The plan actually consists of a number of components. The first sets out a set of guidelines and general principles derived from the objectives of having science and technology lead future economic development, enabling China during the plan period to 'leapfrog' into positions of leadership in the emerging science-based industries, and establishing a capability for 'indigenous innovation' (*zizhu chuangxin*, also translated as 'independent' or 'homegrown' innovation). The latter has led to considerable confusion inside China and abroad since, in its ambiguity, it has been construed by some as echoing self-defeating techno-nationalist notions of self-reliance (*zili gengsheng*) from the Maoist period – when Chinese research and innovation activities were largely cut off from the international community and experienced significant retardation as a result. In explicating the concept, however, the plan points to *zizhu chuangxin* as having three components: genuinely 'original innovation' (*yuanshi chuangxin*), 'integrated innovation' (*jicheng chuangxin*, or the fusing together of existing technologies in new ways), and 're-innovation' (*yinjin xiaohua xishou zaichuangxin*),which involves the assimilation and improvement of imported technologies.

A second part of the plan identifies its priority areas and programmes. As seen in Table 10.3, these include 11 broad 'key areas' pertaining to national needs and eight areas of 'frontier technology'. Within these, the plan identifies a series of priority projects. For instance, under the 'new materials' area, the plan includes work on smart materials, high temperature superconducting technology, and efficient energy materials.

In addition to the priority areas, the MLP identifies a series of large national 'mega-programmes' in engineering and science. As discussed further below, the inclusion of these programmes has been one of the more controversial aspects of the plan, although it does reflect the legacy of science planning in China, especially the continuing influence of the '*liangdan yixing*' model on Chinese thinking. The plan also calls for an expansion of basic research, including foci on the development of new disciplines and interdisciplinary areas, science frontiers, and fundamental research in support of major national strategies, as well as the mega-programmes identified in Table 10.3.

Table 10.3   Areas and programmes identified in China's MLP

Key areas (11)	Mega engineering programmes (16) (*)
Energy	Core electronic components, high-end generic chips, and basic software
Water and mineral resources	Extra large scale IC manufacturing and technique
Environment	New generation broadband wireless mobile telecommunication
Agriculture	Advanced numeric controlled machinery and basic manufacturing technology
Manufacturing	Large-scale oil and gas exploration
Transportation	Large advanced nuclear reactor
IT industry and modern services	Water pollution control and treatment
Population and health	Genetically modified organism new variety breeding
Urbanization and urban development	Drug innovation and development
Public securities	AIDS, virus hepatitis, and other major diseases control and treatment
National defence	Large aircrafts
	High definition observation system
	Manned aerospace and lunar exploration
Frontier technology (8)	**Mega science programmes (4)**
Biotechnology	Protein science
Information	Quantum research
New materials	Nanotechnology
Advanced manufacturing	Development and reproductive biology
Advanced energy	
Ocean	
Laser	
Aerospace and aeronautics	

Source: State Council, *Nation's Medium to Long-Term Plan for the Development of Science and Technology.*
(*) The MLP only identifies 13 mega engineering programmes.

A third section of the plan deals with ongoing reforms in the research and innovation systems and the further development of the national innovation system. This section highlights important objectives pertaining to the continued reform of a large number of government research institutes, changes in the management of S&T, and the need to encourage Chinese enterprises to assume a leading role in the nation's innovation system, including policies to promote industrial research in companies and supports for small and medium-sized enterprises. This new emphasis on the central role of enterprises reflects growing concerns that China's corporate entities are not generating enough output in terms of new intellectual capital that results in new, commercially-viable products and services.

One recent survey, for instance, indicated that Chinese enterprises are not taking research and development seriously. According to the survey report submitted to an ongoing meeting of the Standing Committee of China's top advisory body, the National Committee of the Chinese People's Political Consultative Conference (CPPCC), only 25 per cent of the large and medium-sized enterprises have established research and development departments. The report points out that in Harbin, capital of northeast China's Heilongjiang Province, only 8.3 per cent of large and medium-sized enterprises say R&D investment accounts for more than 5 per cent of their sales revenue, and another 14.1 per cent say R&D investment has reached 3 per cent of their sales revenue. The report blames the system of performance appraisal operating in state-owned enterprises, noting that it emphasizes increasing the value of state-owned assets but lacks criteria to appraise the technological innovation of enterprises. The survey also refers to China's fiscal input on science and technology, only 10 per cent going for supporting scientific and technological innovations of enterprises (Xinhua News Agency, 2006).

The final sections of the plan, as well as an accompanying document, deal with the introduction of a policy framework for the implementation of the plan, including preferential tax policies, policies for high-technology industry zones, the assimilation of foreign technology, and the strengthening of intellectual property protection for Chinese innovators. They also include important policies to strengthen and diversify the funding for science and technology, make expenditures more efficient, and develop the nation's human resources for science and technology, including the cultivation of world-class senior experts, an expanded role for scientists and engineers in industry, policies to recruit talents from abroad, and reforms in education to support the goals of greater creativity and innovation.

10.4   Debates over the MLP

As noted above, the process of developing the plan involved many participants from across China's S&T community, it touched many local and national interests, and was not without conflict. One important issue concerned the critical relationship between 'indigenous innovation' and technology imports. Chinese economists involved in the planning argued strongly that at China's current level of economic development and comparative advantage, the MLP should focus on maintaining China's status as the world's leading manufacturing base and that the most cost-effective way to upgrade China's technological capabilities for that purpose would be to continue to encourage technology transfers from multinational corporations. The technical community, in general, rejected this neoliberal thinking, arguing that China could no longer expect to receive core technologies from the international sources, for the reasons noted above, and that, overall, China's technological gains from MNCs were disappointing. China certainly cannot ignore the reasoning advanced by the economists, but given the large financial and policy resources being committed to 'indigenous innovation' in all its manifestations, the advocates of a strategic science and technology policy in the technical community clearly have carried the day.

A second issue, as noted above, involved the selection of several mega-programmes and the continued relevance of the *'liangdan yixing'* model. While most Chinese planners would acknowledge that the world has changed dramatically since the days of the *liangdan yixing* programmes, the notion of centrally selected R&D objectives and centrally mobilized resources to support those objectives seems to have special appeal within Chinese political culture. In essence, the fundamental debate has been whether this characteristic of the political culture makes for good science and creative innovation.

At a more pragmatic level, however, the issue has been whether the commitment of substantial amounts of funding to large national projects is a wise use of resources and whether, in particular, the expenditures are the best mechanism to move Chinese science and technology to international frontiers. The debate has been sparked, in part, by criticisms about the effectiveness of such a national programmes as the State High-Tech Research and Development Program (the '863 Program') and the State Basic Research and Development Program (the '973 Program'). Since most national programmes are controlled by the Ministry of Science and Technology (MOST), which controls approximately 15 per cent of the government's R&D expenditures, the doubts expressed about national programme effectiveness have inevitably also been taken as criticism of the ministry, which is thought to champion national programmes as a way to enhance its budget and overall importance, while also meeting national goals.

Criticisms of the national programmes, and by implication the inclusion of so many mega-programmes in the MLP, surfaced in 2004. In July of that year, a group of prominent US-based Chinese life scientists who were attending a symposium in Beijing communicated to Premier Wen Jiabao their belief that the funding of the biosciences in the 863 and 973 programmes was biased and inefficient, lacking in transparency, and too often subject to the preferences of MOST officials rather than scientists (Cyranoski, 2004). They expressed concern that work under the MLP also could be put at risk if changes were not made. Additional criticism appeared in a well-publicized article in 'China Voice II', a Chinese-language supplement to *Nature*, in November 2004 by US-based neuroscientists Yi Rao, Bai Lu and senior life scientist Chen-lu Tsou of the Chinese Academy of Sciences. They argued for changing the ways mega-programmes are organized and funded, and suggested that the ministry be dissolved, or at least have a reduction in its control over research funding (Yao, Lu, and Tsou, 2004).

In the same *Nature* supplement, Mu-ming Poo, another US-based scientist who also serves as director of the Institute of Neuroscience of the Chinese

Academy of Sciences in Shanghai, drew an analogy between the 'big science'/'little science' distinction and the operation of a planned economy in comparison with a market economy (Poo, 2004). Also focusing on problems with the 863 and 973 programmes, Poo argued that the pursuit of mega-programmes diverts resources from programmes supporting bottom-up, investigator-driven projects, which often produce more original research. In his view, large national projects have channelled funds to mediocre laboratories, often on the basis of personal connections and limited peer review. The result has been little impact on the direction of research or the productivity of the participating laboratories, in Poo's opinion.

Dissents of this sort clearly have not won support of the political leadership, given the important role of the mega-programmes in the MLP. In fact, at least one such programme – lunar exploration – started way ahead of the initiation of the planning process as China's first Chang'e 1 satellite is now orbiting around the Moon. It will be interesting to see, however, how these large projects are organized and administered. Most of the large engineering programmes will not be run principally by MOST, although the research programmes in 'frontier technologies' and the mega science programmes will be. A cynical interpretation of the MLP might be that it represents only a repackaging of existing MOST programmes and other national programmes, such as the National Key Technologies Program, administered by the National Development and Reform Commission (NDRC) and other agencies. The 'frontier technologies' programme, for instance, is constituted by the same project areas as the 863 Program, except that 'advanced manufacturing' replaces 'automation', while nanotechnology, a field focused by the 973 Program, is now part of MLP's mega-science programmes, administered by MOST.

10.5   Trends in implementation

Since the initiation of the plan, the Chinese government has taken a number of measures, or supporting policies, to provide for its implementation. Some 99 such policies, which fall into the categories of taxation, IPR protection, human resources, and building a culture of innovation, and finance, are anticipated. Responsibility for the development of the supporting policies has been allocated to a number of central government bodies – NDRC, the state-owned Assets Supervision and Administration Commission (SASAC), and the Ministries of Science and Technology, Education, Finance, and so on – and to specific individuals within those bodies who are charged with building interministerial cooperation to produce policy documents by specific dates. However, getting the 99 policies in play has been more challenging than originally thought as only roughly one-half of these have been announced as of the end of 2007, behind the schedule stipulated by the State Council.

Given the weak tradition of innovation in Chinese industry and the goal of bringing Chinese industrial enterprises to the forefront of the national innovation system, it is not surprising that many of these supporting policies focus on that goal. Thus, we see the introduction of a variety of incentive programmes – generous tax deductions and exemptions, accelerated depreciation schedules for equipment and facilities, and a variety of soft loans from China's policy banks. In addition, enterprises will be given preferential treatment in applying for funding from such programmes as the 863 Program and the State Infrastructure Program if they collaborate with institutions of learning.

In support of the effort to promote 'indigenous innovation' and reward the development of Chinese intellectual property and technical standards, government agencies led by the Ministry of Science and Technology are preparing catalogues for the accreditation of 'innovative products', or those having new proprietary intellectual property, a proprietary brand name, having advanced technical features in comparison with international benchmarks, that have marketability and perhaps can be a substitute for an import. Products so accredited as 'innovative' will be included in a national catalogue and will be given preferential consideration in government procurement and in government-supported national projects, and the companies producing them will be accorded the status of a high-tech enterprise (and thus be eligible for other policy preferences). If an enterprise is granted an international patent, it will receive a subsidy from the government for the costs of maintaining such patent (MOST, 2007b).

Implementing policies now taking effect – led by NDRC, and the Ministries of Science and Technology and Education – also include major financial commitments for new or upgraded facilities. These include infrastructure for 12 major national projects in particle physics, astronomy, ocean exploration, remote sensing, aviation, geophysics, materials safety, protein analysis, and space-based weather monitoring. They also include the building of 30 new national laboratories, and new support for the National Key Laboratories programme which will involve the creation or upgrading of 300 national key labs.

One of the more interesting aspects of the MLP implementation is the encouragement of China's wealthy local governments to invest more in science and technology. Spending on science and technology by provincial governments constituted slightly more than 39 per cent of total governmental expenditure on science and technology in 2005 (up from 29 per cent in 1995), and has been rising rapidly during the last two years.[2] Eight of China's wealthier provinces now spend considerably more on science and technology as a percentage of total government spending than the national average of 2.08 per cent (Shanghai – 4.78 per cent, Zhejiang – 3.95 per cent, Guangdong – 3.66 per cent, Beijing – 3.55 per cent, Tianjin – 2.62 per cent, Liaoning – 2.32 per cent, Fujian – 2.29 per cent, and Jiangsu – 2.13 per cent).[3]

This surge in local government spending on science and technology, thus, is leading to major new funding sources for R&D and to new cooperative projects between the central government and local governments. For instance, the Chinese Academy of Sciences efforts to cultivate relations with local governments have led to the establishment of a number of new institutes involving substantial local funding, including those in Shenzhen (Advanced Technology), Suzhou (Nano-technology and Nano-bionics), Qingdao (Biomass Energy and Bioprocess Technology), Yantai (Coastal Zone Research for Sustainable Development), Ningbo (Materials Technology and Engineering), Xiamen (Urban Environment), and Guangzhou (Biomedicine and Health). In addition, local governments are being enlisted to help support major new 'big science' facilities such as the new Spallation Neutron Source national laboratory in Dongguan, Guangdong province, a key project of the MLP which will involve both national government spending, with land and operating expenses being provided by the province (MOST, 2007a; Hao and Jia, 2007).

While the MLP implementation has been 'slow out of the gate', science mega-programmes have been moving forward. In nanotechnology, for example, 29 projects have been selected with a total fund of RMB262 million being allocated. As each project presumably is evolved researchers from more than one institution, funding intensity – less than RMB5 million per year on average – is hardly significant and it is arguable whether there is a resource concentration factor in the arrangement. It also remains to be seen whether the projects selected will contribute to China's leapfrog into international research frontiers. Equally concerned are how scientists working on different projects collaborate with each other to generate synergy and what benchmark will be used to evaluate the first two-year performance and determine their continuous funding (in fact, one would also be interested in knowing how these projects have been selected and whether scientists were on an equal footing in the process). Although the projects are supposed to be basic research oriented under the MLP, some are about the application of certain nanotechnology, and there is one more question as to how they are related to other MOST administered programs related to nanotechnology, especially the 863 Program and the Torch Program that is focused on high-tech industrialization, presumably some are led by the same chief scientists.

10.6   Governance and accountability

The significant expansion of government funding for research and innovation promised by the MLP is raising new concerns about the performance of the research system and whether or not national resources are being used wisely. While China still is a long way from democratic accountability, there is no doubt that criticisms of the system, and frequent reports of fraud and other types of misconduct in the technical community, including the recent scandal involving the 'China chip' (the development of which was funded by MOST, Ministry of Education, NDRC, and the Shanghai Municipal government),[4] are raising questions in the National People's Congress, the Ministry of Finance, and elsewhere about the public administration of science in China and the management problems of government agencies. To its credit, MOST responded quickly to recent cases of misconduct and instituted a package of new evaluation and budgeting procedures intended to monitor research more closely and prevent and punish fraud and other forms of deviant behaviour in scientific research. But it remains to be seen whether these procedures are enforced to safeguard academic integrity.

As we have seen, the effective implementation of the plan will require complex interministerial cooperation at the central government level between MOST, NDRC, the mission-oriented ministries, the National Natural Science Foundation of China, and the Chinese Academy of Sciences, as well as the Ministry of Finance and SASAC, and also intergovernmental cooperation between the central government and the provinces and cities – which are each supposed to work out their own 'local' plan for S&T development. While there clearly will be a lot more money for science and technology flowing through China, the persistence of institutional and management problems do raise questions about the achievement of 'quality spending'.

To improve management at the central government level, especially for the mega-engineering projects, MOST has proposed online management systems for tracking the involvement of technical experts (to avoid conflicts of interest), for monitoring the performance of researchers, and for facilitating applications for funding (Hao and Gong, 2006). While MOST will seek to ensure its own continuing role in science policy and national research coordination, it is clear that major responsibilities for the implementation of the plan reside elsewhere – principally with NDRC – and it will be interesting to see whether the challenges of implementing the MLP successfully will engender a series of new administrative arrangements.

For instance, governance problems arise with the emphasis placed on the 'public goods' aspects of the plan, especially with regard to the sciences and technologies pertaining to the environment, resources, and public health. Research institutes in these areas have been subject to disruptive reforms in recent years, and there are, again, problems of bureaucratic fragmentation and lack of coordination among the relevant ministries. It would not be surprising to see some major ministerial reorganizations in the areas of agriculture, forestry, water resources and land use management in an effort to streamline bureaucratic functions and provide greater organizational coherence for an expanded 'public goods' research agenda.

Problems of these sorts have given rise to discussions of creating a supraministerial 'Office of Science and Technology Policy' at the State Council level, as some scientists from both inside and outside China suggested in the 2004 discussion, noted above, to improve interministerial coordination and provide science advice to a leadership which will be facing many new technical issues resulting from the plan's implementation. It is quite conceivable that with the implementation of the MLP, pressures will grow for the creation of a policy coordination and administrative mechanism of this sort.

10.7   Conclusion

Given both its breadth and depth, the MLP is likely to generate major impacts on Chinese science and technology over the course of the next 15 years. At the very least, if China reaches its spending goals for research and development, the major enhancement of facilities and equipment now being made will have made it a global scientific centre. Of course, spending alone does not guarantee scientific distinction and technological prowess. The drafters of the MLP certainly recognize this point and have sought to encourage ongoing institutional and cultural change as a means to remove obstacles to the achievement of its goals.

Yet a number of intriguing challenges remain. Even though the MLP does have many strikingly different features from earlier plans, the influence of the *liangdan yixing* experience is still evident in the prominence of state-directed mega-programmes in the MLP, and one wonders whether this approach to the acquisition of national innovative capabilities is the right one for the early twenty-first century.

Conversations with science policy specialists in China often keep returning to the question of the proper role of the state in scientific and technological development in the current era. In China, of course, this remains something of a conundrum in the first instance because of the ideological legacy of socialism and central planning. But, following more than two decades of reform, the political economy of science and technology has become much more complex, especially with the important roles that MNCs play in research and innovation in China, and because of the growth of new technically progressive Chinese firms, many of which have no, or only tenuous, relations with the state. In one sense, the roles played by these two dynamic sectors suggest that the course of Chinese scientific and technological development over the coming 15 years cannot accurately be defined, simply, as state-led innovation. On the other hand, China's policy framework and processes clearly privilege the institutions of the state and make it more difficult for the MNC and new technology enterprise sectors, as stakeholders in the national innovation effort, to be represented. Thus, the programme for state-led innovation runs the risk of marginalizing two of the more technically dynamic sectors in the society, especially if the pursuit of *zizhu chuangxin* assumes a more techno-nationalist stance towards MNCs.[5]

Efforts to promote state-led innovation must also face underlying cultural factors. Again, the state is aware that much needs to be done to generate the growth of a 'culture of innovation' in China, but it remains unclear whether this can be accomplished by state action. The cultural problem has been linked to the persistence of Confucian values stressing obedience to hierarchy and the ways in which such values come to shape the educational experiences of Chinese youth. It has also been linked to state policies which encourage technical entrepreneurs to spend more of their time and energy cultivating good relations with officials rather than developing and marketing innovative products (Stevenson-Yang and DeWoskin, 2005). Others have called attention to the absence of skills and comfort levels for managing technology in a rapidly changing environment, especially when international and cross cultural cooperation is called for.

The MLP, thus, can be thought of as a grand experiment, and many observers inside and outside of China view it that way. It is, in the first instance, of relevance to debates that have gone on in many countries for some time about the utility of national, state-directed programmes of innovation in contrast to the belief that decentralized, market-responsive approaches are far more successful. Most students of innovation, of course, recognize a proper role for the state in promoting new knowledge and techniques, but determining what is 'proper' remains contentious, and varies from country to country. But the question does require that mechanisms for assessing and evaluating the costs and benefits of state action be employed. Market forces in China are beginning to help in these tasks, as is the increasing commitment to formal research evaluation, and the growing concerns expressed about government accountability also show promise. The lack of full, democratic accountability, on the other hand, deprives China of an important source of information and feedback as to whether that proper role is being approached.

A second interesting question for which the MLP will provide insights is finding the proper balance between indigenous efforts at research and innovation and involvement with global technology flows and knowledge development. In spite of strong techno-nationalist themes in the discourse over the MLP in China, it is inconceivable that Chinese science and technology could have progressed to current levels without the productive engagement it has had with foreign universities and research centres, and foreign corporations, over the past 25 years. Chinese leaders appear to sense that the terms of this engagement may be changing, as noted above. Nonetheless, it is also true that the globalization of research and innovation continues apace, and it is unlikely that any country can progress without involvement in it. China's leaders seem to recognize this point and have gone to great lengths to remind their foreign counterparts that the MLP is not designed to insulate China from transnational cooperation and significant participation in the world's new emerging global knowledge system. China's growing S&T relations with the EU and Russia, as well as its established ties with the US and Japan, seem to demonstrate this point quite well. The real challenge will be whether China's scientific community, as presently constituted, can prepare and orient itself to work effectively within the parameters of the new collaborative, network-oriented cross-border knowledge communities and alliances that lie increasingly at the heart of the global innovation system today.

The implementation of the MLP will also help us better understand the role of science and technology in national development. With a per capita GDP of only $1,700 in 2005, and with a vast peasant population, China by these measures remains a developing country. By any number of indicators of scientific activity, however, it is not. China ranked fourth in international science and technology publications, surpassing Germany, France, Italy, and Canada, whose economies are more developed. China has a relatively comprehensive science and technology system with indigenous R&D in the life sciences, nano-science, space technology, and other fields which are internationally important, if not among the world's most advanced. Its pool of scientists and engineers devoted to research and development (about 1 million) is second only to the USA, and China is about to surpass the USA in the conferring of doctoral degrees in science and engineering. Resources for scientific and technological development of these sorts encouraged the initiation of the MLP, and become a test as to whether they can be mobilized and organized in such a way as to accelerate economic and social development.

In spite of the great promise of science and technology in China, we have seen that there are also daunting problems to overcome. In addition to those noted above, we also should recognize that the talent issue could become a bottleneck impeding China's becoming an 'innovation-oriented nation'. China lost an entire generation of scientists and engineering and other professionals to the Cultural Revolution, which has been compounded by the outflow of students and researchers abroad in the reform and open-door era. At first glance, the numbers lost to the brain drain do not appear to be that significant – some 1 million Chinese have gone overseas, some 300,000 have returned, and, meanwhile, China has turned out at least 1 million postgraduates and more than 30 million college graduates over the 1982–2005 period. But those who remain overseas often include the best and brightest. The government has tried to lure them back by initiating various incentive measures providing opportunities to lead China's scientific and educational enterprises; but such measures have not been particularly successful. While many professionals have taken advantage of the opportunity that China's booming economic growth has brought, first-rate academics have been hesitant, fearing that their careers could be jeopardized by returning to a Chinese research environment which has not yet been improved to the point where it can accommodate first-class work. In addition, some of the returnees, such as Chen Jin, have tainted the image of the overseas students as a whole, which deters those who have considered moving back.

China's demographic changes could also affect the prospects for the MLP. China's college-aged population (18 to 22 years old) is set to decline after

2010, and after 2015 it will be outnumbered by those in the 55 to 64 bracket who are approaching retirement. Thus, China will be facing the ageing of its population before becoming a high-income country, a fact which will affect the long-term supply of scientists and engineers and other professionals, and also will call for the diversion of societal resources to support those who have become less productive.

As with so many other aspects of Chinese life, such as the deteriorating environment, China is in a race to acquire the knowledge and wealth necessary to solve or ameliorate its problems before they become overwhelming. The MLP represents a strategic orientation towards this race as well as the ever-present struggle to ensure the country's long-term competitiveness in the face of the rapid and dramatic changes happening in the world of science and technology. Viewed from these perspectives, the implementation and impact of the MLP will certainly be worth watching for those interested in the future trajectory of international science and technology development.

Notes

1. An earlier and shorter version of this paper appeared in *Physics Today* (December 2006), 38–43. Partial support for this work came from a US National Science Foundation grant (OISE-0440422).
2. In nanotechnology, local government spending is expected to exceed that of national government during the current 11th Five Year Plan period.
3. In absolute terms, Guangdong leads all provinces, following by Shanghai, Zhejiang, Beijing, and Liaoning (MOST, 2007c).
4. Chen Jin of Shanghai Jiaotong University was caught falsifying a so-called China chip – a digital signal processor (DSP), supposedly designed and manufactured in China, by scratching the logo off the original chips and replacing a 'Made in China' label. Although the funds from various government ministries have been taken back, Chen was not prosecuted for his cheating, nor were those involved in evaluating and assessing his projects pursued for their responsibility (Hao, 2006a, 2006b).
5. In some ways, of course, MNCs have enjoyed policies which have provided privileges not available to Chinese firms, a situation which has generated resentment in some quarters in China. On the other hand, some of the rhetoric and objectives of the MLP suggests a considerably less welcoming orientation towards foreign investors in the future, and measures to promote Chinese technical standards incorporating Chinese intellectual property, could lead to a far more discriminatory policy environment for MNCs.

References

Cyranoski, David (2004) 'Biologists Lobby China Government for Funding Reform', *Nature*, 430 (26 July): 495.
Hao Xin (2006) 'Scientific Misconduct: Invention of China's Homegrown DSP Chip Dismissed as a Hoax', *Science*, 312 (19 May): 987.

Hao Xin (2006) 'Scientific Misconduct: Scandals Shake Chinese Science', *Science*, 312 (9 June): 1464–6.

Hao Xin and Gong Yidong (2006) 'China Bets Big on Big Science', *Science*, 311 (17 March): 1548–9.

Hao Xin and Jia Hepeng (2007) 'Research Facilities: China Supersizes its Science', *Science*, 315 (9 March): 1354–6.

Kostoff, R.N., M.B. Briggs, R.L. Rushenberg, C.A. Bowles and M. Pecht (2006) 'The Structure and Infrastructure of Chinese Science and Technology', *DTICTechnical Report Number ADA443315* (http://www.dtic.mil/). Defense Technical Information Center (Fort Belvoir, VA).

Ministry of Science and Technology (2007) *Science and Technology Newsletter*, 466 (20 February).

Ministry of Science and Technology (2007) *Science and Technology Newsletter*, 467 (28 February).

Ministry of Science and Technology (2006) *S&T Statistical Databook 2006*. Available online at http://www.most.gov.cn/eng/statistics/2006/index.htm (accessed 29 March 2007).

Poo, Mu-ming (2004) 'Big Science, Small Science' (in Chinese), *Nature*, 432, China Voice II (18 November 2004): A18–A23.

Qiu, Jane (2007) 'Chinese Law Aims to Quell Fear of Failure', *Nature*, 449 (6 September): 12.

Stevenson-Yang, Anne and Ken DeWoskin (2005) 'China Destroys the IP Paradigm', *Far Eastern Economic Review*, 168(3): 9–18.

Suttmeier, Richard P., Xiangkui Yao, and Alex Zixiang Tan (2006) *Standards of Power? Technology, Institutions, and Politics in the Development of China's National Standards Strategy*, Seattle, WA: National Bureau of Asian Research.

Xinhua News Agency (2006) 'Chinese Enterprises Failing to Fund research, Advisory Body Told', BBC Monitoring Asia Pacific – Political, 6 July.

Yao, Yi, Bai Lu, and Chen-Lu Tsou (2004) 'Fundamental Transition from Rule-by-Man to Rule-by-Merit: What Will Be the Legacy of the Medium to Long-Term Plan of Science and Technology?' (in Chinese), *Nature*, 432, China Voice II (18 November): A12–A17.

11
Outsourcing of R&D to Asia: A Case for Continental Specializations?

Albert-Jan Abma, Henny J. van der Windt, Nicolien F. Wieringa and Menno P. Gerkema

11.1 Introduction

This chapter discusses the future research and development (R&D) portfolio of multinational enterprises, with respect to organization, strategy and allocation. Current foreign investments in R&D show that the rise of science and technology in India and China has inspired multinational companies to experiment with new transnational innovation strategies. Offshore outsourcing[1] of R&D was out of the ordinary in the 1980s and before; it was generally believed that R&D could be successfully performed by western companies only, and that the research activities should be located near a firm's headquarters (Mol, 2005). Today, we know that these two presumptions have become outdated, as several chapters in this volume attest. Many research-intensive companies with headquarters in OECD countries cooperate with local research institutes in India or China, or have even started completely new, decentralized R&D departments in these countries. Every year, more companies start up R&D enterprises in Asian countries, or decide for extra investments in current initiatives. Superficially, it might seem an unavoidable trend that R&D will finally move from the West to the East, but is this true and what should be the underlying driving forces? This chapter examines the phenomenon of offshore outsourcing of R&D, with a focus on the dynamics of innovation processes at the firm level.

We focus on the concept of outsourcing of R&D in relation to the factors that affect the sources and success of innovation. Will the outsourcing of core R&D activities become a dominant R&D strategy for multinationals in the long run? Our primary focus is on outsourcing of R&D from OECD countries to low-wage countries in Asia. By lowering the costs of R&D, multinationals have the possibility to choose new innovation strategies. As a result, outsourcing of R&D will affect many actors worldwide. National governments, both in Asian and in OECD countries, should react on the reallocation and reorganization of R&D and reconsider their national innovation policies. As a consequence, universities need to rethink their policy too and should ask themselves how their education of scientists and engineers would best fit the new scientific and technological infrastructure.

Our major point is that the outsourcing of R&D has both positive and negative effects upon a firm's innovativeness. It depends upon the nature of a company how the effects are balanced and whether the outsourcing of R&D will be successful in the long run. We characterize product development by two general kinds of activities, alternately sequenced in a cyclical process of production of 'basic' knowledge and strategic deliberation. We argue that outsourcing can strengthen activities concerning the production of basic knowledge; however, it can negatively affect the strategic deliberations of the firm. To optimize the complete value chain of product development, outsourcing of R&D should be combined with smart strategies for integration of R&D into the firm's strategic management functions.

Finally, we will discuss the strategic consequences of this scenario at the governmental and educational level. We will show why the outsourcing of R&D might lead to continental specialization between Asian countries and OECD countries with respect to certain R&D activities. We suggest Asia can become a centre for execution of (basic) scientific research, while R&D activities in OECD countries should focus on the management of scientific knowledge production and the integration of science into the strategic choices. This scenario has far reaching consequences for many actors.

11.1.1   Outsourcing and the current technology policy in OECD countries

Until the beginning of the twenty-first century it was generally thought that R&D activities were inappropriate for outsourcing to foreign countries, in contrast to industrial production and marketing activities (Mol, 2005). Of all the activities concerning new product development, R&D was thought to be too risky to be moved to low-wage countries. As a pillar of local economies in the West it was though that R&D-intensive industry strengthened local economies by spin-off activities. As a result, the technology policy in OECD countries aimed at initiating and sustaining R&D to stimulate economic activity in a broad way.

Today, the effectiveness of such technology policy is being called into question. Recent empirical research on the Dutch manufacturing industry, for example, shows that R&D intensity did not correlate to higher levels of spin-off activities. On the contrary, R&D intensity was a positive predictor for raising outsourcing levels, suggesting that firms in R&D-intensive industries increasingly relied on partnership relations with outside suppliers (Mol, 2005). While multiplier effects of R&D on local economies appear of limited value, preservation of R&D activities in itself is still important.

The outsourcing of R&D activities may have negative consequences for the local scientific and technological infrastructure. Many authors refer to the concept of 'regional innovation systems'[2] to emphasize the importance of the interrelationships between R&D and academic research, governmental organizations, supplier firms and educational institutions (Gunasekara, 2006). Within this theoretical concept, the loss of one partner will weaken the system as a whole. For example, if R&D departments leave the OECD countries, graduates in science and technology will face increasing difficulties in finding a job; consequently, fewer students would be willing to study science and technology and the drop in student numbers can finally lead to lower funding of universities. The direct interrelationships between R&D and the governmental and academic environment make clear why preservation of R&D is an important policy item in OECD countries.

The outsourcing of R&D to low-wage countries began in the 1980s. Grandstrand et al. (1993) reported that multinational corporations were experimenting carefully with worldwide geographical decentralization of R&D. Since then, the outsourcing of R&D to low-wage countries has emerged, as will be illustrated later. Since the 1990s Asian countries like China, Korea and India have shown tremendous scientific growth compared to western countries (King, 2004). Some analysts claim India and China would become leading R&D centres of the world. According to Mashelkar, former Director-General of the Indian Council of Scientific & Industrial Research (CSIR), India can become the world's number-one knowledge production centre by 2020 if it plays its cards right (Mashelkar, 2005).

OECD countries have responded on increasing outsourcing by policy plans aimed at even further strengthening of the local scientific and technological infrastructure. First, the United States started, relatively successfully, with generous public funding of high-quality academic research (Pavitt, 2000). This strategy has been followed by the European Union leading to the current policy goal to become, technologically, the most competitive region in the world ('European Research Area') (EU, 2004). However, the important question remains whether or not this technology policy will be effective in the long run. Will OECD countries keep their competitive advantage while the scientific infrastructure of Asian countries keeps growing? And, will strengthening of high-quality research remain the right policy goal in case outsourcing of R&D leads to a new, global production strategy of multinational firms?

11.1.2 Problem statement

The objective of this chapter is to explore in which circumstances the outsourcing of R&D may become a favourable strategy for firms. Our focus will be on the dynamics of innovation processes within companies. The aim is to explore current innovation routines and to focus on the key factors for success and failure. Subsequently, it will be discussed how the implementation of outsourcing of (parts of) R&D will affect these key factors. The hypothesis holds that the outsourcing of R&D may offer new opportunities to deal with the management puzzle of optimizing the process of technological innovation. We will discuss the following questions:

1. Which (groups of) activities are characteristic for technological innovation?
2. How does (partial) outsourcing affect the success factors of current strategies for new product development?
3. Which parts of the innovation cycle have already been moved to Asia?
4. How can outsourcing of R&D lead to new transnational production strategies?
5. How should governments in OECD countries react to outsourcing trends by changes in technology policy and the education system for scientists and engineers?

11.2   Theoretical framework

In order to understand the dynamics of technological innovation, we have adopted a multidisciplinary approach. Innovation is interpreted as a cyclical process wherein technological, socio-cultural and commercial factors interact closely. Multidisciplinary approaches for understanding the dynamics of innovation can be found in works that discuss the management of innovation. Several theoretical models for managing technological innovation have been suggested. The models can be classified, roughly, into two groups (depending on the paradigmatic position of the author involved): models that presume a so-called 'technology push' as the driving force for innovation (TP) and models where a 'market pull' (MP) is seen as a starting point.

In its simplest form a TP model describes a linear management process for product development that starts with scientific research or technological enquiry and ends with the launch of a commercial product. The several R&D activities involved are presented schematically in Table 11.1. In a certain way MP models are complementary to TP models. Where TP models start with disclosing technology, MP models begin with an identification of customers' needs. Table 11.1 presents an overview of MP models, based on research work of Spivey et al. (as cited in Bishop, 2004).

The stepwise approach of both TP and MP models is a powerful tool to gain insights into the complex innovation processes that take place. However, there is one major activity that remains implicit in both models, namely the ongoing strategic deliberation during innovation. To start a new phase after finishing the former one implies a strategic decision on one or more management levels. As a result, the innovation cycle is not only a line of certain research activities, but also a succession of decision points.

The principle of a succession of decision points is schematically visualized by the so-called innovation funnel of Wheelright and Clark (as cited in

Table 11.1 General comparison of TP models with MP models

Technology-push models	Market-pull models
Disclosing technology	(No analogous step)
Linking technology with needs	Initial screening
	Preliminary market assessment
Assessing technology	Preliminary technical assessment
Matching technology with functional needs	Detailed market study
Refining technology for specific needs	Business/financial analysis; Product development; In-house testing
Preparing to launch into the users' world	Customer tests; Test market; Trial production; Pre-commercialization
Managing a technology over its life cycle, especially at introduction	Production start-up
	Market launch

Source: Bishop (2004).

Ganguly, 1999). Although the funnel model of Wheelright and Clark is, in fact, an example of a MP model, the general idea also holds for TP models. The shape of the funnel begins with the wide end where many innovative ideas are born. Like a fish in a funnel, an innovative idea has to pass several gates. In order to pass a gate ideas have to fulfil certain selection criteria. According to Wheelright and Clark, ideas should first be tested on feasibility, then the firm should examine how the innovation can be realized (capability phase) and subsequently the innovation should be managed (implementation phase). Every next gate the selection criteria become more severe; in the end few ideas appear apt to pass all the gates and come to a commercial launch. The overall process of selection of ideas shows the dynamics of innovation as a chain of well defined decision points ('gates') that require strategic deliberation.

Inspired by Wheelright and Clark, but at a more abstract level, we propose a new, simple model that represents technological innovation as a cyclical process of two alternating kind of activities: production of 'basic' knowledge and strategic deliberation (Figure 11.1).

With this model for innovation we understand technological innovation as an ongoing story with no well-described starting point. 'Basic' knowledge is interpreted as specialized knowledge generated within a certain research discipline (for example, science, technology or marketing). In the process of product development this specialized knowledge has to be transmitted into a new context that goes beyond the borders of the disciplines involved. In this wider context, strategic deliberation takes place through integrating different kinds of basic knowledge with considerations for corporate strategy. Strategic deliberation corresponds to decision making for passing the gates

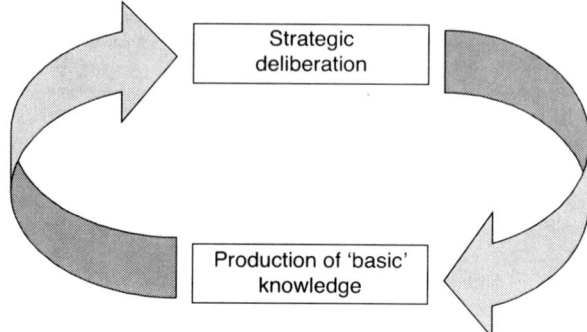

Figure 11.1   Technological innovation represented as a cyclic alternation of two kinds of activities

of the funnel. When an idea has passed the gate new specialized knowledge needs to be generated before one can decide for passage to the next gate. It depends on the underlying driving force for innovation such as which disciplines need to be involved at which time. In this way the model describes both TP and MP models.

The activity of strategic deliberation needs more explanation. Strategic deliberation should be understood as a process of integrating knowledge. Kaufmann and Tödtling have described the mechanism of integrating knowledge (Kaufmann and Tödtling, 2001) by using the concept of self-referential systems in social theory.[3] They consider integration of knowledge as a process of 'border-crossing', where a certain kind of information is being extracted from one knowledge system and is placed into another. 'Border crossing' is more than just exchange of data. The knowledge generated within one information system is not immediately meaningful within the new system and requires more or less translation, with all possible risks of other interpretations. Therefore strategic deliberation is a complicated activity that is crucial for successful product development.

In summary, we have proposed an abstract model for describing technological innovation to transcend the differences between TP and MP models and to underline the importance of two kinds of activities in technological innovation: production of basic knowledge and strategic deliberation. The central idea behind this abstract model is that technological innovation can only become successful when both kinds of activities are balanced and rightly performed. In the next section we will discuss the phenomenon of outsourcing of R&D with respect to the two-phase model by using empirical data and theoretical notions presented in the literature.

11.3 Outsourcing of R&D as a strategy for improving product development

In this section we will explore how the outsourcing of R&D can affect the two major types of activities in product development: 'production of basic knowledge' and 'strategic deliberation'. First, the consequences for strategic deliberation are being discussed, followed by an analysis of the changes in the production of basic knowledge.

11.3.1 Strategic deliberation: integration of knowledge is being challenged

Activities concerning strategic deliberation appear to be critical for the success of new product development. One important integration step is the search for synergy between the department of marketing and R&D. Griffin and Hauser have described 15 empirical studies presenting evidence that adequate interaction among marketing and R&D enhances new product success substantially (Griffin and Hauser, 1996). These studies are mostly based on empirical research among 100–200 firms or projects (1,500–3,000 cases in total). With respect to the proposed two-step innovation model, integration of multidisciplinary knowledge is a crucial activity that can make the difference between success and failure of innovation.

In the context of innovation, we discern three ways for integrating knowledge. The first method for integration of knowledge is working in multidisciplinary project teams (Ganguly, 1999). According to Ganguly, it is becoming more and more exceptional in industry that individual scientists work in the isolation of their laboratory on some fundamental scientific problem. Nowadays business-related scientific projects are conducted by multidisciplinary teams of scientists, marketing managers, marketing researchers and so on. This multidisciplinary approach has proven to be more productive, while also being cost effective, although the assembly and activation of those teams is an intense and time-consuming exercise. Once activated, such dedicated teams have been found to raise creativity, energy and output to levels which frequently surpass all expectations and prior experience (Ganguly, 1999). It can be expected that outsourcing of R&D departments will make it more difficult to continue to work with multidisciplinary teams in product development.

Secondly, the integration of knowledge is also important with respect to the problem of co-ordination between different organizational levels of firms. This kind of integration is one step beyond the interaction between, for example, engineers and marketing experts in multidisciplinary teams, and requires congruence between higher and lower management. According to Grant, this process of fine-tuning is especially important in the context of new product development. He describes five levels of integration that are mutually related in a hierarchy of knowledge integration: single task capabilities, specialized capabilities, activity-related capabilities, broad functional capabilities and cross-functional capabilities (Grant, 1996). New product development belongs to the top of the integration hierarchy (cross-functional capabilities), since it includes integration of many broad functional capabilities (for example, operational capability, R&D design capability, marketing & sales capability). Activities at the top of the integration hierarchy are most complex and most difficult to achieve as they suppose proper integration at all lower levels. When R&D activities do not match strategies at higher levels of management, product development will fail. Again we think that outsourcing of R&D makes integration of knowledge more complicated.

Thirdly, the integration of knowledge is important for organizational learning and the transfer of 'tacit knowledge'. The argument is based on the widely used distinction between explicit knowledge (or 'information') on the one hand and 'tacit' knowledge (or 'know-how') on the other. Information is defined as easily codifiable knowledge that can be transmitted without loss of integrity once the syntactical rules required for deciphering are known. By comparison, know-how involves knowledge that is tacit, 'sticky', complex, and is difficult to codify (Dyer and Nobeoka, 2000). The main point of Dyer and Nobeoka is that tacit knowledge can only be transmitted during personal interaction and is much more difficult to integrate than explicit information. They conclude that integration of tacit knowledge should be understood as a learning process and requires an appropriate learning context. According to Dyer and Nobeoka, only a few firms are able to creative an adequate learning context together with external partners. They mention the Toyota group of companies as a successful organization due to an open attitude towards organizational learning. Knowledge could diffuse more quickly within Toyota's production network than in competing carmaker networks, because an advanced knowledge-sharing network had been developed. Probably, it takes many years to achieve a network of participants with institutionalized routines for learning. However, the case of Toyota shows that outsourcing of R&D can be combined with preservation of a high quality learning context necessary for product development. Probably only a few companies have the capabilities to become successful.

In summary, we conclude that strategic deliberation is an important, but hazardous activity as a part of the proposed two-phase innovation model. Grant (1996) considers the integration of specialized knowledge to be the most important capability a company needs in order to prosper in dynamically-competitive environments (Grant, 1996). He has collected evidence, that in cases of unstable market conditions and increasing intensity and diversity of competition, firms establish their long-term strategies on organizational capacities rather than on served markets. He argues that if knowledge is a critical input into all production processes, and if efficiency requires that it is created and stored by individuals in specialized form, and if production requires the application of many types of specialized knowledge, then the primary role of the firm is the integration of knowledge (Grant, 1996). When firms decide that they are going to outsource R&D, they challenge an important success factor for integration of knowledge. Outsourcing of R&D is a risky enterprise that should be used with care. It depends on the firm's qualities whether the strategy of outsourcing will be successful.

11.3.2   Production of knowledge: strengthening of scientific and technological research

The strategy of the outsourcing of R&D often aims to find locations where a better performance of scientific research can be achieved. According to the two-phase model, the outsourcing of R&D should improve the production of basic knowledge. In order to achieve this R&D should be recognized as a self-referential system.

It is important to note that R&D workers participate in more than one self-referential system at the same time. According to Kaufmann and Tödtling, the difference between academic researchers and researchers from R&D results only from interaction with other systems. Compared to academic researchers, company or contract researchers also have to consider the business system's ways of operation. This results, for example, in different routines for disclosure of knowledge and the reward systems. 'Pure' scientists focus on publications while company and contract researchers focus on patents and commercially useful results. For researchers involved in more than one system it can be very difficult to manage the conflicting interests of making R&D results public versus restricting access through patents or secrecy. Related to this, industrial research has a more pronounced interest in applied short-term research, and it is usually more flexible and willing to engage in interdisciplinary R&D than university science which tends to be more rigid (Kaufmann and Tödtling, 2001). One could presume that the production of scientific knowledge will improve when R&D is performed more autonomously.

Furukawa and Goto (2006) have provided some evidence that pharmaceutical firms stimulate their scientists to participate in the system of scientific research in order to enhance innovation. They conducted an empirical analysis of the role of corporate scientists in Japanese pharmaceutical companies using data on published papers and patent applications. They found that scientists with the highest publication performance scores did not apply for a considerably greater number of patents than other researchers in their companies. Instead, they found that 'core scientists' had a positive effect on the number of patent applications filed by their co-authors. Furukawa and Goto conclude that these scientists probably play an important role as central conduits for the in-flow of knowledge from outside their companies. Apparently, a company needs scientists that primarily take care of the level of scientific research within the firm in order to guarantee the success of innovation on the whole.

According to Kaufmann and Tödtling, it is the exchange of formerly unrelated information between systems that reinforces innovativeness. Crossing the border between different systems stimulates changes in the systems in general. In the particular case of industry–science interaction this might, among other things, result in product innovation (Kaufmann and Tödtling, 2001). Interaction between systems is thus interpreted as a success factor for innovation. We might conclude, perhaps paradoxically, that innovation is overall better off when R&D is not only awarded for their added value for the process of strategic deliberation, but also for their contribution to science as a self-referential system. When people involved in the innovation process can have stronger participation in the relevant self-referential systems, the interaction may be more substantial and can enhance the overall process of technological innovation.

The outsourcing of R&D can enhance scientific research in a number of different ways. First of all, R&D can be located near so-called scientific 'hot-spots'. Recent research undertaken by Papanastassiou and Pearce shows that the outsourcing of R&D can be driven by the desire to get into better contact with key scientific institutes and to disclose state-of-the-art knowledge at an early stage (Papanastassiou and Pearce, 2005). The research is based on an analysis of foreign investments in different kinds of R&D in the United Kingdom. Firms seem willing to activate strong local scientific inputs for longer-term and speculative research programmes. A second way to improve the production of scientific research is to move R&D to places where adequately skilled personnel are abundant. According to Reddy, companies have faced shortages of research in the OECD countries in the 1990s, especially in the field of electronics and biotechnology, compelling them to use more geographically dispersed researchers, including some in developing countries (Reddy, 1997). The outsourcing of R&D can thus be a global strategy to recruit talent and ensure high-quality R&D in the long run. Thirdly, R&D can be outsourced to low-wage countries. On the one hand this strategy can be interesting for saving costs. On the other hand, the strategy allows firms to get more research power for the same price. Since more research power can lead to more innovativeness, this outsourcing strategy can enhance a firms performance in highly competitive markets.

In summary, we conclude that the production of basic knowledge, as a part of the two-phase innovation model, consists of the activities of people who participate in specialized, self-referential systems. The outsourcing of R&D can be an interesting strategy to let the science system work more effectively, and can help to improve the knowledge input in new product development.

11.4 Empirical exploration of current strategies for the outsourcing of R&D

Now that we have developed a theoretical notion on the positive and negative effects of the outsourcing of R&D on transnational production strategies, it is important to verify our conclusions in practice. What are the experiences of multinational firms with the outsourcing of R&D to Asia? In this chapter we report our initial findings. For further research, of course, more advanced methodology is needed. This section begins with a description of some trends in the outsourcing of R&D to Asian countries in general. Subsequently, we explain our explorative approach, leading to a list of cases of outsourcing of R&D to India and China. Finally, we discuss the cases critically with respect to our two-phase model for new product development.

11.4.1 Trends in the outsourcing of R&D to Asian countries

To get a proper understanding of trends in the outsourcing of R&D to Asian countries, it is important to take a closer look at the type of activities involved in decentralized R&D. Following Grandstrand et al. (1993), we discern two kinds of driving forces for the outsourcing of R&D: 'demand-oriented' and 'supply-oriented' forces. Demand-oriented outsourcing often refers to small R&D departments that are necessary for certain, demand-driven reasons. For example, firms with foreign manufacturing subsidiaries need local technical support laboratories. Or, as is the case for clinical tests, government regulations sometimes reinforce the obligations to set up local adaptive R&D. Supply-oriented outsourcing often takes place when foreign circumstances can improve the quality of R&D. For example, the possibility of performing R&D in cooperation with local customers can lead to product improvement. A firm can acquire foreign R&D laboratories to incorporate specialized knowledge and know-how. One can also start a new decentralized R&D laboratory in order to 'tap into' a scientific infrastructure abroad. And, last but not least, cost and productivity of R&D vary among nations over time.

The internationalization of R&D since the 1960s can be understood as a shift towards demand-oriented outsourcing (Gassmann, as cited in Heiman, 2005). Decentralization started during the 1960s and 1970s and when sales activities were moved into target markets and production activities was moved to low-wage countries. Technical centres were founded to support production plants. Later on, decentralized R&D departments became increasingly involved in design and product development, often closely related to the capacities of the local production facilities. Since the late 1980s multinationals have outsourced basic research projects to joint ventures and leading universities and institutes, often located in close proximity. R&D activities became more oriented towards international markets and international centres of knowledge, while the competences and the strength of R&D facilities increased. According to Hirisch-Kreinsen (as cited in Heiman, 2005), in the late 1990s foreign production facilities and foreign R&D facilities were increasingly separated. R&D became an internationally independent function, especially in large companies. For increased efficiency, the number of R&D facilities was then reduced to fewer key R&D centres during the 1990s, while remaining centres were increasingly integrated within transnational R&D strategies.

The establishment of R&D in Asian countries has started (besides some precursors, for example, Astra Zeneca in India, see below) since the early 1990s. We present some data from China and India, but also other Asian countries like Malaysia or Taiwan could have been taken into account. In China, first, new, international R&D initiatives were launched with exponential growth from 34 R&D units in early 2000 to 82 in 2002 to 199 in 2004 (von Zedtwitz, 2004). Other authors report 700 international R&D units in 2006 (AWT, 2006). The enterprises are generally organized in three different ways (von Zedtwitz, 2004). Firstly, R&D is performed in independent, but fully owned laboratories. Secondly, R&D activities are conducted under a branch of a Chinese operation or under a joint venture construction. Thirdly, there is cooperation with Chinese research universities and research institutes. According to von Zedtwitz, a large number of foreign R&D sites are engaged in research and technology development and should be considered as an increasingly important source and provider of global technology. It may be clear that the outsourcing of R&D to China has become supply driven.

India also shows an enormous growth of international R&D departments. The establishment of R&D centres by global firms was conspicuously lacking prior to the liberalization of the Indian economy in 1991. While the nation had a strong pool of highly trained engineers and scientists, India had a lax intellectual property protection regime. Since India signed the GATT agreement in 1993, at least 60 international R&D departments and 25 technological alliances started until 2000, mostly based in the software industry, but also in the biotechnological and pharmaceutical industry (Bowonder and Richardson, 2000). According to a recent publication more than 100 multinationals have started one or more R&D departments in India (AWT, 2006).

In summary, it can be concluded that India and China have become favourable locations for the outsourcing of R&D. Foreign R&D has established, for supply-driven reasons, positive contribution to the multinationals' research performance.

11.4.2 Exploration of individual international R&D departments in Asia

It is interesting to take a closer look at the currently established international R&D departments. Reddy has described some international research centres in India in more detail. The Swedish pharmaceutical company Astra Zeneca, for example, has established the Astra Research Centre India (ARCI) in Bangalore in 1985. Astra considered the department as a centre for creative applied research. Its main objectives were: the pursuit of scientific research leading to the discovery of new diagnostic procedures, novel therapeutic products, and targets for rational drug design for diseases afflicting large populations in both developing and developed countries. Personnel consist of scientists, of mainly Indian origin, who were expected to interact closely with home research groups of Astra, the Indian Institute of Science and universities in India and abroad (Reddy, 1997).

Texas Instruments India (TI-India) is another case described by Reddy. The primary driving forces behind the location of R&D in India were to gain access to R&D personnel as well as to have long-term strategic presence in the Asia-Pacific region. TI-India is treated as a global resource centre, which means that the R&D output has to be used worldwide. The R&D activities include development of computer assisted design (CAD) software systems using integrated circuits (IC) design; software for applications other than CAD, IC design of application specific memory products. The initial focus of the Indian partner was on differentiated products including digital signal processors, memories and mixed ICs. Now, one is increasingly paying attention to the area of methodology development; TI-India is one of the four R&D centres established to pursue R&D of this nature (Reddy, 1997).

In Table 11.2 we present the findings of our initial empirical exploration. The aim of our approach is to explore to what extent R&D departments in Asian countries contribute to multinationals' entire R&D agenda. For some companies we have compared the numbers of scientists and engineers in the home country with the numbers in Asian departments. Data are estimated by comparison of different websites and refer to the period 2006–07.

Table 11.2 shows that the size of an individual R&D department can be substantial sometimes, but that, on the whole, outsourcing is limited to a relatively small fraction of the company's global R&D staff. Cases from the software industry and the electronics industry show the highest levels of outsourcing, with decentralized R&D departments up to almost 3,000 scientists and engineers (Motorola) and outsourcing levels of about 10 per cent of the total R&D staff. The fraction of decentralized R&D seems even lower in the biotechnology, consumer goods and pharmaceutical sectors, in which decentralized laboratories do often not exceed the number of 100 staff members. Novo Nordisk is an exception, with about 10 per cent of their R&D being outsourced to China. The data presented should be used with care. The firms presented are only multinationals. Further, data presented in Table 11.2 can become outdated soon, since many firms are at the moment investing in R&D in Asian countries.

However, the interesting point is that the outsourcing of R&D to Asian countries shows huge differences between companies in the same sector. Motorola and Nokia have about the same size of global R&D staff, but have

Table 11.2   Current R&D sizes of foreign R&D departments in India and China

Sector	Company (home country)	R&D size home country in numbers of scientists and engineers (location)	R&D size at decentral R&D in numbers of scientists & engineers (location)
Biotechnology	Amgen (USA)	12.500 (worldwide)	No department in 2006, planned in 2007
Biotechnology	Monsanto (USA)	1900 (USA)	50 (Bangalore, India)
Consumer goods	Proctor & Gamble (USA)	9000 (worldwide)	100 (Beijing, China)
Consumer goods	Unilever (The Netherlands/ United Kingdom)	800 (Vlaardingen, the Netherlands); 700 (Port Sunlight, UK)	70 (Mumbai, India)
Electronics	Philips (The Netherlands)	15.000 (Eindhoven, the Netherlands)	800 (Shanghai, China)
Pharmaceuticals	Astra Zeneca (UK/Sweden)	5500 (at all 3 locations in Sweden, the R&D headquarter included)	70 (Bangalore, India)
Pharmaceuticals	Bayer (Germany)	1600 (at R&D headquarters, Wuppertal, Germany)	100 (Beijing, China)
Pharmaceuticals	Novo Nordisk (Denmark)	3100 (worldwide)	570 (Tianjin, China)
Software industry	Motorola (USA)	25.000 (worldwide)	2800 (6 centres in India)
Software industry	Nokia (Finland)	20.000 (worldwide)	500 (6 centres in China)

completely different outsourcing levels (2,800 and 500 respectively). The same point holds for the pharmaceutical industry. In Sweden alone Astra Zeneca has more R&D personnel employed than Novo Nordisk has worldwide, but Novo Nordisk has about eight times more scientists employed in Asia. Apparently, companies have different strategies for outsourcing of R&D.

	Demand-oriented outsourcing	Supply-oriented outsourcing
Quality of production of basic knowledge	—	+ +
Quality of strategic deliberation	+	— —

Figure 11.2   Global overview of the suggested consequences of outsourcing

11.4.3   Critical discussion of current international R&D activities in Asia

In this subsection we discuss the current trends in the outsourcing of R&D to India and China with respect to the suggested two-phase model for technological innovation. The consequences for the success of product development will depend upon the kind of outsourcing involved. Figure 11.2 shows schematically how the production of basic knowledge and strategic deliberation are affected by both demand-oriented outsourcing and supply-oriented outsourcing.

Demand-oriented outsourcing has slight negative effects on the basic production of knowledge, but improves strategic deliberation. Demand-oriented R&D facilities are often small – for example, laboratories for technical support and clinical trials lead to dissipation of research power at the headquarters and are therefore unfavourable for the production of scientific knowledge. On the other hand, the proximity to production plants, local authorities or local customers can enhance the integration of local knowledge with research activities and can contribute positively to the quality of strategic deliberation. Especially when local people are involved in R&D, a multinational firm can overcome the differences in culture and language more easily and will have better communication with local manufacturing department, suppliers and governments.

Supply-oriented outsourcing has probably opposite effects on technological innovation. Since the flourishing of R&D is the major driving force behind decentralization, we presume the quality of knowledge production should be improved compared with activities in the home country. As explained in section 11.3 it is important to give scientists and engineers the opportunity to focus on research and to create an interesting climate for research activities. Relatively large R&D departments are to be expected, since firms should strive for a certain 'critical mass' for successful R&D. We also presented several arguments why supply-oriented outsourcing is unfavourable for strategic deliberation. As R&D activities are performed at a considerable distance from higher management and marketing departments, integration of knowledge becomes more complicated. In the case of Asian countries linguistic and cultural differences will also probably have negative effects on the process of integration of knowledge with the home departments.

The recently established R&D centres in India and China illustrate that the current outsourcing of R&D is mostly supply driven. As can be concluded from Figure 11.2, this is a rather risky strategy, with large revenues for the production of basic knowledge, but also serious drawbacks for strategic deliberation. In this situation one might expect that a firm is interested in the outsourcing of R&D as long as risks remain acceptable. From this point of view one can explain why decentralized international R&D departments in Asia do not exceed 10 per cent of a firm's global staff. It also explains why outsourcing levels are highest in the software industry and electronics. As mentioned in section 11.3, firms in these sectors are confronted with shortages of scientists and engineers in OECD countries and have to find alternative solutions. Novo Nordisk is a remarkable case, with relatively high investments in R&D in low-wage countries, in contrast to the other cases of the pharmaceutical industry (see chapter by Kjersem and Gammeltoft, this volume). It remains to be seen if the outsourcing of R&D proves to be a good strategy for Novo Nordisk in the long run, but apparently, short-term experience seems promising. Perhaps this firm has found proper strategies for dealing with the difficulties concerning strategic deliberation.

In summary, we conclude that many multinational firms have started R&D centres in India and China. At this moment decentralized R&D departments are often small compared to centres in the home country, but this situation can change. Probably, the outsourcing of R&D is still an experiment for most companies. It is still unsure whether improvement in knowledge production is possible with acceptable risks for the quality of strategic deliberation. We think future trends will strongly depend on three factors. Firstly, it will be important for Asian countries to create and preserve the right political and juridical climate for innovation. Secondly, macro-economic factors like the shortage of scientists in OECD countries can force firms to intensify the strategy of outsourcing of R&D to Asia. Thirdly, and that is the central point of this chapter, the success of outsourcing depends on the way a firm deals with the problem of strategic deliberation. All factors should be explored in more detail in order to be able to predict future trends.

11.5   Discussion and conclusions

The results of the analysis in this chapter can be discussed in relation to possible future developments. As explained below, we expect that the strategy of the outsourcing of R&D might be increasingly successful in the near future, despite the relatively small investments of most firms at this moment. On a global scale we think that outsourcing of R&D can lead to a continental specialization in R&D activities. In this section we discuss this future scenario from three points of view: the firm level, the governmental level, and the educational level. From each perspective we present arguments that apply for the continental specialization of parts of R&D, and, simultaneously, we discuss strategies for the key players – firms, governments and universities. Finally, we summarize this section and come to a final conclusion.

11.5.1   Firm level: a balanced strategy for transnational product development?

The success of the outsourcing of R&D, as described in section 11.3, depends strongly on the way in which a firm is able to optimize the production of basic knowledge, without harming the process of strategic deliberation. We expect that this need for a balanced transnational innovation strategy can perhaps be fulfilled in two ways. A firm can decide to outsource only less risky forms of R&D, like pre-competitive research and contract research. Pre-competitive research refers to activities such as 'embodiment of knowledge' or 'disclosing technology' that have been described by TP models for managing innovation (section 11.2) as early stages of product development. In this stage of the innovation process the integration of knowledge is less critical since interaction of R&D with marketing department and higher management is less frequent. The same holds for contract research, which is often done when strategic decisions have been taken and an application needs to be worked out. Also in this stage of technological innovation interaction with marketing and corporate management is less important, although managers of R&D should clearly have in mind the objectives for research and have these discussed before with representatives of marketing and corporate management.

We expect that activities other than pre-competitive research and contract research are more difficult to be outsourced. TP and MP models both describe activities like 'matching technology with functional need', 'market scanning' and 'refining technology' as important tasks in product development, that require close interaction between marketing, corporate management and R&D. Current solutions like the establishment of multidisciplinary teams will probably no longer hold, despite modern communication techniques. If a firm is willing to outsource R&D activities other than pre-competitive and contract research, solutions should be found for sustaining the quality of strategic deliberation. Perhaps intermediates are needed to facilitate transfer of knowledge between R&D personnel, marketing experts and corporate management.

The idea of intermediates, as suggested above, is perhaps the second opportunity for firms to develop a balanced transnational production strategy. We think the intermediates should be specialists in the integration of specialized knowledge from the different systems involved in technological innovation: science, technology, marketing and corporate management. They do not necessarily need to be a specialist in one of the systems involved, but are experts in transmitting knowledge from one system to the other and back. If successful, the idea of specialists in strategic innovative deliberation can evolve into a new self-referential system. This extra system in innovation processes not only guarantees the quality of strategic innovative deliberation within a firm, but also offers scientific researchers the opportunity to focus more on participation in the science system and hence improve the production of basic knowledge. As a result, technological innovation can be enhanced from two sides simultaneously. It is interesting to explore whether this system has already evolved somewhere, since it can also be an attractive routine without the outsourcing of R&D.

If the two suggested solutions for the problem of strategic deliberation prove to be successful, outsourcing can become a very attractive strategy for multinationals. As one part of this development the cost price of scientific research can drop significantly, and firms get the opportunity to develop new models for managing R&D. Trends in current R&D like the preference for MP models and the shift away from expensive basic scientific research can perhaps stop and can be redirected into the start-up of large low-cost laboratories in Asian countries. If so, the TP model for managing innovation may become popular again. This trend has enormous consequences and we believe that the likelihood of the scenario also depends on synergy with trends at the governmental and educational level.

11.5.2   Governmental level: stimulating the search for a new competitive advance?

On a macroeconomic level, new transnational strategies for product development can lead to changes in the local scientific infrastructures in both Asia and the OECD countries. We think the need for experts in strategic innovative activities may offer countries the opportunity to attract a new kind of highly skilled employment. On a global scale we think that OECD countries will probably have the optimal location for activities concerning facilitating strategic deliberation. Specialists in integrating knowledge should have frequent contact with as many participants of the innovation process as possible. Multinational headquarters are mostly located in the West and probably will not move to low-wage countries. We expect that new products will often be developed for western markets first where purchasing power is more apparent. As proximity to markets is almost a necessary condition for market departments, these will also remain in the West. Strategic innovative deliberation will be located in OECD countries so long as headquarters and marketing departments do not move.

It might be possible that the two identified central functions within the innovation process can develop further into separate entities. While OECD countries can become a centre for management of scientific research and integration of science into business, Asian countries could provide the world of basic scientific and technological knowledge. The success of this development of continental specialization will depend strongly on the way specialists in strategic innovative deliberation find methods for improving transmission of specialized knowledge between the systems involved in technological innovation. It also depends on the competition between Europe, the US and Asian countries for the best scientific infrastructure, the development of new (Asian) markets and, of course, the kind of products and the scientific disciplines involved. Further research is needed to assess the likelihood of continental specialization.

Governments in OECD countries have to choose between two policy strategies: fighting against outsourcing, or stimulating the creation of a new competitive advance for their country. At this moment it seems to us that OECD governments prefer the first strategy. For example, the European Union strives for becoming, technologically, the most competitive region of the world by the policy goal that expenditures in R&D should rise to 3 per cent of GNP in 2010. To reach that goal, approximately 1.2 million additional research personnel will be needed between 2003 and 2010, including 700,000 additional researchers (EU, 2004). This strategy aims at preserving or even strengthening the current position in the global market for R&D activities. It remains to be seen whether this strategy proves to be successful. Anyway, we conclude that this strategy does not take into account the new economic opportunities from the outsourcing of R&D can give. We believe it can be interesting for governments of OECD countries to develop a technology policy aimed at stimulation of competitive advance in new markets for R&D.

11.5.3   Educational level: educating a new kind of scientists?

In this final subsection we discuss the consequences of the outsourcing of R&D in relation to the education of scientists and engineers. Most curricula in science aim at educating specialists in the production of scientific knowledge. This goal is perfectly right for graduates working at universities, but it does not fit the future demand for scientists, when R&D activities are being outsourced. To revise the current education of scientists and engineers, we believe that it is still important for science students to become familiar with scientific research and to be able to disclose and interpret the latest scientific insights. Additionally, they need competences that are scarcely trained in most curricula yet, like translating knowledge from one system to another, integration of knowledge of different systems and management of scientific research.

We expect that there will be educational conflicts in terms of the priority and time devoted to pure science on the one hand and strategic aspects on the other. We believe that the revision of the educational system will be of major importance for the future of science education at universities in OECD countries. When graduates have difficulties in finding a job, the science education programme may become less attractive for aspirant students and, subsequently, student numbers will drop. If the numbers reach below a certain critical mass, the continuation of the educational programme is under pressure. More possibilities for an educational programme aiming at strategic decision making will be important for a proper fit with future jobs in transnational R&D, but at the same time it may also be a prudent strategy to maintain the quality of the current educational program for pure scientists.

11.5.4 Final conclusion

In the introduction to this chapter we described our contribution as an analysis of the dynamics of innovation processes at the firm level. From this point of view we defined the success of innovation as a proper balance between two kinds of activities: knowledge production and strategic deliberation. Outsourcing of R&D from OECD countries to Asian countries like China and India often disturbs the current balance in the firms' innovation routines. It depends upon the situation whether a firm will succeed in finding a new balance that contributes to the optimization of the firms' innovation processes on the whole. Macroeconomic developments, like the national knowledge infrastructure in OECD countries and the competences of available research and management personnel, are also important factors. Depending upon the choices of firms, national governments and universities make in the near future, the trend of offshore outsourcing can develop in different ways. It is far too early to conclude that the current trend towards outsourcing of R&D from OECD to Asian countries is unavoidable. Relevant actors in OECD countries could become prepared for new transnational innovation strategies. In that case, offshore outsourcing can also lead to partial offshore outsourcing of R&D and new continental specializations.

Notes

1. Some authors distinguish 'outsourcing' from 'offshoring'. According to Van Gorp et al. (2006), the offshoring of activities should involve a foreign location, whereas outsourcing is executed on the domestic market and involves a third party. Offshore outsourcing, then, is the establishment of activities in foreign countries with a third party. In our chapter we prefer to use the term 'outsourcing' when we make reference to the establishment of new R&D activities, because we want to emphasize the fact that activities are performed by a new party, namely non-OECD countries (for example, in Asia).

2. The concept of regional innovation systems is defined by many authors in different ways. Gunasekara (2006) has clustered these definitions at the supranational, national, sectoral technological, local and regional levels. In our chapter we do not use regional innovation systems as a central theoretical concept. We just want to note that we think that the economic importance of R&D will be underestimated if the notion of regional innovation systems is not taken into account.

3. The concept of self-referential systems holds, basically, that a social system behaves independently, for example with respect to interpreting external influences, organizing its internal structure, and how the system behaves in the context of its environment (Kaufmann and Tödtling 2001). Kaufmann and Tödtling have suggested that science is a self-referential system, as are both 'business' and 'policy'. Any social system is distinguishable by different modes of interpretation, decision rules, objectives, and specific communication standards.

References

AWT (2006) *Bieden en binden. Internationalisering van R&D als beleidsuitdaging. Adviesraad voor het Wetenschaps- en Technologiebeleid*, Rijswijk, The Netherlands: Quantis.

Bishop, G.L. (2004) *A Comprehensive Model for Technology Push Product Development*. Brigham Young University. URL: http://contentdm.lib.byu.edu/ETD/image/etd394.pdf. Accessed on 8 January 2008.

Bowonder, B. and P.K. Richardson (2000) 'Liberalization and the Growth of Business-led R&D: the Case of India', *R&D Management*, 30(4): 279–88.

Dyer, J.H. and K. Nobeoka (2000) 'Creating and Managing a High-performance Knowledge-sharing Network: the Toyota Case', *Strategic Management Journal*, 21: 345–67.

EU High Level Group on Human Resources for Science and Technology in Europe (2004) *Increasing Human Resources for Science and Technology in Europe*. July. URL: http://ec.europa.eu/research/conferences/2004/sciprof/pdf/final_en.pdf. Accessed on 8 January 2008.

Furukawa, R. and A. Goto (2006) 'The Role of Corporate Scientists in Innovation', *Research Policy*, 35: 24–36.

Ganguly, A. (1999) *Business-driven Research and Development: Managing Knowledge to Create Wealth*. London: Macmillan Press Ltd.

Gorp, D. van, P.K. Jagersma en M. Ike'e (2006) *Offshoring in the Service Sector: A European Perspective*. NRG Working Paper Series, vol. 06-06. Nijenrode Business Universiteit. URL: http://www.nyenrode.nl/download/NRG/workingpapers/NRG06-06.pdf.Accessed on 8 January 2008.

Grandstrand, O. et al. (1993) 'Internationalization of R&D – a Survey of Some Recent Research', *Research Policy*, 22: 413–30.

Grant, R.M. (1996) 'Prospering in Dynamically-competitive Environments: Organizational Capability as Knowledge Integration', *Organization Science*, 7(4): 375–87.

Griffin, A. and J.R. Hauser (1996) 'Integrating R&D and Marketing: A Review and Analysis of the Literature', *Journal Product Innovation Management*, 13: 191–215.

Gunasekara, C. (2006) 'Academia and Industry: the Generative and Developmental Roles of Universities in Regional Innovation Systems', *Science and Public Policy*, March: 137–50.

Heiman, P. (2005) *Foreign-owned R&D Facilities in China, England, Germany and Sweden: An Analysis of Regional Entry and Integration Behavior*. University of Augburg,

Germany. URL: http://deposit.ddb.de/cgi-bin/dokserv?idn=976393433, Accessed on 8 January 2008.

Kaufmann, A, and F. Tödtling (2001) 'Science–Industry Interaction in the Process of Innovation: the Importance of Boundary-crossing Between Systems', *Research Policy*, 30: 791–804.

King, D.A. (2004) 'The Scientific Impact of Nations: What Different Countries get for Their Research Spending', *Nature*, 430: 311–16.

Mashelkar, R.A. (2005) 'India's R&D: Reaching for the Top', *Science*, 307: 1415–17.

Mol, M.J. (2005) 'Does Being R&D Intensive Still Discourage Outsourcing? Evidence from Dutch Manufacturing', *Research Policy*, 34: 571–82.

Papanastassiou, M. and R. Pearce (2005) 'Funding Sources and the Strategic Roles of Decentralised R&D in Multinationals', *R&D Management*, 35(1): 89–100.

Pavitt, K. (2000) 'European Funding: Why European Union Funding of Academic Research Should be Increased: a Radical Proposal', *Science and Public Policy*, 27(6): 455–60.

Reddy, P. (1997) 'New Trends in Globalization of Corporate R&D and Implications for Innovation Capability in Host Countries: a Survey from India', *World Development*, 25: 1821–37.

von Zedtwitz, M. (2004) 'Managing Foreign R&D Laboratories in China', *R&D Management*, 34(4): 439–52.

12
Science, Technology and Innovation Dynamics in India and China: Some Concluding Thoughts

Govindan Parayil and Anthony P. D'Costa

12.1 Introduction

The thematic focus of this volume has been the emerging new innovation dynamics in Asia with a particular emphasis on India and China. As clearly spelt out in the introductory chapter, our objective was not to present a strict comparison of India's and China's R&D and high-technology achievements. Rather, our objective was to highlight the dynamic role played by Indian and Chinese firms and other economic actors such as the government, universities and research institutions in forging this nascent dynamics within the global innovation scene. We have given a comprehensive account of what we mean by the term 'new innovation dynamics' in the introductory chapter. Our objective has been to look at the emerging picture as well as the similarities and differences, broadly, in the evolution and structuration of a new innovation dynamics by conceptually linking the innovation systems, institutions, policies, S&T infrastructure and output, MNC investments in R&D and so on in these Asian giants.

12.2 India and China and the emerging innovation dynamics

India and China are termed 'high-tech hopefuls' (Economist, 2007) in an age where scientific and technological advances are seen not a sole preserve of the developed world, as argued by several authors in this volume. A handful of emerging nations such as South Korea, Singapore, Taiwan, India, China, Brazil, South Africa, and Malaysia have captured the imagination of academics and the business press because of these nations' emergence as competitive players in the global high-tech field. However, it was India and China that drew the greatest attention of academics, policy makers and business analysts and managers to their high-tech ambitions due to their large-scale engagement with contemporary global innovation dynamics.

Why and how India and China became such important players in the globalization of R&D, among other topics of importance in innovation dynamics, are the sorts of questions that the various chapters in this volume have attempted to answer. Those who follow a path-dependent route to examine the historical dimensions of global political economy would find this rather puzzling. How did these nations that were struggling to catch up (during their early postcolonial developmental phase up until the 1980s) with the advanced industrialized nations, which are considered the 'traditional' players in R&D-based growth and development, suddenly emerge to compete with and even challenge the monopoly of the established players?

It is an enduring question for historians, economists, and others interested in the past of Indian and Chinese science and technology to ask why these giants, which had been such pre-eminent powers in science and mathematics since ancient times, had almost suddenly stopped making advances in these domains and virtually stopped innovations in industry and economy while the Europeans had just begun to pick up from where the Indians and Chinese had stopped sometime in the fourteenth and fifteenth centuries. Although an interesting question in itself to delve into, a question to which numerous authors had attempted to provide answers, our primary concern in this volume has been to focus on the present status and future possibilities of science, technology, research and innovation in China and India.

Although India and China emerged independent developmental states after the Second World War under different historical circumstances, they inherited similar developmental challenges. After centuries of colonial rule and exploitation, most of their people were living in acute poverty and material deprivation. Industrial development was slow and agriculture suffered centuries of neglect. As is well known, under different political dispensations, both countries attempted to re-build their shattered economies. India under Nehru decided to pursue national development in a democratic framework with a mixed economy model where both the state and markets were to play along to the tune of Five-Year Plans. China under Mao decided to go full steam on a collectivized socialist economy model by forcing the peasants out of their farms and into factories. Both countries muddled along for decades, poverty and deprivation continued to haunt the majority of their people because of technological stagnation, low productivity and stunted economic growth. Both countries invested early on in heavy industries and built large dams and industries, but enterprises under state control in both countries performed rather poorly. It was only in the late 1970s that China decided on a new course of capitalist modernization by opening up its economy for outside investment and international trade. India was late in coming to do the same and embarked on economic liberalization and opening its economy to the forces of globalization since the mid-1980s but more forcefully since the 1990s.

What attracted MNCs to invest in R&D activities was not because these two giants had excelled in inventing and developing new technologies and knowledge that they could leverage and exploit, but because they were becoming better at diffusing outside knowledge in their high-tech sectors such as IT and biotech with the help of their highly educated human resources. The high levels of investment that India and China made in higher education in science, engineering and technology created a large pool of competent human resources fit for the emerging knowledge economy. Their enterprising and highly motivated scientists, engineers and managers who emigrated to find better professional opportunities abroad are now available to be utilized within these countries on value-adding innovative activities. The decision of MNCs such as Microsoft, IBM, GE, Texas Instruments, Hewlett Packard, Siemens, Alcatel and numerous others to set up R&D centres came out of this new optimism that the engineers and scientists that India and China annually produced at some of their best universities and institutes of technologies could be employed at these centres along with returning engineers and managers who were eager to take up jobs and assignments in their native lands. Among others, Chapter 9 by D'Costa and Kobayashi not only underscores the importance of human resources to innovation capability but also suggests that countries such as Japan, who are slow to tap into these resources, are likely to face considerable competitive pressures in the future.

It has been observed that investment in R&D by MNCs around the world continued to increase in recent years, with a large proportion of this going to India and China. The world's top 1,250 firms invested US$510 billion in R&D in 2006 – an increase of 10 per cent over the 2005 figure (Willman, 2007). Global competition is cited as the prime reason for this increased spending (Cookson, 2007) and firms in China and India represented some of the sharpest increase in R&D spending. This is particularly important because almost all Fortune 500 companies have already begun investing in R&D in China and India for many years as the authors in this volume have shown.

As Liu and Lundin showed in Chapter 2, the Chinese system of innovation is directed largely by the state, with large government research institutes (GRIs) taking a leading role. This is in stark contrast to the Indian model of innovation where GRIs have yet to play a leading role in influencing innovation dynamics as argued by D'Costa (Chapter 4) and Chaturvedi and Chataway (Chapter 6). There are other shortcomings in India, which lacks an ecosystem for innovation such as not enough PhDs in engineering, the poor quality of university research, less emphasis on non-IT sectors and so on. So while it was mostly market forces led by both foreign and local MNCs and local startups that played key roles in forging India's high-tech innovation, these same forces are also, in the absence of coherent innovation strategy, distorting national capabilities.

Liu and Lundin show that China's state-led planned innovation model is slowly giving way to a more market-driven innovation in China. China's open-door policy to foreign direct investment in the high-tech sector during its reform period in tandem with the diversification and decline of state-run enterprises led to the rise of market-driven factors in forging China's innovation dynamics. Indigenous innovations driven by Chinese firms as well as open innovations driven by foreign firms are interesting aspects of the emerging innovation dynamics in China. However, it is the central government that dictates the policies and strategies for S&T development and expansion in China as Cao, Suttmeier and Simon argue in Chapter 10. The situation is paradoxical since earlier centrally-driven innovation attempts by India met with mixed results, while China's current more decentralized system to generate innovations is also producing mixed outcomes. What this suggests is that there is a fine balance between open-ended and state-led system of innovation and thus tensions between the desire to build indigenous technological capabilities and multinational-dominated innovation systems.

The internationalization of R&D into China, according to Sylvia Schwaag Serger (Chapter 3), could be explained by two factors: (1) government stipulations for FDI require foreign MNCs to conduct some R&D functions in China in a manner akin to a quid pro quo for technology for China in return for market opening; and (2) a significant and growing share of R&D activities by MNCs in China could be explained by a unique combination of China's market size, strategic importance and attractive human capital. One important finding of Schwaag Serger (Chapter 3), Kjersem and Gammeltoft (Chapter 7), Abma et al. (Chapter 11) and others is that increasing MNC investment in R&D in China and India is not conducted at the expense of R&D activities in Europe and North America, rather it benefits both parties. This is an important finding to allay protectionist tendencies in Europe and North America.

A critical examination of the costs and consequences of the internationalization of R&D into India and China is extremely important in an analysis of the normative implications of the new innovation dynamics unfolding in these countries. By analysing the biotechnology and pharmaceutical sectors of India and China within the context of the increasing globalization of R&D by MNCs in these countries, Thomas (Chapter 5) shows that the newly emerging innovation dynamics benefits only those who have profited from economic globalization. While local MNCs were focusing their attention on providing products for the burgeoning middle classes as well as developing products for exporting to rich countries, as shown by both Thomas and Chaturvedi and Chataway (Chapter 6), foreign MNCs were investing in R&D to develop new drugs and biochemical products as well as using the local populations for clinical trials of new drug combinations and therapeutic techniques by taking advantage of low costs, lax regulatory standards and the ready availability of patients for clinical trials and complying physicians. One key development that helped MNCs to take advantage of the new innovation environment is the compliance of these countries to a uniform standard of intellectual property rights mandated by the TRIPS Agreement under the WTO. What is happening under these neoliberal conditions is the slow and steady erosion of price controls on all sorts of essential medicines and staples that used to protect the poor and economically marginalized populations. Notwithstanding the benefits of internationalization of R&D in developing countries, we cannot afford to ignore the social costs of the new innovation dynamics, an issue that is broached by several chapters in this volume.

12.3 Some concluding thoughts

By providing a detailed account of the new innovation dynamics in India and China, we have attempted to raise several questions that will require further detailed investigations. We hope that others will pick up from where we have left off by continuing discussions on these and other issues within the field of innovation studies and economic development. What sort of challenges and opportunities will the globalization of R&D to Asia pose to OECD countries and other emerging and developing countries? Do India's and China's R&D policies offer useful lessons for other emerging nations in Asia, Africa and Latin America? Rather than focusing only on the macroeconomic aspects of the challenges and benefits of the globalization of R&D, we need to focus on important micro- and meso-level issues pertaining to this important topic within the dynamics of innovation in Asia.

References

Cookson, Clive (2007) 'Global Competition Sparks Spending Spree', *Financial Times*, London, 29 October. Accessed at http://www.ft.com, accessed 18 December 2007.

Economist, The (2007) 'High-tech Hopefuls: A Special Report on Technology in India and China', 10 November.

Willman, John (2007) 'US Leads Way as Global R&D Spending Rises 10%', *Financial Times*, London, 12 November.

Index

292

Breinigsville, PA USA
16 March 2011
257793BV00005B/25/P